Monoclonal Antibodies

Special Issue Editor
Christian Klein

MDPI • Basel • Beijing • Wuhan • Barcelona • Belgrade

MDPI

Special Issue Editor
Christian Klein
Roche Pharmaceutical Research & Early Development
Switzerland

Editorial Office
MDPI
St. Alban-Anlage 66
Basel, Switzerland

This edition is a reprint of the Special Issue published online in the open access journal *Antibodies* (ISSN 2073-4468) from 2017–2018 (available at: http://www.mdpi.com/journal/antibodies/special_issues/moloclonal_antibodies).

For citation purposes, cite each article independently as indicated on the article page online and as indicated below:

Lastname, F.M.; Lastname, F.M. Article title. *Journal Name* **Year**, *Article number*, page range.

First Editon 2018

ISBN 978-3-03842-875-6 (Pbk)
ISBN 978-3-03842-876-3 (PDF)

Cover image courtesy of Christian Klein.

Table of Contents

About the Special Issue Editor

Christian Klein is a Distinguished Scientist, Head of Oncology Programs and Department Head Cancer Immunotherapy Discovery at the Roche Innovation Center Zurich, specialized in the discovery, validation and preclinical development of antibody-based cancer immunotherapies and bispecific antibodies. During his 16 years at Roche, he has made major contributions to the development and approval of obinutuzumab, the preclinical development of currently nine clinical stage bispecific antibodies/antibody fusion proteins, and the development of Roche's proprietary bispecific antibody platforms, e.g., the CrossMAb technology and immunocytokine and T-cell bispecific antibody platforms. He is author of >100 publications and reviews in the field of antibody engineering and cancer therapy and co-inventor on >120 patent applications. After completing his diploma in biochemistry at the University of Tübingen and his dissertation in biochemistry at the Technical University in Munich, he completed his habilitation in biochemistry at the Ludwig-Maximilians University Munich 2017 and acts as external lecturer there.

Preface to "Monoclonal Antibodies"

Monoclonal antibodies are established in clinical practice for the treatment of various diseases including cancer, autoimmunity, metabolic and infectious diseases. Over the last 20 years, monoclonal antibodies have established themselves as therapeutics and various so-called "blockbuster" drugs are in fact antibodies. Currently, ca. 80 monoclonal antibodies have been approved in Europe and the US and several hundred of them are currently in early and advanced clinical trials. Notably, in the last decade, the field has significantly advanced, and, nowadays, a large proportion of antibodies in development is made up by engineered antibodies including bispecific antibodies, antibody drug conjugates and novel antibody-like scaffolds. This Special Issue on "Monoclonal Antibodies" includes original manuscripts and reviews covering various aspects related to the discovery, analytical characterization, manufacturing and development of therapeutic and engineered antibodies.

The collection starts with a number of reviews. Cho and colleagues review the state-of-the-art in therapy of multiple myeloma where antibody-based immunotherapies are changing the current treatment paradigm, and Wang-Lin and Balthasar summarize pharmacokinetic and pharmacodynamic considerations important to consider for the treatment of bacterial infections by monoclonal antibodies. Finally, Fülöp and colleagues review the role of complement activation in infusion reactions associated with the application of monoclonal antibodies and the potential use of complement factor H for its prevention.

A first series of original articles describes novel monoclonal antibodies for potential diagnostic or therapeutic application. Rashidian and colleagues describe a novel rabbit monoclonal antibody, MRQ-67, that specifically recognizes the R132H mutation of Isocitrate dehydrogenase 1 (IDH1) which is prevalent in diffuse astrocytomas, oligodendrogliomas, and secondary glioblastomas but not the wildtype IDH1. MRQ-67 is able to identify neoplastic cells in glioma tissue specimens, and can be used as a tool in glioma subtyping. Zhang and colleagues have identified novel monoclonal antibodies against the plasmodium falciparum Circumsporozoite Protein that is a major and immunodominant protective antigen on the surface of plasmodium sporozoites. These antibodies are specific for the central repeat region and mediate protection against challenge with sporozoites. Finally, Rocha and colleagues generated antibodies directed against novel epitopes of the Dengue nonstructural protein 1 (NS1) which is a multi-functional glycoprotein essential for viral replication and modulation of host innate immune responses and represents a surrogate marker for infection. These antibodies are able to differentiate Dengue and Zika virus infections and may contribute to the development of novel diagnostic tools. In a series of three articles, Strube and colleagues describe approaches useful for the manufacturing and analytical characterization of monoclonal antibodies. A first article by Schmidt et al. describes aqueous two-phase extraction (ATPE) as a method to capture monoclonal antibodies using a combined harvest and capture step during the downstream process. A subsequent article by Kornecki et al. focuses on the characterization and classification of host cell proteins (HCPs) and how to categorize and avoid them in the manufacturing process. Finally, Zobel-Roos et al. propose a process analytical approach allowing for controlled automation of the downstream process by inline concentration measurements based on UV/VIS spectral analysis. In the same area, Radhakrishnan and colleagues show how time-dependent media supplementation by MnCl2 can be used to control the glycosylation profile of antibodies. Castellanos and colleagues use small-angle scattering (SAS) combined with size-exclusion multi-angle light scattering high-performance liquid chromatography and molecular modeling to characterize antibody—antigen complexes in solution.

Lastly, two articles deal with engineering monoclonal and bispecific antibodies. Tam and colleagues have identified a set of novel mutations in the Fc-portion of antibodies that abrogate immune effector function of the respective antibodies. Such Fc-mutations are essential for the development of antibody therapeutics where simultaneous FcgR activation is undesired for the mechanism of action, e.g., for T-cell bispecific antibodies. Dheilly and colleagues engineered novel CD47-CD19 bispecific antibodies based on low affinity CD47 inhibitory antibodies. The corresponding CD47-CD19 bispecific antibody inhibited tumor growth in vivo and induced a long lasting anti-tumor immune response that could be further enhanced in combination with chemotherapy or PD-1/PD-L1 checkpoint blockade.

This collection of articles is of value to readers working in the field of monoclonal and therapeutic antibodies.

Christian Klein
Special Issue Editor

Article

Monoclonal Antibodies against *Plasmodium falciparum* Circumsporozoite Protein

Min Zhang [1,2,3], **Rajakumar Mandraju** [1,4], **Urvashi Rai** [1], **Takayuki Shiratsuchi** [1,5] and **Moriya Tsuji** [1,*]

[1]   HIV and Malaria Vaccine Program, Aaron Diamond AIDS Research Center, Affiliate of The Rockefeller University, New York, NY 10016, USA; zhanmin@iu.edu (M.Z.); mandraju@gmail.com (R.M.); urvashi.rai@gmail.com (U.R.); Shiratsuchi.Takayuki@hq.otsuka.co.jp (T.S.)

[2]   Department of Pathology, New York University School of Medicine, New York, NY 10016, USA

[3]   Department of Pharmacology and Toxicology, Indiana University School of Medicine, Indianapolis, IN 46202, USA

[4]   Department of Immunology, UT Southwestern Medical Center Dallas, TX 75390, USA

[5]   Otsuka Pharmaceutical Co., Ltd., Osaka 540-0021, Japan

*   Correspondence: mtsuji@adarc.org; Tel.: +1-212-448-5021

Received: 10 May 2017; Accepted: 1 August 2017; Published: 23 August 2017

Abstract: Malaria is a mosquito-borne infectious disease caused by the parasite *Plasmodium* spp. Malaria continues to have a devastating impact on human health. Sporozoites are the infective forms of the parasite inside mosquito salivary glands. Circumsporozoite protein (CSP) is a major and immunodominant protective antigen on the surface of *Plasmodium* sporozoites. Here, we report a generation of specific monoclonal antibodies that recognize the central repeat and C-terminal regions of *P. falciparum* CSP. The monoclonal antibodies 3C1, 3C2, and 3D3—specific for the central repeat region—have higher titers and protective efficacies against challenge with sporozoites compared with 2A10, a gold standard monoclonal antibody that was generated in early 1980s.

Keywords: *Plasmodium falciparum*; circumsporozoite protein; CSP; monoclonal antibody; 2A10; 3C1; 3C2; 3D3

1. Introduction

In 2015, there were 214 million new cases of malaria (range 149–303 million) and an estimated 438,000 malaria deaths (range 236,000–635,000) worldwide [1]. Malaria is a mosquito-borne disease caused by the protozoan parasite, *Plasmodium* spp. Malaria is transmitted among humans by the bite of female mosquitoes of the genus *Anopheles*. The battle against malaria has been fought using a wide range of interventions, including insecticide-treated bed nets, indoor residual spraying, effective medicines, and vaccine [2–5]. However, emerging antimalarial drug resistance and insecticide resistance threaten malaria control and public health [6–8]. The only approved malaria vaccine is RTS,S/A01 (trade name Mosquirix) to date. RTS,S/A01 represents it's composed of *P. falciparum* CSP repeat region (R), T-cell epitopes (T) fused to the hepatitis B surface antigen (S) and assembled with un-fused copies of hepatitis B surface antigen, and a chemical adjuvant (AS01) is added to increase the immune system response. The efficacy of RTS,S/AS01 against all episodes of severe malaria is approximately 50% in young children in Africa [9–11]. A completely effective vaccine is not yet available for malaria. The novel vectored immunoprophylaxis, an adeno-associated virus-based technology to introduce effective antibody genes in mammalian host, has been added to currently available tools to control malaria [12]. A highly efficient neutralization antibody is one of the essential components of the vectored immunoprophylaxis [12]. Sporozoites are the infectious form of the parasites inside mosquito salivary glands. The circumsporozoite protein (CSP) is a major protein on the surface of *Plasmodium* sporozoites and an immunodominant protective antigen in irradiated sporozoites [13]. The overall structure of CSP is conserved among *Plasmodium* species, consisting of a species-specific central tandem repeat region flanked by conserved N-terminus and C-terminus [14]. The N-terminus is proteolytically processed during sporozoite invasion into host cells, unmasking the C-terminal cell-adhesive domain [15,16]. The C-terminus contains a thrombospondin repeat domain and T cell epitopes. The central repeat region, which is composed of approximately 30 tandem repeats of asn-ala-asn-pro (NANP), corresponds to highly immune-dominant B-cell epitopes [17,18].

The transmission of malaria from mosquito to mammalian host can be prevented by antibodies against CSP, such as the monoclonal antibody (mAb) 2A10 [12,19]. The mouse mAb 2A10 is directed against the central repeat region of *P. falciparum* CSP (PfCSP) [12,20–22]. The mouse mAb 2A10 is a useful tool for the study of PfCSP in a mouse model. Delivery of adeno-associated virus expressing 2A10 into mice results in long-lived mAb expression and protection from sporozoite challenge. Vectored immunoprophylaxis provides an exciting new approach to the urgent goal of effective malaria control [12]. However, the mice expressing the CSP-specific mAb 2A10 lower than 1 mg/mL could not be completely protected [12]. Thus, highly potent CSP-specific antibodies are desired for the immunoprophylaxis to control this infectious disease. Here, we report a generation of novel and potent CSP-specific antibodies against PfCSP. In addition, we characterized the mAbs' subclasses, titers, and protections for sporozoite challenge. Importantly, the protective efficacies of 3C1, 3C2, and 3D3 were found to be better than the reference mAb 2A10.

2. Materials and Methods

2.1. Expression and Purification of Recombinant PfCSP

PfCSP coding sequence without glycosylphosphatidylinositol (GPI) anchor (GenBank: M19752.1) was amplified using Phusion® high fidelity DNA polymerase (Cat#M0530S, New England Biolabs, Ipswich, MA, USA) with specific primers containing *EcoR* I and *Not* I restriction enzyme recognition sites. The PCR product was purified using Qiagen PCR cleanup kit (Qiagen, Germantown, MD, USA). Both the PCR product and pET20b vector were digested with restriction endonucleases *EcoR* I and *Not* I (New England Biolabs) according to the manufacturer's protocol. After gel purification, the digested PCR product was ligated into the linearized pET20b vector using Roche rapid DNA ligation kit (Cat. No. 11635379001, Roche, Branford, CT, USA), and then transformed into Top10F' chemically competent *E. coli.* (Invitrogen, Grand Island, NY, USA) and plated onto Luria-Bertani (LB) agar plates containing ampicillin. A single colony was picked from the plate and inoculated into LB broth plus ampicillin. The recombinant plasmid was purified from the overnight culture using Qiagen plasmid purification kit. The purified plasmid was validated by DNA sequencing and transformed into the BL21(DE3) strain for protein expression. When the culture reached an optical density (OD, 600 nm) of 0.5–0.6, PfCSP expression was induced using IPTG (1 mM) at 20 °C. Then the overnight culture was pelleted by centrifugation and lysed with lysozyme buffer and followed by sonication. Lysate was cleared by centrifugation and the His-tagged PfCSP was purified using Ni^{2+}-affinity chromatography (Qiagen, Germantown, MD, USA).

PfCSP purification: 25 mL of nickel nitrilotriacetic acid (Ni-NTA) agarose beads were loaded onto a 22 mL phenyl sepharose column (Pharmacia/Pfizer, New York, NY, USA), washed and equilibrated by 200 mL of His Elution Buffer (50 mM TRIS hydrochloride (Tris-HCl) (pH 8.0), 300 mM imidazole, 50 mM NaCl, 0.1 mM ethylenediaminetetraacetic acid (EDTA), and 1 mM phenylmethane sulfonyl fluoride (PMSF) and 500 mL of His Binding Buffer (50 mM Tris-HCl (pH 8.0), 5 mM imidazole. 100 mM NaCl, 0.1 mM EDTA, and 1 mM PMSF). Then the clarified lysate from 1 L culture was added to the column and washed with 250 mL of His Binding Buffer followed by 500 mL of His Wash Buffer (50 mM Tris-HCl (pH 8.0), 20 mM imidazole. 300 mM NaCl, 0.1 mM EDTA, and 1 mM PMSF). Then, the bound protein was eluted with 20 × 15 mL of His Elution Buffer (50 mM Tris-HCl (pH 8.0), 200 mM imidazole. 300 mM NaCl, 0.1 mM EDTA, and 1 mM PMSF). Proteins were resolved on sodium dodecyl sulfate polyacrylamide gel electrophoresis (SDS-PAGE) followed by Coomassie Brilliant Blue staining. Tris-HCl, EDTA, and PMSF are from Sigma-Aldrich, St. Louis, MO, USA.

2.2. Generation of Hybridomas

The recombinant PfCSP was shipped to Green Mountain Antibodies, Inc. (Burlington, VT, USA) for the immunization of mice, followed by the fusion to generate monoclonal antibodies. Briefly, mice were primed with 50 μg of PfCSP emulsified with complete Freund's adjuvant, followed by weekly immunization of 50 μg of PfCSP emulsified with TiterMax® (Sigma-Aldrich, St. Louis, MO, USA) and SAS® (Sigma-Aldrich) (alternate week). One week after administering seven doses of immunization, the lymph node was isolated. B cells were purified from using anti-B220 magnetic-activated cell sorting (MACS), and then fused with a mouse myeloma cell line. Cloning was achieved by limiting dilution. After re-cloning, positive clones that secrete immunoglobulin G (IgG) against the full-length PfCSP were selected by enzyme-linked immunosorbent assay (ELISA) (Table 1).

2.3. ELISA Assay

The ELISA plates were first coated with peptides representing PfCSP N-terminal, the central repeat, or C-terminal regions (10 μg/mL) and then blocked with 3% bovine serum albumin (BSA) in phosphate buffered saline with Tween-20 (PBST) (Table 2). MAbs were 10-fold diluted (0.1–1000 ng/mL) and added to the plates and incubated for 1 h. After washing the plates, horseradish peroxidase (HRP)-conjugated goat anti-human IgG Fc Fragment was added. One hour later, tetramethylbenzidine (TMB) High Sensitivity Substrate was added, and ODs were read at 450 nm. Peptide AIAWAKARARQGLEW was used as a negative control. The mAb 2A10 was used as a positive control [19,23].

2.4. Immuno-Fluorescence Assay

1×10^4 salivary gland PfCSP/Py sporozoites were loaded on MP biomedical multi-test glass slides (MP Biomedicals, Santa Ana, CA, USA). PfCSP/Py is an infectious *P. yoelii* parasite bearing a full length of *P. falciparum* circumsporozoite protein (25). After air drying at room temperature, the slides were fixed with 4% paraformaldehyde for 10 min at room temperature, and then blocked with 3% BSA in PBST. The mAbs were two-fold diluted from 1.31 mg/mL to 5 ng/mL, and added to the PfCSP/Py sporozoites-coated wells on the slides for 45 min. After washing with PBS containing 0.05% Tween-20 three times, the slides were incubated with Alexa Fluor 594 conjugate goat anti-mouse IgG (H + L) antibody. One hour later, the slides were washed and mounted in PBS containing 50% glycerol and 1% (*w/v*) *p*-phenylenediamine to reduce bleaching.

2.5. Sporozoite Neutralization Assays

In vitro neutralization assays were conducted by pre-incubating 2×10^4 PfCSP/Py sporozoites with 100 μg mAb on ice for 45 min, and then adding to 1×10^5 Hepa1-6 cells. Forty-two hours post infection, liver stage parasite burden wear measured by quantitative polymerase chain reaction (qPCR) of *P. yoelii* 18S rRNA as previously described [24]. Mouse glyceraldehyde 3-phosphate dehydrogenase (GAPDH) was used as an internal control. In vivo neutralization assays were conducted by pre-incubating 50 PfCSP/Py sporozoites dissected from infected mosquito salivary glands with 5 or 50 μg mAb on ice for 45 min, and then intravenously injecting into BALB/c mouse. The presence of parasite in blood was determined by Giemsa staining of the blood smear of the recipient mouse.

2.6. Giemsa Stain

Starting three days after sporozoite challenge, a drop of blood was collected from the mouse tail vein for thin blood smears on pre-cleaned glass slides. Thin blood smears were fixed with absolute methanol and then stained with diluted Giemsa stain (1:20, v/v) for 20 min. % parasitemia (% of parasitized red blood cells among total red blood cells) were examined with a 100× oil immersion objective under the microscope.

3. Results

3.1. Generation of Hybridomas

PfCSP was expressed and purified from *E. coli.* (Figure 1), and then immunized BALB/c mice. The immune spleen cells from the mice producing anti-PfCSP antibodies were fused with myeloma cells, and six hybridoma cell lines (2D4, 3C1, 3C2, 3D3, 4C1, 4C6) were cloned. The mAbs 4C6 2D4 3D3 were identified as belonging to subclass IgG1. The mAbs 3C2 and 4C1 were isotyped as IgG2b class. The mAb 3C1 belonged to subclass IgG3 (Table 1).

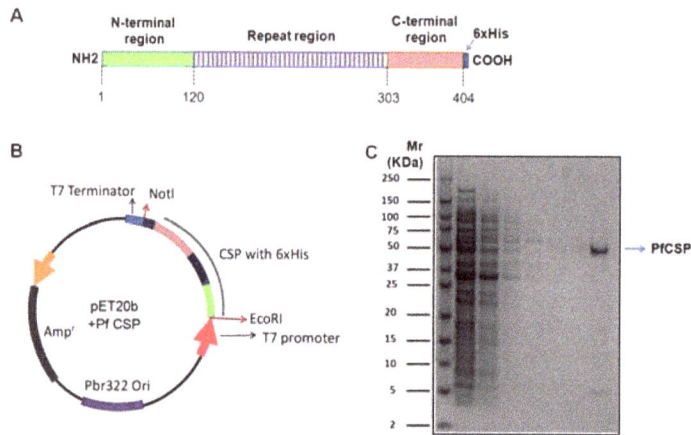

Figure 1. Expression and purification of a recombinant *P. falciparum* circumsporozoite protein (PfCSP). (**A**) Schematic representation of the recombinant PfCSP. Pf CSP coding sequence excluding C-terminal glycosylphosphatidylinositol (GPI) anchor, composed of N-terminal, central repeat, and C-terminal regions, was fused with 6XHis tag at its C-terminus, and cloned into pET20b vector (Stratagene, La Jolla, CA, USA); (**B**) Schematic representation of the PfCSP expression plasmid in this study. The full length of PfCSP without GPI anchor was cloned into pET20b vector between *EcoR* I and *Not* I; (**C**) Expression and purification of a recombinant PfCSP from *E. coli*. The recombinant PfCSP was expressed in BL21 (DE3), and then purified by Ni-Affinity Chromatography. Lane 1, Protein marker; Lane 2, crude extract; Lane 3, flow through; Lane 4–7, washes; Lane 8, elute. Data are representative from three independent experiments.

Table 1. The titers of the PfCSP-specific mAbs *.

Name of the mAb	Titer (IFA)	Titer (ELISA)	Subclass
2A10	40 ng/mL	10 ng/mL	IgG2a
4C6	80 ng/mL	1 μg/mL	IgG1
2D4	80 ng/mL	500 ng/mL	IgG1
3C2	10 ng/mL	1 ng/mL	IgG2b
4C1	328 μg/mL	200 ng/mL	IgG2b
3C1	5 ng/mL	5 ng/mL	IgG3
3D3	10 ng/mL	2 ng/mL	IgG1

* PfCSP: *P. falciparum* circumsporozoite protein; IgG: immunoglobulin G; IFA: immunofluorescence assay; mAb: monoclonal antibody.

3.2. Specificity of Anti-PfCSP mAbs

The specificity of the mAbs has been explored by measuring their reaction with peptides covering PfCSP N-terminal, the central repeat, and C-terminal regions (Table 2). The mAbs 2D4, 4C1, and 4C6 recognized the PfCSP C-terminal region. The mAbs 3C1, 3C2, and 3D3 recognized the PfCSP central repeat region (Figure 2).

Table 2. Synthetic peptides representing PfCSP.

Peptide ID #	Sequence	Position
1	MMRKLAILSVSSFLF	N-terminus
2	SSFLFVEALFQEYQC	N-terminus
3	QEYQCYGSSSNTRVL	N-terminus
4	NTRVLNELNYDNAGT	N-terminus
5	DNAGTNLYNELEMNY	N-terminus
6	LEMNYYGKQENWYSL	N-terminus
7	NWYSLKKNSRSLGEN	N-terminus
8	SLGENDDGNNEDNEK	N-terminus
9	EDNEKLRKPKHKKLK	N-terminus
10	HKKLKQPADGNPDP	N-terminus
11	NANPNVDPNANPNVD	Repeats
12	NPNVDPNANPNVDPN	Repeats
13	NVDPNANPNANPNAN	Repeats
14	NPNANPNANPNANPN	Repeats
15	NANPNANPNANPNAN	Repeats
16	NANPNANPNANPNVD	Repeats
17	NPNVDPNANPNANPN	Repeats
18	NANPNANPNKNNQGN	Repeats
19	NNQGNGQGHNMPNDP	C-terminus
20	MPNDPNRNVDENANA	C-terminus
21	ENANANSAVKNNNNE	C-terminus
22	NNNNEEPSDKHIKEY	C-terminus
23	HIKEYLNKIQNSLST	C-terminus
24	NSLSTEWSPCSVTCG	C-terminus
25	SVTCGNGIQVRIKPG	C-terminus
26	RIKPGSANKPKDELD	C-terminus
27	KDELDYANDIEKKIC	C-terminus
28	EKKICKMEKCSSVFN	C-terminus
29	SSVFNVVNSSIGLIM	C-terminus
30	IGLIMVLSFLFLN	C-terminus
31	AIAWAKARARQGLEW	Negative Control Peptide

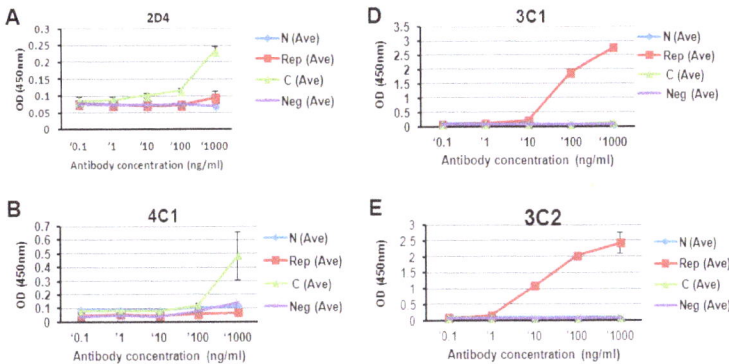

Figure 2. *Cont.*

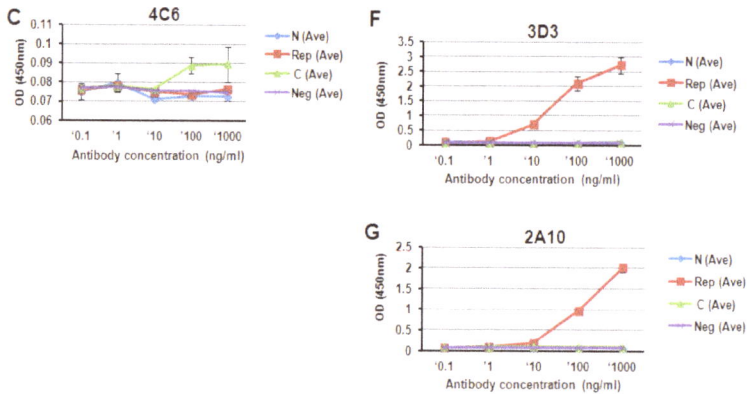

Figure 2. Specificity of anti-PfCSP monoclonal antibodies (mAbs) by enzyme-linked immunosorbent assay (ELISA). Peptides representing PfCSP N-terminal, central repeat, and C-terminal regions were used to evaluate specificity of anti-PfCSP mAbs (Table 1). (**A**), 2D4; (**B**), 4C1; (**C**),4C6; (**D**), 3C1; (**E**), 3C2; (**F**), 3D3; (**G**), 2A10. The mAb 2A10 was used as a positive control. ELISA was performed in duplicate. Data are representative of three independent experiments. OD, Optical density.

3.3. Titration of the PfCSP-Specific mAbs

The antibody titer was tested by enzyme-linked immunosorbent assay (ELISA) and immunofluorescence assay (IFA) (Table 1 and Figure 3). ELASA using peptides covering PfCSP showed that the titers of mAbs recognizing the PfCSP central repeat region were higher than those recognizing the PfCSP C-terminal region. The titers of the three mAbs recognizing the PfCSP central repeat were higher than the control 2A10 (3C2 > 3D3 > 3C1 > 2A10). IFA using the *Plasmodium* sporozoites expressing PfCSP [25] also showed that the titers of the mAbs recognizing the PfCSP central repeat were higher than those recognizing the PfCSP C-terminal region. The titer of the three mAbs recognizing the PfCSP central repeat were higher than the control 2A10 (3C1 > 3D3 = 3C2 > 2A10).

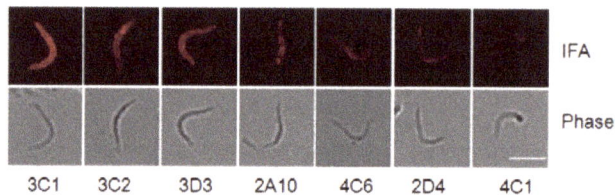

Figure 3. Immunofluorescence assays. PfCSP/Py (a *P. yoelii* parasite bearing *P. falciparum* circumsporozoite protein) salivary gland sporozoites [25] were incubated with 160 ng/mL mAbs, except 4C1 at 328 μg/mL, followed by incubation with Alexa Fluor 594 goat anti-mouse IgG (H + L) *antibody.*

3.4. Protection of the PfCSP mAbs against PfCSP/Py Sprozoite Challenge

We then examined the protection of the PfCSP mAbs against malaria sporozoite challenge in vitro and in vivo. For the malaria sporozoite challenges, we used the highly infectious hybrid PfCSP/Py sporozoite, which is based on rodent *P. yoelii* parasite and its CSP is replaced by the full-length of CSP from *P. falciparum* [25]. We found that mAb 3C1, 3C2, and 3D3 significantly inhibited the parasite development in Hepa 1–6 cells compared with 2A10, which is an effective mouse mAb specific for the PfCSP central repeat [19,23] (Figure 4). This was in agreement with the in vivo neutralization assay (Table 3 and Figure 5). Fifty μg of 3C1, 3C2, and 3D3 completely protected the mice from PfCSP/Py sporozoite challenge. The protective effect of 3C1, 3C2, and 3D3 were better than the previously generated mAb 2A10. Even 5 μg of 3C1 partially protected the challenged mice compared to the mAb 2A10.

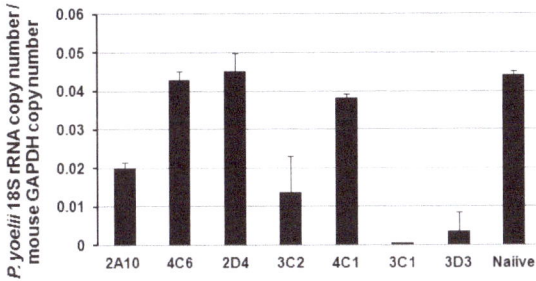

Figure 4. In vitro neutralization assay. 2×10^4 PfCSP/Py sporozoites were pre-incubated with 100 μg of each mAb, and then added to Hepa1-6 cells. Forty-two hours post infection, liver stage parasite burden wear measured by *P. yoelii* 18S rRNA/mouse glyceraldehyde 3-phosphate dehydrogenase (GAPDH). Naive mouse serum was used as control. The in vitro neutralization assay was performed in triplicate. Data are representative of two independent experiments.

Table 3. In vivo neutralization assay.

Amount of mAb	5 μg				50 μg			
Days Post Challenge	Day 3	Day 4	Day 5	Day 14	Day 3	Day 4	Day 5	Day 14
2A10	0 *,#	4	5	5	0	0	3	3
4C6	0	5	5	5	0	1	4	4
2D4	0	4	5	5	0	0	3	3
3C2	0	1	5	5	0	0	0	0
4C1	0	5	5	5	0	3	4	4
3C1	0	1	4	4	0	0	0	0
3D3	0	2	5	5	0	0	0	0
Naiive	0	5	5	5	0	5	5	5

* Five mice per group. # The number of infected mice.

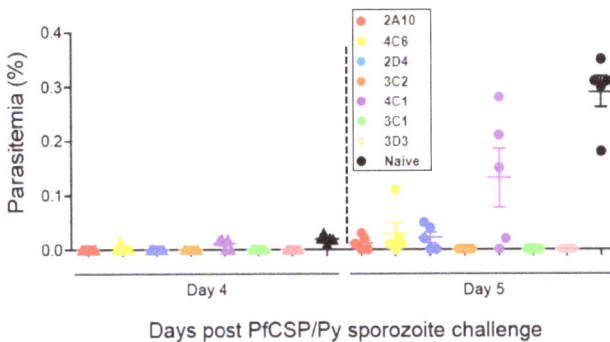

Figure 5. Parasitemia of mice in the neutralization assay. Fifty PfCSP/Py sporozoites were incubated with 50 μg mAbs followed by i.v. injection into BALB/c mice (five mice per group). Parasitemia were counted by Giemsa stains of mouse tail blood followed by microscopy. Data are parasitemia of the mice four and five days post challenge. Naive mice were i.v. injected with 50 PfCSP/Py sporozoites as positive controls.

4. Discussion

The CSP consists of the N-terminal flanking region, the central region that contains repetitive immunodominant B-cell epitopes, and the C-terminal flanking region that contains multiple T-cell epitopes. The N-terminus of the CSP is proteolytically processed during the sporozoite invasion into host cells [15,16]. This may explain why we did not obtain specific antibodies against the N-terminus. The abundant NANP repeats present within the central region are likely to contribute to the high neutralization efficacies of the mAbs against PfCSP central repeat region, as previously published [26,27]. In fact, mAbs, which recognize the PfCSP central repeat region, have been shown to exert a potent neutralization activity against the sporozoites [26,27]. All the mAbs 3C1, 3C2, 3D3, and 2A10, recognize the central repeat region of the PfCSP. Fifty μg of the novel mAbs 3C1, 3C2, and 3D3 completely protected the mice from PfCSP/Py sporozoites challenge; while the reference mAb 2A10 only partially protected the mice. A likely explanation is that the native structure of the central repeats of the CSP is not in a random coil state, and the repeat region is predicted to form a rod-like structure [28]. It is speculated that these mAbs recognize structurally different epitopes coded by NANP repeat, resulting in different protection efficacies. The titers (3C1 > 3D3 = 3C2 > 2A10) of these novel mAbs determined by IFA using a whole malaria parasite (sporozoite), as an antigen, corroborate their protection efficacies in vitro (3C1 > 3D3 > 3C2 > 2A10), as well as in mice (3C1 > 3D3 = 3C2 > 2A10). These indicate that mAbs having higher titers against the native from of the CSP expressed by sporozoites exert higher protection efficacies.

Synthesized peptides and sporozoites were used in ELISA and IFA, respectively, to determine the antibody titers. Sporozoites express a native form of the PfCSP, whereas synthesized peptides represent the primary structure of PfCSP. B-cell epitopes are typically classified as either linear epitopes or conformational epitopes, which constitute the spatially folded amino acids and lie far away in the primary sequence. The difference seen by ELISA and IFA may reflect the structural properties of unique B-cell epitopes recognized by our mAbs.

Over the past few years, there has been growing interest in use of vectored immunoprophylaxis to protect hosts from HIV. Vectored immunoprophylaxis is based on adeno-associated virus (AAV) as a vehicle for generating the existing anti-HIV neutralizing antibodies in humans [29,30]. Recently vectored immunoprophylaxis has been utilized for other diseases including malaria and colorectal cancer [12,31]. This new tool requires potent neutralizing antibodies. Although human monoclonal antibodies against PfCSP have been generated, only one mouse mAb against PfCSP, 2A10, has been used as a gold standard mAb for more than three decades. It is noteworthy that a few new mouse mAbs against PfCSP, which we generated in this study, are found to be more potent than 2A10. Therefore, we believe it is important to assess the characteristics of these newly generated mAbs before humanizing them for the purpose of clinical applications, such as a vectored immunoprophylaxis, in the future. Moreover, the mouse mAbs generated in this study are useful tools for the study of PfCSP in a mouse model.

5. Conclusions

In summary, here we report a generation of novel mAbs specific against the CSP from *P. falciparum*. The mAbs 2D4, 4C1, and 4C6 recognize the C-terminal region of PfCSP. The mAbs 3C1, 3C2, and 3D3 recognize the central repeat region of PfCSP, and their titers and protection efficacies are higher than 2A10, which has been widely used as a gold standard antibody against PfCSP.

Acknowledgments: This work was supported by grants from NIH AI073658 and AI081510 (both to M.T.).

Author Contributions: M.T. conceived and designed the experiments; M.Z., R.M., U.R., and T.S. performed the experiments and analyzed the data; M.Z. and M.T. wrote the paper.

Conflicts of Interest: The authors declare no conflict of interest.

References

1. World Health Organization. *World Malaria Report*; World Health Organization: Geneva, Switzerland, 2015.
2. Cheah, P.Y.; White, N.J. Antimalarial mass drug administration: Ethical considerations. *Int. Health* **2016**, *8*, 235–238. [CrossRef] [PubMed]
3. Ochomo, E.; Chahilu, M.; Cook, J.; Kinyari, T.; Bayoh, N.M.; West, P.; Kamau, L.; Osangale, A.; Ombok, M.; Njagi, K.; et al. Insecticide-treated nets and protection against insecticide-resistant malaria vectors in western Kenya. *Emerg. Infect. Dis.* **2017**, *23*, 758–764. [CrossRef] [PubMed]
4. Mashauri, F.M.; Manjurano, A.; Kinung'hi, S.; Martine, J.; Lyimo, E.; Kishamawe, C.; Ndege, C.; Ramsan, M.M.; Chan, A.; Mwalimu, C.D.; et al. Indoor residual spraying with micro-encapsulated pirimiphos-methyl (Actellic (R) 300CS) against malaria vectors in the Lake Victoria basin, Tanzania. *PLoS ONE* **2017**, *12*, e0176982. [CrossRef] [PubMed]
5. Matuschewski, K. Vaccines against malaria-still a long way to go. *FEBS J.* **2017**. [CrossRef] [PubMed]
6. Alout, H.; Labbe, P.; Chandre, F.; Cohuet, A. Malaria vector control still matters despite insecticide resistance. *Trends Parasitol.* **2017**, *33*, 610–618. [CrossRef] [PubMed]
7. Antony, H.A.; Parija, S.C. Antimalarial drug resistance: An overview. *Trop. Parasitol.* **2016**, *6*, 30–41. [PubMed]
8. Barik, T.K. Antimalarial drug: From its development to deface. *Curr. Drug Discov. Technol.* **2015**, *12*, 225–228. [CrossRef] [PubMed]
9. Rts, S.C.T.P. Efficacy and safety of RTS,S/AS01 malaria vaccine with or without a booster dose in infants and children in Africa: Final results of a phase 3, individually randomised, controlled trial. *Lancet* **2015**, *386*, 31–45.
10. Agnandji, S.T.; Lell, B.; Fernandes, J.F.; Abossolo, B.P.; Methogo, B.G.; Kabwende, A.L.; Adegnika, A.A.; Mordmuller, B.; Issifou, S.; Kremsner, P.G.; et al. A phase 3 trial of RTS,S/AS01 malaria vaccine in African infants. *N. Engl. J. Med.* **2012**, *367*, 2284–2295. [PubMed]
11. Kazmin, D.; Nakaya, H.I.; Lee, E.K.; Johnson, M.J.; van der Most, R.; van den Berg, R.A.; Ballou, W.R.; Jongert, E.; Wille-Reece, U.; Ockenhouse, C.; et al. Systems analysis of protective immune responses to RTS,S malaria vaccination in humans. *Proc. Natl. Acad. Sci. USA* **2017**. [CrossRef] [PubMed]
12. Deal, C.; Balazs, A.B.; Espinosa, D.A.; Zavala, F.; Baltimore, D.; Ketner, G. Vectored antibody gene delivery protects against *Plasmodium falciparum* sporozoite challenge in mice. *Proc. Natl. Acad. Sci. USA* **2014**, *111*, 12528–12532. [CrossRef] [PubMed]
13. Kumar, K.A.; Sano, G.; Boscardin, S.; Nussenzweig, R.S.; Nussenzweig, M.C.; Zavala, F.; Nussenzweig, V. The circumsporozoite protein is an immunodominant protective antigen in irradiated sporozoites. *Nature* **2006**, *444*, 937–940. [CrossRef] [PubMed]
14. Ferguson, D.J.; Balaban, A.E.; Patzewitz, E.M.; Wall, R.J.; Hopp, C.S.; Poulin, B.; Mohmmed, A.; Malhotra, P.; Coppi, A.; Sinnis, P.; et al. The repeat region of the circumsporozoite protein is critical for sporozoite formation and maturation in *Plasmodium*. *PLoS ONE* **2014**, *9*, e113923. [CrossRef] [PubMed]
15. Coppi, A.; Pinzon-Ortiz, C.; Hutter, C.; Sinnis, P. The *Plasmodium* circumsporozoite protein is proteolytically processed during cell invasion. *J. Exp. Med.* **2005**, *201*, 27–33. [CrossRef] [PubMed]
16. Coppi, A.; Natarajan, R.; Pradel, G.; Bennett, B.L.; James, E.R.; Roggero, M.A.; Corradin, G.; Persson, C.; Tewari, R.; Sinnis, P. The malaria circumsporozoite protein has two functional domains, each with distinct roles as sporozoites journey from mosquito to mammalian host. *J. Exp. Med.* **2011**, *208*, 341–356. [CrossRef] [PubMed]
17. Nardin, E.H.; Oliveira, G.A.; Calvo-Calle, J.M.; Castro, Z.R.; Nussenzweig, R.S.; Schmeckpeper, B.; Hall, B.F.; Diggs, C.; Bodison, S.; Edelman, R. Synthetic malaria peptide vaccine elicits high levels of antibodies in vaccinees of defined HLA genotypes. *J. Infect. Dis.* **2000**, *182*, 1486–1496. [CrossRef] [PubMed]
18. Stoute, J.A.; Slaoui, M.; Heppner, D.G.; Momin, P.; Kester, K.E.; Desmons, P.; Wellde, B.T.; Garcon, N.; Krzych, U.; Marchand, M. A preliminary evaluation of a recombinant circumsporozoite protein vaccine against *Plasmodium falciparum* malaria. *N. Engl. J. Med.* **1997**, *336*, 86–91. [CrossRef] [PubMed]
19. Anker, R.; Zavala, F.; Pollok, B.A. VH and VL region structure of antibodies that recognize the (NANP)3 dodecapeptide sequence in the circumsporozoite protein of *Plasmodium falciparum*. *Eur. J. Immunol.* **1990**, *20*, 2757–2761. [CrossRef] [PubMed]

20. Charoenvit, Y.; Mellouk, S.; Cole, C.; Bechara, R.; Leef, M.F.; Sedegah, M.; Yuan, L.F.; Robey, F.A.; Beaudoin, R.L.; Hoffman, S.L. Monoclonal, but not polyclonal, antibodies protect against Plasmodium yoelii sporozoites. *J. Immunol.* **1991**, *146*, 1020–1025. [PubMed]

21. Persson, C.; Oliveira, G.A.; Sultan, A.A.; Bhanot, P.; Nussenzweig, V.; Nardin, E. Cutting edge: A new tool to evaluate human pre-erythrocytic malaria vaccines: Rodent parasites bearing a hybrid *Plasmodium falciparum* circumsporozoite protein. *J. Immunol.* **2002**, *169*, 6681–6685. [CrossRef] [PubMed]

22. Foquet, L.; Hermsen, C.C.; van Gemert, G.J.; Van Braeckel, E.; Weening, K.E.; Sauerwein, R.; Meuleman, P.; Leroux-Roels, G. Vaccine-induced monoclonal antibodies targeting circumsporozoite protein prevent *Plasmodium falciparum* infection. *J. Clin. Investig.* **2014**, *124*, 140–144. [CrossRef] [PubMed]

23. Zavala, F.; Cochrane, A.H.; Nardin, E.H.; Nussenzweig, R.S.; Nussenzweig, V. Circumsporozoite proteins of malaria parasites contain a single immunodominant region with two or more identical epitopes. *J. Exp. Med.* **1983**, *157*, 1947–1957. [CrossRef] [PubMed]

24. Bruna-Romero, O.; Hafalla, J.C.R.; Gonzalez-Aseguinolaza, G.; Sano, G.; Tsuji, M.; Zavala, F. Detection of malaria liver-stages in mice infected through the bite of a single Anopheles mosquito using a highly sensitive real-time PCR. *Int. J. Parasitol.* **2001**, *31*, 1499–1502. [CrossRef]

25. Zhang, M.; Kaneko, I.; Tsao, T.; Mitchell, R.; Nardin, E.H.; Iwanaga, S.; Yuda, M.; Tsuji, M. A highly infectious *Plasmodium* yoelii parasite, bearing *Plasmodium falciparum* circumsporozoite protein. *Malar. J.* **2016**, *15*, 201. [CrossRef] [PubMed]

26. Potocnjak, P.; Yoshida, N.; Nussenzweig, R.S.; Nussenzweig, V. Monovalent fragments (Fab) of monoclonal antibodies to a sporozoite surface antigen (Pb44) protect mice against malarial infection. *J. Exp. Med.* **1980**, *151*, 1504–1513. [CrossRef] [PubMed]

27. Gysin, J.; Barnwell, J.; Schlesinger, D.H.; Nussenzweig, V.; Nussenzweig, R.S. Neutralization of the infectivity of sporozoites of *Plasmodium* knowlesi by antibodies to a synthetic peptide. *J. Exp. Med.* **1984**, *160*, 935–940. [CrossRef] [PubMed]

28. Plassmeyer, M.L.; Reiter, K.; Shimp, R.L., Jr.; Kotova, S.; Smith, P.D.; Hurt, D.E.; House, B.; Zou, X.; Zhang, Y.; Hickman, M.; et al. Structure of the *Plasmodium falciparum* circumsporozoite protein, a leading malaria vaccine candidate. *J. Biol. Chem.* **2009**, *284*, 26951–26963. [CrossRef] [PubMed]

29. Balazs, A.B.; Ouyang, Y.; Hong, C.M.; Chen, J.; Nguyen, S.M.; Rao, D.S.; An, D.S.; Baltimore, D. Vectored immunoprophylaxis protects humanized mice from mucosal HIV transmission. *Nat. Med.* **2014**, *20*, 296–300. [CrossRef] [PubMed]

30. Balazs, A.B.; Chen, J.; Hong, C.M.; Rao, D.S.; Yang, L.; Baltimore, D. Antibody-based protection against HIV infection by vectored immunoprophylaxis. *Nature* **2012**, *481*, 81–84. [CrossRef] [PubMed]

31. Hu, S.; Dai, H.; Li, T.; Tang, Y.; Fu, W.; Yuan, Q.; Wang, F.; Lv, G.; Lv, Y.; Fan, X.; et al. Broad RTK-targeted therapy overcomes molecular heterogeneity-driven resistance to cetuximab via vectored immunoprophylaxis in colorectal cancer. *Cancer Lett.* **2016**, *382*, 32–43. [CrossRef] [PubMed]

antibodies

MDPI

Article

Functional, Biophysical, and Structural Characterization of Human IgG1 and IgG4 Fc Variants with Ablated Immune Functionality

Susan H. Tam *, Stephen G. McCarthy *, Anthony A. Armstrong, Sandeep Somani, Sheng-Jiun Wu, Xuesong Liu, Alexis Gervais, Robin Ernst, Dorina Saro, Rose Decker, Jinquan Luo, Gary L. Gilliland, Mark L. Chiu * and Bernard J. Scallon

Janssen Research & Development, LLC, 1400 McKean Road, Spring House, Ambler, PA 19477, USA; aarmst12@its.jnj.com (A.A.A.); ssomani@its.jnj.com (S.S.); swu4@its.jnj.com (S.-J. W.); xliu8@its.jnj.com (X.L.); agervais@its.jnj.com (A.G.); rernst2@its.jnj.com (R.E.); dsaro@its.jnj.com (D.S.); rdecker5@its.jnj.com (R.D.); jluo@its.jnj.com (J.L.); ggillila@its.jnj.com (G.L.G.); bhscallon@gmail.com (B.J.S.)

* Correspondence: susanhtam@gmail.com (S.H.T.); sgmcc0711@gmail.com (S.G.M.); mchiu@its.jnj.com (M.L.C.); Tel.: +1-610-299-7994 (S.H.T); +1-610-836-2569 (S.G.M.); +1-610-651-6862 (M.L.C.)

Received: 19 July 2017; Accepted: 21 August 2017; Published: 1 September 2017

Abstract: Engineering of fragment crystallizable (Fc) domains of therapeutic immunoglobulin (IgG) antibodies to eliminate their immune effector functions while retaining other Fc characteristics has numerous applications, including blocking antigens on Fc gamma (Fcγ) receptor-expressing immune cells. We previously reported on a human IgG2 variant termed IgG2σ with barely detectable activity in antibody-dependent cellular cytotoxicity, phagocytosis, complement activity, and Fcγ receptor binding assays. Here, we extend that work to IgG1 and IgG4 antibodies, alternative subtypes which may offer advantages over IgG2 antibodies. In several in vitro and in vivo assays, the IgG1σ and IgG4σ variants showed equal or even lower Fc-related activities than the corresponding IgG2σ variant. In particular, IgG1σ and IgG4σ variants demonstrate complete lack of effector function as measured by antibody-dependent cellular cytotoxicity, complement-dependent cytotoxicity, antibody-dependent cellular phagocytosis, and in vivo T-cell activation. The IgG1σ and IgG4σ variants showed acceptable solubility and stability, and typical human IgG1 pharmacokinetic profiles in human FcRn-transgenic mice and cynomolgus monkeys. In silico T-cell epitope analyses predict a lack of immunogenicity in humans. Finally, crystal structures and simulations of the IgG1σ and IgG4σ Fc domains can explain the lack of Fc-mediated immune functions. These variants show promise for use in those therapeutic antibodies and Fc fusions for which the Fc domain should be immunologically "silent".

Keywords: Fc engineering; silent effector function; IgG1; IgG2; IgG4; IgG sigma; developability; pharmacokinetics; crystal structure

1. Introduction

The biological functionality of therapeutic antibodies depends on the interactions of two regions of the protein with components of its external environment: the antigen binding region (Fab) interacting with an antigen, and the fragment crystallizable (Fc) region interacting with components of the immune system. The Fc region of the immunoglobulin (IgG) antibody, which is the focus of this study, can have interactions with Fc gamma receptors (FcγR) and first subcomponent of the C1 complex (C1q) that mediate antibody-dependent cellular cytotoxicity (ADCC), complement-dependent cytotoxicity (CDC), antibody-dependent cellular phagocytosis (ADCP), induction of secretion of mediators, endocytosis of opsonized particles, as well as modulation of tissue and serum half-life through interaction with the neonatal Fc receptor (FcRn) [1,2]. Numerous publications have reviewed the application of enhanced

Fc effector function to increase biologic activity [3–6]. In addition, the coupling of the Fab and Fc regions can impact the therapeutic window for safety and efficacy of antibodies and Fc fusion proteins [3,7].

Fc-mediated effector functions are best avoided for some applications such as systemic neutralization of cytokines, targeting cell surface antigens on immune cells, or when engineering bispecific molecules to bring target diseased cells within proximity of effector immune cells to provide a more specific immune receptor engagement [3,8,9]. In each of these cases, it is best not to stimulate unwanted cell and tissue damage or risk undesired effector cell activation, immune cell depletion, or FcγR cross-linking that might induce cytokine release through engagement of Fc-mediated effector functions [3]. An important consideration in such biological processes is that the complexity of FcγR functional properties is increased by the varying densities of activating and inhibitory receptors on the different effector cell populations [10]. Likewise, since the threshold of activation can be variable with different patients, it would be prudent for safety considerations to develop antibodies with a more silent Fc framework. Thus, development of completely silent Fc domains can be critical for biologics that do not require FcγR or C1q mediated effector functions [11].

When an antibody with no effector function is required, there are different approaches one may take to generate a molecule with the desired properties. Unfortunately, results of some strategies often come with liabilities to the molecular profile. For instance, Fab or F(ab′)$_2$ fragments can be generated; however, such molecules have shorter half-lives in patient sera. Chemical modifications can extend the half-life of such molecules, but can also bring potential risks with toxicities [12]. Another strategy has been to eliminate the N-linked glycosylation at residue Asparagine 297 (European Union (EU) numbering) [13–16]; however, this can reduce antibody solubility and stability. Another approach employs mutagenesis of specific Fc amino acid residues to specifically influence effector functions [17].

An example of this approach is illustrated in the first marketed therapeutic antibody (Orthoclone OKT3, a murine IgG2a) in which two mutations in the lower hinge (L324A/L235A, referred to as AA) were introduced to mitigate the induction of cytokine storm [18,19]. Also, because FcγRs are highly selective in subclass specificity and affinity [20,21], another approach may be to move the Fab domains onto Fc regions which elicits less effector function such as human IgG2 (huIgG2) or IgG4 (huIgG4) [22]. In addition, swapping among the sequences of the four human IgG (huIgG) subtypes has been used to design more silent Fc domains [3,4,23–25] that have resulted in such variants as huIgG2/4 [26], huIgG2m4 [27], and L234F/L235E/P331S (FES) [28]. Notably, Vafa and co-workers employed multiple strategies to develop a huIgG2 variant, termed huIgG2 sigma (IgG2σ) that showed undetectable Fc-mediated effector function and C1q binding [29]. In utilizing such strategies for silencing Fc effector function, it needs to be recognized that there is some potential for huIgG2 subtype molecules to form heterogeneous isoforms which can be a challenge in the generation of a homogeneous product [30–32].

Although huIgG4 has weak binding affinity to most FcγRs except for the high affinity receptor FcγRI, it does retain the ability to induce phagocytosis by macrophages (expressing FcγRI, FcγRIIa, and FcγRIIIa) and possibly activate monocytes when in an immune complex due to activating FcγRs on specific immune cells. Recent additional approaches to generate antibodies with no effector function have included disruption of proline sandwich motifs [33], and incorporation of asymmetric charged mutations in the lower hinge or constant heavy chain domain 2 (C$_H$2) domain [34]. Because development of antibodies with silent Fc domains continues to be important for various therapies, and because the threshold of activation may be different for each patient or disease population, efforts are on-going to obtain the most silent Fc variants which will have improved safety and good manufacturing qualities.

We describe here the functional and structural characteristics of three novel silent Fc designs: huIgG1 sigma (IgG1σ), which is a variant of huIgG1, and huIgG4 sigma1 (IgG4σ1) and huIgG4 sigma2 (IgG4σ2), which are variants of huIgG4. The effector functions of these silent Fc variants are compared to those of previously described constructs such as huIgG1 L234A/L235A (AA) [25], huIgG4 S228P/L234A/L235A (PAA) [19,35], and huIgG1 L234F/L235E/P331S (FES, a triple mutant being employed in a clinical anti-interferon receptor antibody [36]). We also present a comparison of serum half-lives in mice and cynomolgus monkeys, an evaluation of potential immunogenicity, and an assessment of biophysical stability. Crystal structures and molecular modeling were carried out to understand the mechanism for lack of interactions between the IgG variants and the Fc gamma receptors. The aim of the studies presented here is to provide data on the biological, biophysical, and structural properties of the huIgG1σ, huIgG4σ1, and huIgG4σ2 along with other commonly used silent Fc formats to enable development of better quality antibody therapeutics.

2. Results

2.1. Design of Silent Fc HuIgGs

Silent human IgG Fcs were designed by uniquely combining the sequence from previously characterized variants having mutations in the hinge and C_H2 regions [29]. Mutations present in these variants are listed in Figure 1 and Table 1. All sequences described follow the EU numbering system [37,38].

Both huIgG1σ and the huIgG2σ constructs contain a total of seven mutations in their hinge and C_H2 domains (Figures 1 and 2; Table 1). However, only six of the mutations are common to both huIgG2σ and the huIgG1σ variant described here. HuIgG1σ uniquely includes an L234A mutation, and huIgG2σ has a V309L mutation. HuIgG1 naturally has a leucine at that position.

Compared to huIgG4 PAA, huIgG4σ1 and huIgG4σ2 have two and three additional mutations, respectively, for a total of five and six mutations (Table 1). While modeled after huIgG2σ, these huIgG4 variants do not have the full set of sigma mutations present in huIgG2σ because the huIgG4 sequence naturally contains four of those residues.

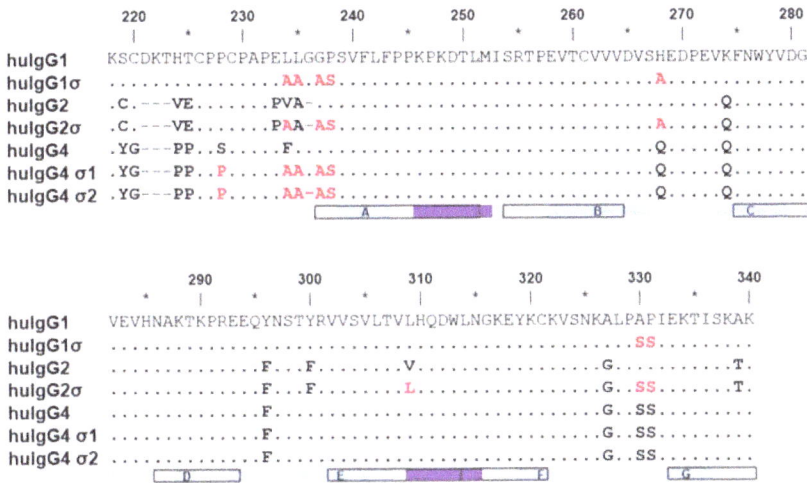

Figure 1. Alignment of hinge and constant heavy chain domain 2 (C_H2) domain amino acid sequences of wild-type human immunoglobulin G1 (IgG1), IgG2 and IgG4 as well as their sigma variants. The alignment above uses EU numbering. Residues identical to wild-type IgG1 are indicated as dots; gaps are indicated with hyphens. Sequence is given explicitly if it differs from wild-type IgG1 or from the parental subtype for σ variants. In the latter case, sequence is colored red. Open boxes beneath the alignment correspond to International Immunogenetics Information System (IMGT) strand definitions (labeled) [39]. Light blue and purple boxes beneath the alignment correspond to the strand and helix secondary structure assignment for wild-type IgG1 in the Protein Data Bank (PDB) 3AVE (chain A) [40]. Residues 267–273 form the BC loop and 322–332 form the FG loop.

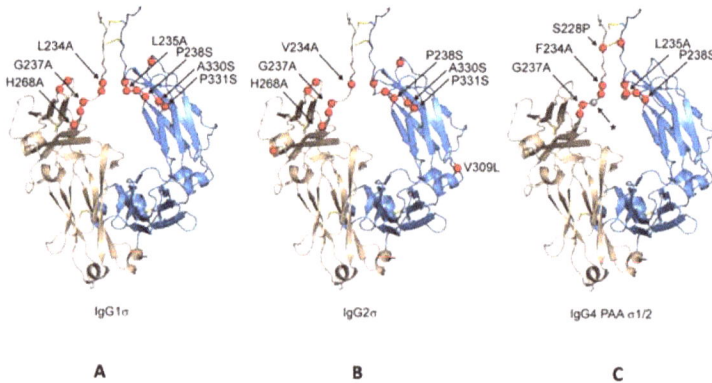

A   **B**   **C**

Figure 2. Schematic indicating positions of mutations on generic Fc structure. Positions mutated relative to parental subtype are depicted as red spheres on a structure of IgG1 Fc having an idealized hinge (wheat and blue, cartoon) for IgG1σ (**A**), IgG2σ (**B**), and IgG4σ1/2 (**C**). For clarity, the sites of mutation are labeled on only one of two equivalent Fc chains. IgG4σ2 differs from IgG4σ1 by the deletion of G236 (position indicated with a small, grey sphere and labeled with an asterisk).

Table 1. List of mutations in the hinge and constant heavy chain domain 2 (C_H2) regions of different silent immunoglobulin (IgG) designs.

	IgG1 AA	IgG1σ	IgG1 FES	IgG2σ	IgG4 PAA	IgG4σ1	IgG4σ2
Core hinge					S228P	S228P	S228P
Lower hinge	L234A	L234A	L234F	V234A	F234A	F234A	F234A
	L235A	L235A	L235E		L235A	L235A	L235A
							ΔG236
		G237A		G237A		G237A	G237A
		P238S		P238S		P238S	P238S
C_H2 domain of Fc		H268A		H268A			
				V309L			
		A330S		A330S			
		P331S	P331S	P331S			

Δ Indicates deletion of residue. European Union (EU) numbering is used for the amino acid substitutions.

2.2. FcγR Binding and Competition

Since effector functions such as ADCC and ADCP are induced by the binding of antibody Fc with the various FcγRs on the surface of different immune cell types, it is crucial to minimize and if possible to eliminate this engagement. Testing for binding of the assorted Fc variants to human FcγRs was determined using an AlphaScreen competitive binding assay. The AlphaScreen competition strategy involves the following: biotin-labeled huIgG is captured on streptavidin donor beads; His-tagged FcγRs are captured on nickel acceptor beads; unlabeled competitor (test) antibodies are applied as serial dilutions. There is a reduction in (maximal) signal when competition takes place. To provide an avidity component comparable to that created by multiple antibodies bound to a cell surface, the test antibodies were cross-linked with anti-F(ab′)$_2$. This method increases the sensitivity of the assay to aid in confirming low-level binding.

Competitive binding analyses of Ab1 (anti-tumor necrosis factor (TNF)α huIgG1) samples to the different FcγRs are shown in Figure 3. On CD64 receptor (FcγRI), Ab1 huIgG1 wild-type (WT) binds well with 50% inhibition at approximately 1 µg/mL. HuIgG1 AA and huIgG1 FES show low level binding at 500 µg/mL, the highest concentration tested. However, the huIgG1σ, huIgG4σ1, and huIgG4σ2 variants show no indications of binding, even at concentrations up to

1 mg/mL (Figure 3A). On CD32a (huFcγRIIa-H131 high and huFcγRIIa-R131 low affinity variants), CD32b (huFcγRIIb), and CD16a (huFcγRIIIa), the huIgG1 AA and huIgG1 FES variants showed varying degrees of binding to these receptors, yet the sigma (σ) variants showed complete lack of binding compared to the huIgG1 wild-type (WT) (Figure 3B–E). HuIgG4 PAA was not included in this test panel due to reagent constraints, but huIgG4 PAA has been reported to have reduced, but not eliminated binding to FcγRs [29,33]. These collective data demonstrated that complexes formed from the IgG1σ and IgG4σ variants did not bind to FcγRs even at high concentrations.

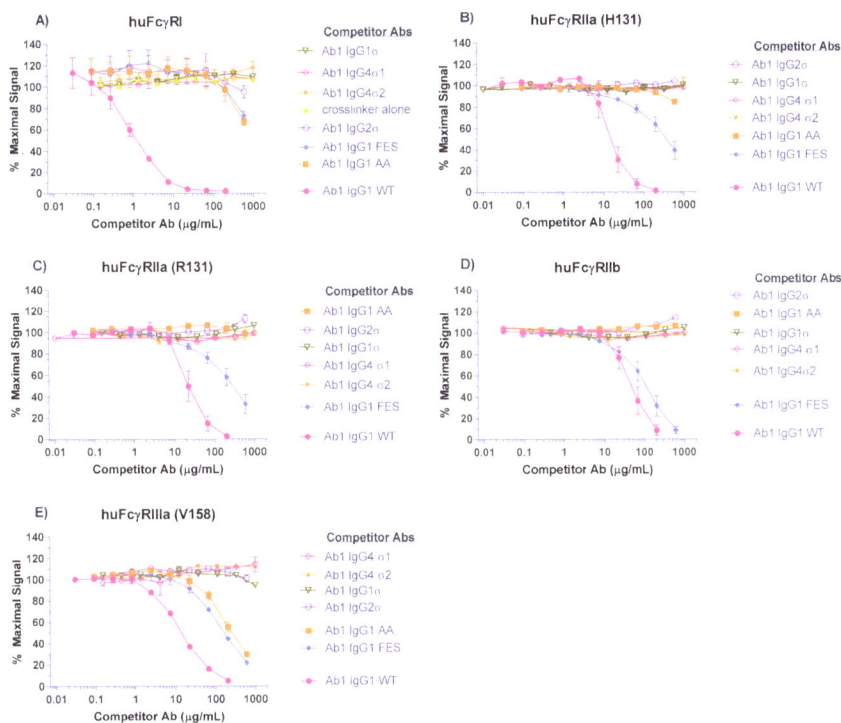

Figure 3. Testing for interactions of cross-linked huIgG variants with Fcγ receptors. Binding of huIgG Ab1 (anti-TNFα) molecules to human FcγRs was assessed using AlphaScreen bead assays in a competitive format. To increase sensitivity of the assays, test samples were cross-linked to introduce binding avidity by using a goat F(ab′)₂ anti-huIgG F(ab′)₂-specific fragment in 1:1 molar ratio with the test antibodies. Cross-linked test antibodies at the designated concentrations were co-incubated with biotin-labeled huIgG Fc fragment (to avoid binding to cross-linker), the respective His-tagged FcγRs, nickel chelate acceptor beads, and streptavidin donor beads. Plates were read on the EnVision multi-label plate reader, and data plotted with GraphPad Prism v6.0 software. Shown are binding of cross-linked huIgG variants to: (**A**) huFcγRI, (**B**) huFcγRIIa-H131, high affinity allotype, (**C**) huFcγRIIa-R131, low affinity allotype, (**D**) huFcγRIIb, (**E**) huFcγRIIIa-V158, high affinity allotype. Non-binding of cross-linker alone is shown with the high affinity huFcγRI. All points represent the mean of duplicate samples ± range. The plot labels refer to IgG1 AA—human IgG1 L234A/L235A; IgG1 FES—human IgG1 L234F/L235E/P331S; and IgG1 WT—wild type human IgG1.

2.3. Target Mediated Effector Function In Vitro

2.3.1. Antibody-Dependent Cell-Mediated Cytotoxicity (ADCC)

When the Fc region of an antibody binds to FcγR on immune cells, cytotoxic factors are released causing ADCC and death of target cells, whose membrane-surface antigens have been bound by the specific antibody. To compare cytotoxic potential of the silent Fc variants, K2 cells (T72-18, TNFα SP2/0) expressing the Δ1–12 variant of human TNFα were used as target cells and human peripheral mononuclear cells (PBMCs) were the immune effector cells in ADCC assays [41]. Figure 4 shows the combined data from six independent experiments that used PBMC effector cells from 15 different donors. Figure 4A shows that Ab1 huIgG1σ and huIgG2σ have no cytotoxic activity compared to huIgG1 AA, huIgG1 FES variants and the positive control Ab1 huIgG1 WT. Figure 4B shows also that Ab1 huIgG4σ1 and huIgG4σ2 have no cytotoxic activity compared to the huIgG4 PAA and huIgG1. At concentrations up to 1 mg/mL, Ab1 huIgG1σ, huIgG2σ, huIgG4σ1 and huIgG4σ2 showed minimal to no killing of target cells. In particular, the sigma variants displayed negligible cytotoxic activity at levels below that of the negative control Ab2 (anti-F glycoprotein of Respiratory Syncytial Virus a huIgG1) which did not bind to TNFα. In this assay with numerous donors, Ab1 huIgG1 AA was the least silent Fc variant; HuIgG1 AA, huIgG1 FES, and huIgG4 PAA showed intermediate cytotoxic levels, and the most ADCC-silent molecules were the huIgG1σ and huIgG4σ variants.

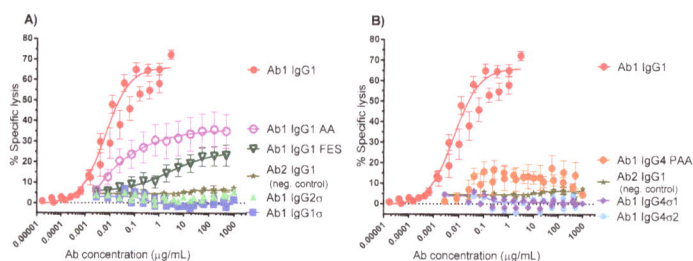

Figure 4. Antibody-dependent cellular cytotoxicity (ADCC) assays (combined data from six experiments). Titrating amounts of test antibodies (Abs) were added to K2 target cells, followed by addition of human PBMC immune effector cells. The extent of target cell lysis was quantitated after 2 h. Samples were tested in duplicate for each individual experiment. In the combined data shown here, 15 donors were tested, and each point represents the mean value ± standard error of the mean. % Specific lysis is shown in (**A**) for IgG1σ compared to IgG1 AA, FES, and IgG2σ; and in (**B**) for IgG4σ1 and IgG4σ2 compared to IgG4 PAA (S228P/F234A/L235A). Ab1 IgG was used as the positive control and Ab2 IgG1 (neg. control) was the negative control.

2.3.2. Antibody-Dependent Cellular Phagocytosis (ADCP)

Binding of the Fc region to FcγRs on certain immune cells such as macrophages can cause cell death by phagocytosis. ADCP relies on macrophages to bind and devour target cells following antibody binding. To further evaluate the lack of effector function with the silent Fc variants, ADCP flow cytometric assays were performed using macrophages derived from primary human monocytes as effector cells and K2 cells as target cells. Macrophages and K2 cells were labeled with separate fluorescent dyes and co-cultured for 5 h in the presence of different concentrations of test antibody. Quantitation of antibody-opsonized target cells was measured by dual-label flow cytometry. Representative data from two experiments are shown in Figure 5. Ab1 huIgG1 shows approximately 30% target cell killing by macrophages, whereas Ab1 huIgG4 PAA, Ab1 huIgG4σ1, and Ab1 huIgG4σ2 show low level killing (<5%), comparable with that of Ab2 (non-TNF binding)

IgG1 (Figure 5A). In another study, Ab huIgG1 shows a similar extent of cell killing by macrophages (using a different donor), but Ab1 huIgG4σ1, Ab1 huIgG4σ2 and Ab1 huIgG1σ1 (along with Ab1 huIgG2σ, Ab1 huIgG4 PAA and Ab2 huIgG1 negative control) show minimal K2 target cell phagocytosis relative to positive and negative controls at concentrations up to 1 mg/mL (Figure 5B). These results suggest that the huIgG1σ and huIgG4σ variants would not have Fc-mediated cytotoxic potential when bound to target immune cells.

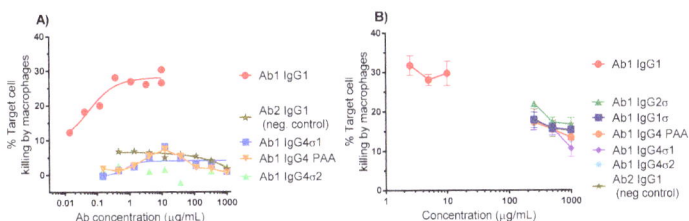

Figure 5. Antibody-Dependent Cellular Phagocytosis (ADCP) (phagocytosis) assays. The extent of phagocytosis was measured using flow cytometry. (**A**) A representative experiment, showing the %Target cell killing by macrophages in a 5 h assay with different concentrations of test antibodies; (**B**) Results using macrophages from a different donor tested in triplicate with a focus on high Ab concentrations (200–1000 µg/mL). % Target cell killing is the number of dual-labeled cells (target cells engulfed by macrophages) divided by total number of target cells × 100. Error bars represent ± standard error of the mean.

2.3.3. Complement-Dependent Cytotoxicity (CDC)

Complement-dependent cell killing occurs when the Fc portion recruits serum complement proteins to the cell bound by the specific antibody; leading to induction of a membrane attack complex and target cell lysis. To determine whether the Fc variants have complement activation capability, CDC assays were performed using K2 target cells, rabbit complement and a panel of Ab1 huIgG Fc variants. As shown in Figure 6 using Ab1 variants, CDC results indicate that huIgG1σ, huIgG4 PAA, huIgG4σ1 and huIgG4σ2, and huIgG2σ samples have minimal activity, like that of the negative control Ab2 huIgG1 at concentrations up to 500 µg/mL. In contrast to the negative control, the positive control Ab1 huIgG1 has measurable target mediated specific killing activity at less than 0.1 µg/mL. Relative to these controls, Ab1 huIgG1 AA and huIgG1 FES show intermediate levels of cytotoxicity. Data indicate that Ab1 huIgG1σ, huIgG4 PAA, huIgG4σ1, huIgG4σ2 and huIgG2σ lack specific CDC-inducing activity.

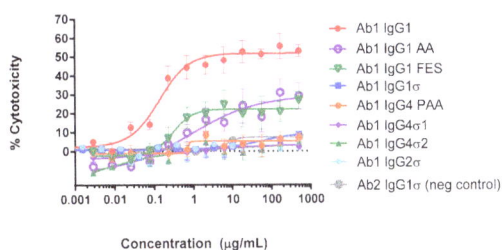

Figure 6. Cell-based CDC assays with IgG Fc variants. K2 mouse myeloma cells stably expressing a transmembrane form of human tumor necrosis factor alpha (TNFα) were incubated with varying concentrations of test antibody and rabbit complement, and then cell viability was measured by detecting lactate dehydrogenase release in the culture supernatant. Error bars represent standard error of the mean from samples analyzed in triplicate. Ab2 IgG1σ does not bind to target cells and is the negative control antibody.

2.4. FcRn Binding

Interaction with neonatal Fc receptor (FcRn) is a critical factor that contributes to sustaining circulating antibody half-life [2], thus better binding to FcRn is expected to give a longer antibody half-life. To determine whether the sigma Fc mutations affect FcRn binding, competition binding by an enzyme-linked immunosorbent assay (ELISA) was performed with varying concentrations of unlabeled silent Fc antibody and a fixed amount of biotin-labeled huIgG1 on plate-bound human FcRn at pH 6. Bound tracer antibody was determined by colorimetric detection after incubation with streptavidin-horseradish peroxidase (HRP) and tetramethylbenzidine (TMB) substrate reagents. ELISA results shown in Figure 7 indicate a slight (less than 2-fold) reduction in binding affinity to FcRn for huIgG1σ relative to the huIgG1 wild-type control. Other tested variants show up to 4-fold reductions in relative affinity. The negative control variant (huIgG1 H435A) of Ab2 appropriately shows no binding to FcRn. While these findings indicate some reduction in relative affinity for the variants tested, all the silent Fc variants can bind FcRn, and huIgG1σ appears to bind FcRn the best.

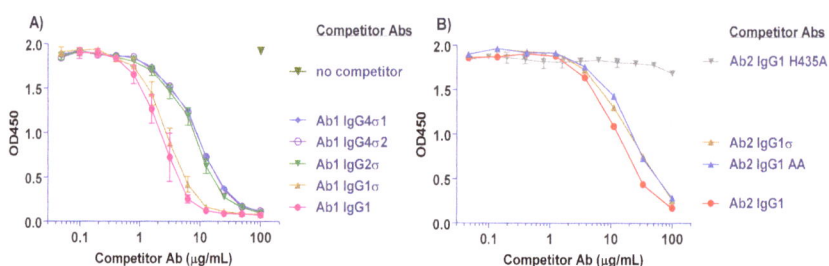

Figure 7. Binding of IgG variants to neonatal Fc receptor (FcRn). In a competition binding format, titrating amounts of antibody were combined with a fixed concentration (4 µg/mL) of biotin-labeled huIgG1. Antibody mixtures were incubated with immobilized human FcRn on a plate for 1 h, washed, detected with streptavidin-horseradish peroxidase (HRP), and developed with tetramethylbenzidine (TMB) substrate. Bound tracer antibody was detected via absorbance at 450 nm measured on a plate reader. (**A**) Ab1 samples and (**B**) Ab2 samples were tested in duplicate and error bars represent mean ± range.

2.5. FcγR Binding Interactions In Vivo

2.5.1. T Cell Immunostimulatory Activity

In vivo, the Fc and FcγR interaction can be evaluated and detected by T cell activation [42]. Human FcγR transgenic mice (FcγR-hu) express all five human FcγRs and have no expression of endogenous mouse FcγRs [43]. When anti-mouse CD3 antibody (Ab4, 2C11) binds to mouse T cells, there is an agonistic effect (presumably due to binding of FcγR on neighboring cells at the same time it is bound to CD3 on T cells) which leads to up-regulation of activation markers such as CD69 and CD25 [42]. An increase in binding of Fc to FcγR correlates with an increased level of T cell activation. This agonist activity was used to evaluate T cell activation of the Fc variants in FcγR-hu mice.

An initial study was done to confirm that T cell activation in the FcγR-hu mice is due to FcγR-Fc binding. Three strains of mice with different gene dosages were used: the 5-transgene FcγR-humanized mice, mice hemizygous for the 5 transgenes, and FcRα chain null mice (FcγR knockouts). Ab4 huIgG1 and Ab4 huIgG1σ were injected into the intraperitoneal cavity (0.5 mg/kg, 10 mL/kg). Approximately 24 h later, mice were euthanized and their spleens removed for T cell activation analyses by flow cytometry. Results demonstrated the dependence on the presence of human FcγR, since very little or no T cell activation using huIgG1 was detected in the null mice, an intermediate level of activation was detected in hemizygous mice, and the most robust activation was noted in the homozygous mice

(Figure 8A). Low to no activation was observed in T cells of huIgG1σ treated mice. Next, homozygous FcγR-humanized mice were used to explore a panel of Ab4 silent Fc variants. Twenty-four hours following the dose of the test antibodies, splenocytes from the mice were isolated and flow analyses performed to determine T cell levels of the early activation marker (CD69), and a late activation marker (CD25).

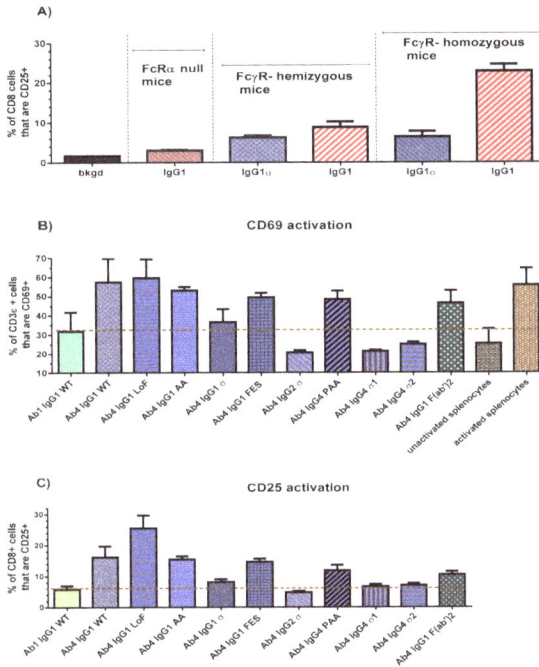

Figure 8. In vivo T cell activation. (**A**) Results of in vivo T cell activation by IgG1σ relative to wild-type IgG1. FcγR-humanized mice, homozygous and hemizygous, and mice lacking FcγRs (null) were injected intraperitoneal with a 0.5 mg/kg dose of either human IgG1σ or wild-type IgG1. After 24 h, spleens were collected from each mouse and the percent of CD8+ cells (as well as CD3ε cells, not shown) that stained positive for CD25 was measured by flow cytometric analyses. (**B**) Comparison of in vivo T cell CD69 activation with panel of silent IgG Fc variants. (**C**) Comparison of in vivo T cell CD25 activation with panel of silent IgG Fc variants. The dashed line represents the level of non-specific activation, derived from the mean value of un-activated splenocytes and splenocytes treated with a control IgG1 WT. Error bars represent the standard error of mean from 3 to 6 samples in both studies.

Comparison of CD69 activation results (expressed as a percentage of CD3ε cells) are shown in Figure 8B for the silent Fc variants, a non-FcγR binding F(ab')$_2$ fragment and a non-T cell binding Ab1 huIgG1. Splenocytes from wild-type IgG (Ab4 huIgG1), low fucose IgG1 (Ab4 huG1 LoF with high affinity binding to CD16), and phorbol myristate acetate treated mice, collectively showed a robust CD69 signal (mean value 58%) compared to un-activated splenocytes (25%) from phosphate- buffered saline (PBS) treated mice. Ab4 F(ab')$_2$ gave a substantial unexpected signal (46%) which suggested that T cells may have a low level of activation merely by bivalent CD3 binding without FcγR engagement. Ab4 huIgG1 induced a higher CD69 activation level than Ab1 huIgG1 WT because it has both CD3 and FcγR engagement. The dashed line in Figure 8B represents a "non-specific activation" level (29%), determined from the average value of untreated splenocytes and splenocytes treated with a non-T cell binding Ab1 huIgG1. Ab4 huIgG1 AA, Ab4 huIgG1 FES and Ab4 huIgG4 PAA produced an increase in

T cell activation relative to the non-specific activation level. However, Ab4 huIgG2σ, Ab4 huG4σ1 and Ab4 huG4σ2 caused a decrease in CD69 activation relative to the non-specific level, and Ab4 huIgG1σ caused an activation level like that of the Ab1 huIgG1 WT which does not bind T cells.

CD25 activation levels of splenic T cells (expressed as a percent of CD8+ cells) from mice treated with the silent Fc variants are shown in Figure 8C. The dashed line represents the level for non-specific activation derived from splenocytes treated with control Ab1 huG1 WT. Consistent with CD69 results, Ab4 huG1 AA, huG1 FES, and huG4 PAA produced an increase in CD25 activation levels that are comparable to that of Ab4 huG1 WT treated mice. As expected, the Ab4 huIgG1 LoF induced a higher CD25 activation signal. Likewise, Ab4 huG1 F(ab')₂ caused a higher CD25 activation than expected (such as in the case of CD69 activation) which may be due to bivalent CD3 binding, clustering, and signaling. Ab4 huIgG1σ, Ab4 huIgG2σ, Ab4 huG4σ1 and Ab4 huG4σ2 induced minimal CD25 activation. These in vivo CD69 and CD25 results indicated that binding of IgG1σ, IgG2σ, IgG4σ1 and IgG4σ2 variants did not activate T cells compared to T cells from mice treated with huIgG WT or huIgG4 PAA.

2.5.2. FcγR-Mediated Anti-Tumor Activity

The B16 metastatic melanoma syngeneic model has been used to assess Fc and FcγR interactions. B16F10 mouse melanoma tumor cells can stimulate lung tumor metastasis, and murine TA99 antibody which targets the gp75 antigen on B16F10 cells, can inhibit tumor cell growth in mice [44,45]. Also, human TA99 antibody can inhibit metastasis, whereas a human IgG1 mutant which does not engage FcγRs has no effect in FcγR-hu mice [43].

This B16 metastatic tumor model was used to test the effect of a huIgG1σ version of Ab5 (anti-grp75) on lung tumor metastasis in FcγR-hu mice. These mice treated with different versions of Ab5: huIgG1, huIgG1 LoF (low fucose with enhanced ADCC, in-house data) and mIgG2a (mouse Ab as positive control) showed a significant reduction in lung tumor metastasis compared to the mice treated with PBS (Figure 9). Lung metastasis foci were decreased by 2-fold in mice treated with Ab5 mIgG2a or Ab5 huIgG1, and 3-fold with Ab5 huIgG1 LoF compared to the PBS treated mice. In contrast, mice treated with huIgG1σ showed no tumor inhibition, and although unexplained, huIgG1σ treated mice showed more tumor metastasis than the PBS treated mice. Results indicate that in vivo antibody-mediated tumor clearance appears to be dependent on binding to FcγRs by huIgG1 or huIgG1 LoF, whereas huIgG1σ which does not bind FcγR is unable to prevent tumor cell growth.

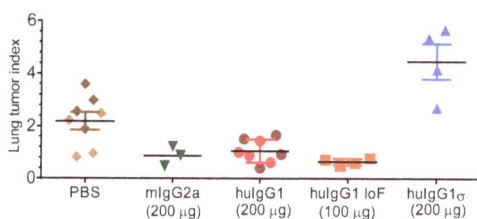

Figure 9. Effect of silent Fc mutations on syngeneic tumor cell metastasis. FcγR-humanized mice were injected IV with B16F10 cells and received either: phosphate-buffered saline (PBS), Ab5 (TA99) mIgG2a, Ab5 huIgG1, Ab5 huIgG1 LoF (low fucose) or Ab5 huIgG1σ (*n* = 4 for groups) on days 0, 2, 4, 7, 9 and 11. On day 21, lungs were harvested and metastasis foci were counted. Data show the mean number of lung metastasis foci expressed as lung tumor index from mice receiving the indicated treatment ± standard error.

2.6. Pharmacokinetics (PK)

PK studies were done to determine whether normal antibody half-life would be altered using antibodies with silent sigma Fc mutations compared to the wild-type huIgG. Human FcRn transgenic mice (Tg32 hemi, 8–10 weeks old, 3–4 per group) were intravenously-injected once with 2 mg/kg doses of the test antibody. Blood was collected at various times up to 35 days post-injection and sera prepared by standard procedure. A Meso Scale Discovery (MSD) assay using anti-human IgG specific reagents was employed for quantitating human IgG in the mouse serum samples. Terminal half-life values were obtained using a one-compartment linear fit using Prism 6.0 software (GraphPad, San Diego, CA, USA).

HuIgG serum concentrations over time for the mice are shown in Figure 10A. PK profiles display a linear decrease in IgG levels from day 7 through 35. HuIgG2σ ($t_{1/2}$ = 5.5 ± 0.5 days) shows a 2-fold shorter half-life than the huIgG1 ($t_{1/2}$ = 11.4 ± 1.7 days). However, PK values for the other Fc variants: huIgG1 AA ($t_{1/2}$ = 7.5 ± 1.1 days), huIgG4 PAA ($t_{1/2}$ = 7.8 ± 3.2 days), huIgG4σ2 ($t_{1/2}$ = 7.9 ± 2.1 days), huIgG4σ1 ($t_{1/2}$ = 8.9 ± 1.3 days), huIgG1σ ($t_{1/2}$ = 9.6 days), huIgG2 ($t_{1/2}$ = 11.2 ± 0.8 days) and huIgG1 FES ($t_{1/2}$ = 9.7 ± 1.1 days) are not significantly different from that of huIgG1. These results indicate that the silent Fc mutations do not alter the PK compared to normal huIgG1, except for huIgG2σ which showed a reduction in half-life.

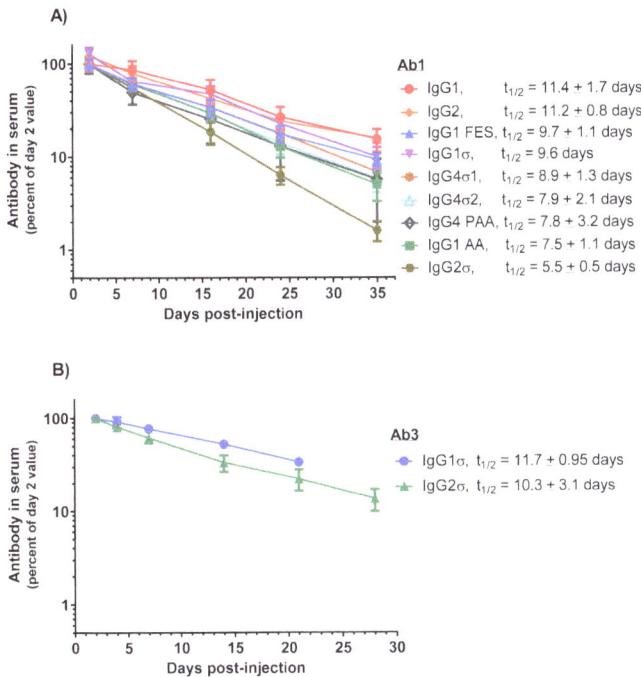

Figure 10. PK comparison of IgG1 WT and a panel of silent Fc variants in mice and cynomolgus monkeys. (**A**) Tg32 human FcRn transgenic mice were IV-injected with 2 mg/kg doses. Serum levels were expressed as a percent of day 2 values, the first timepoint. Data points represent the mean of 3 or 4 mice ± SEM. (**B**) Cynomolgus monkeys were injected intravenously with 1.5 mg/kg dose of an IgG2σ version of Ab3 bispecific antibody and the mean PK profile was compared (in a separate study) with monkeys injected with 1.5 mg/kg of an huIgG1σ version of Ab3. Similarly, IgG1σ and IgG2σ versions of Ab2 × Ab5 antibody were evaluated in the two separate studies.

PK studies in cynomolgus monkeys were conducted also to compare the huIgG1σ and huIgG2σ versions of Ab3, a bispecific antibody (anti-Respiratory Syncytial Virus (RSV)) and glycoprotein gp120 of HIV envelope, huIgG1). Male cynomolgus monkeys (2.5 to 3 kg, 3 per groups) were intravenously injected once with 1.5 mg/kg of the test antibody. Blood samples were collected up to 21 days or 28 days, and PK analyzed as described. Although performed in separate studies, similar half-life values are observed for Ab3 huIgG1σ ($t_{1/2}$ = 11.7 ± 0.95 days) and its huIgG2σ counterpart ($t_{1/2}$ = 10.3 ± 3.1 days); and for Ab2 × Ab5 bispecific (anti-RSV × anti-gp75) huIgG1σ ($t_{1/2}$ = 5.6 ± 0.5 days) and its huIgG2σ counterpart ($t_{1/2}$ = 5.2 ± 0.1 days) (Figure 10B). Considering that Ab2 huIgG1 historically has shown a half-life of 10–12 day in monkeys (Table S1), these monkey data along with the above mouse data, suggest that the huIgG1σ variant (as well as the huIgG2σ) should have a relatively normal serum half-life in humans.

2.7. Immunogenic Potential

To assess the potential immunogenicity of huIgG1 and huIgG4 sigma variants, a T cell epitope analysis was done using the ImmunoFilter™ v2.7 human leukocyte antigen (HLA) class II-peptide binding prediction software (Xencor, Inc., Monrovia, CA, USA) to predict potential immunogenicity in humans. These analyses generate *IScores*, which are weighted, population-relevant values that enable separation of individual 9-mer agretopes into groups by predicted immunogenic risk (PIR). Higher *IScores* indicate a higher PIR. Areas of sequence in the Fc variants which are identical to that of huIgG1 WT Fc sequence are excluded by the application of a tolerance threshold. *IScores* are averaged across all HLA-class II loci for each sequence of interest to create a single average *IScore* value. As shown in Figure 11A, the average *IScores* of the tested variant sequences are like those of their respective wild-type isotype controls. The greatest deviation from wild-type Fc results is a reduction of 1.2% observed with huIgG4 PAA. However, differences observed between the tested variants are inconsequential because all produce average *IScores* of less than 5%. *IScores* of less than 10% are considered to indicate very low risk of immunogenicity with a prediction that less than 10% of the U.S. population would have at least one allele predicted to bind a 9-mer located within the tested sequence. A further breakdown of the *IScore* data into number of predicted agretopes, as well as number of low, medium and high PIR agretopes for each sequence of interest is shown in Figure 11B. Again, individual predicted agretopes show all sequences having similar immunogenic risk profiles as their corresponding wild-type Fc sequences, where predicted agretopes would likely be tolerogenic. None of the Fc variants are predicted to generate any high PIR agretopes (i.e., over 50% of the U.S. population having at least one MHC allele predicted to bind the 9-mer).

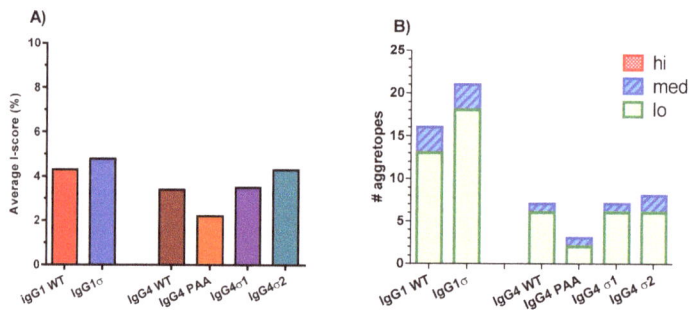

Figure 11. ImmunoFilter™ analyses of the human IgG variants. The amino acid sequences for the different variants were analyzed for predicted immunogenicity *in silico* using ImmunoFilter™ v2.7. (**A**) *IScores* summary which approximated the percent of US population expected to have a human leukocyte antigen (HLA) allele predicted to bind a nine-mer peptide within a sequence of interest. (**B**) The number of individual agretopes and their associated risk potential based on the percent of the population expected to have an allele predicted to bind each agretope (low (lo) = 10–25%, medium (med) = 25–50%, and high (hi) = over 50% of the U.S. population). Resulting *IScores* were averaged across all loci. Higher *IScores* indicate a higher predicted immunogenicity risk.

2.8. Developability Assessment

A panel of biophysical assays was conducted on Abs with Fc variations to assess potential problems in developability that might impact the development and commercial manufacturing of therapeutics. These studies included assessing antibody thermal stability, the stability of highly concentrated samples over time, and stability at low pH. The results are described in the sub-sections below.

2.8.1. Thermal Stability

Differential Scanning Calorimetry (DSC) was used to assess the thermal stability of Ab1 Fc variants. A summary which includes the mid-point of denaturation (Tm) for the transitions is shown in Table 2. HuIgG1 (WT), huIgG1 AA and huIgG2 have similar Tm1 (C_H2 domain) transitions, but huIgG1 FES, huIgG1σ, huIgG2σ, huIgG4 PAA, huIgG4σ1 and huIgG4σ2 show lower Tm1 transition values, suggesting reduced stability of its C_H2 domain attributable to the presence of mutations (Table 1). Ab1 huIgG2 Tm1 transition is 71.6 ± 0.1 °C, while the huIgG2σ variant is 62.0 ± 0.2 °C for Tm1. The Ab1 huIgG4 PAA shows a first transition at 69.5 ± 0.1 °C, while the huIgG4σ1 is at 62.0 ± 0.2 °C and the huIgG4σ2 is at 61.2 ± 0.6 °C, suggesting that the huIgG4 PAA is more thermally stable than its variants. The huIgG4 PAA differences are interesting given that the only sequence differences are in the first two or three residues in the lower hinge (Table 1). The Tm2 and Tm3 indicating the Fab and C_H3 transitions are included for comparison, but generally show normal antibody transition profiles between 69 and 83 °C [46].

Colloidal stability was evaluated over a temperature range by static light scattering (SLS) at two different wavelengths (Table 2). The scattering intensity at 266 nm is more sensitive to fluctuations of smaller aggregates. Measurements of scattering intensity at 473 nm are useful for detection of larger aggregates. The SLS data indicate the onset of aggregation (Tagg) for each antibody sample. Results comparing the antibody variants show that huIgG1 variants have higher aggregation onset temperatures around 68 °C (SLS at 266 nm) compared to the huIgG2 or huIgG4 variants (Tagg = 64–65 °C using SLS at 266 nm). SLS measurements at 473 nm display a similar profile.

Table 2. Thermal stability of the Fc variant antibody panel.

Ab	Fc Variant	Differential Scanning Calorimetry Data			Colloidal Scattering at Two Wavelengths	
		Tm1 (°C)	Tm2 (°C)	Tm3 (°C)	Tagg$_{266nm}$ (°C)	Tagg$_{473nm}$ (°C)
Ab1	IgG1	71.7 ± 0.1	75.2 ± 0.1	82.6 ± 0.1	67.7 ± 0.6	69.7 ± 0.7
Ab1	IgG1 AA	71.7 ± 0.2	75.0 ± 0.1	82.2 ± 0.1	68.0 ± 0.3	69.9 ± 0.5
Ab1	IgG1 FES	64.7 ± 0.0	74.0 ± 0.1	82.5 ± 0.0	ND	ND
Ab1	IgG1σ	60.8 ± 0.1	74.4 ± 0.1	82.7 ± 0.1	68.3 ± 0.7	69.8 ± 0.6
Ab1	IgG2	71.6 ± 0.1	75.9 ± 0.1	ND	65.3 ± 0.3	67.6 ± 0.3
Ab1	IgG2σ	62.0 ± 0.2	74.5 ± 0.1	71.4 ± 0.2	64.0 ± 2.1	66.0 ± 1.7
Ab1	IgG4 PAA	69.5 ± 0.1	73.4 ± 0.1	ND	63.8 ± 1.7	66.0 ± 0.6
Ab1	IgG4σ1	62.0 ± 0.2	70.8 ± 0.1	73.8 ± 0.1	64.1 ± 1.4	66.0 ± 0.5
Ab1	IgG4σ2	61.2 ± 0.6	69.9 ± 0.6	73.2 ± 0.4	63.8 ± 1.8	66.0 ± 0.5

Antibody (Ab) samples are in PBS; values are shown as ± range. Abbreviations: ND, no data; fragment crystallizable, Fc; temperature of melting at midpoint, Tm; IgG1 AA, IgG1 L234A/L235A; IgG1 FES, IgG1 L234F/L235E/P331S; IgG4 PAA, IgG4 S228P/F234A/L235A.

2.8.2. Stability of Concentrated Samples

Samples were concentrated from their initial concentration of 1–2 mg/mL to 40–50 mg/mL and stored at 4 °C. Table 3 summarizes the antibody concentrations and Table 4 summarizes the size exclusion chromatography (SEC) analyses at time of lot release, week 1, week 2, week 3, and week 4 post-release. All molecules show stable antibody concentrations for up to 4 weeks storage at 2–8 °C in PBS. The increasing concentration values observed for all samples with time are attributed to sample evaporation during storage. Aliquots at these time points were diluted and analyzed by SEC. All samples appear visibly clear and SEC analyses results indicate that all molecules at high concentrations consist of greater than 95% monodisperse monomeric antibody.

Table 3. Stability of Concentrated Proteins Stored at 4 °C (concentration of antibodies over time).

Sample ID	Release concentration (mg/mL)	Week 1 (mg/mL)	Week 2 (mg/mL)	Week 3 (mg/mL)	Week 4 (mg/mL)
IgG1	53.9 ± 0.8	53.1 ± 0.3	57.2 ± 0.7	55.1 ± 0.4	55.0 ± 0.3
IgG1 FES	45.5 ± 0.9	48.8 ± 1.2	46.4 ± 0.7	46.7 ± 0.3	46.6 ± 1.5
IgG1σ	41.9 ± 0.6	41.5 ± 0.2	42.0 ± 0.3	42.5 ± 0.9	43.4 ± 0.5
IgG2	41.4 ± 0.1	40.5 ± 0.2	42.0 ± 0.4	42.9 ± 0.2	43.0 ± 0.8
IgG2σ	40.8 ± 0.5	41.3 ± 0.4	40.9 ± 0.6	41.7 ± 1.1	41.8 ± 0.6
IgG4-PAA	48.6 ± 1.5	50.6 ± 0.6	52.5 ± 1.0	53.2 ± 0.2	50.9 ± 1.3
IgG4σ1	50.4 ± 0.4	50.3 ± 0.1	50.0 ± 0.6	50.0 ± 0.2	51.4 ± 0.8
IgG4σ2	51.7 ± 1.1	51.3 ± 0.4	48.4 ± 2.1	53.4 ± 1.6	54.8 ± 0.7

Ab samples are in PBS and shown as mean ± range. Abbreviations: IgG1 AA, IgG1 L234A/L235A; IgG1 FES, IgG1 L234F/L235E/P331S; IgG4 PAA, IgG4 S228P/F234A/L235A.

Table 4. Size exclusion chromatography (SEC) analyses with percent monomer of concentrated proteins.

Sample Name	Release (%)	Week 1 (%)	Week 2 (%)	Week 4 (%)
IgG1	100.0	100.0	99.6	99.4
IgG1 FES	100.0	100.0	100.0	99.0
IgG1σ	100.0	100.0	99.7	99.3
IgG2	98.17	98.0	97.9	97.7
IgG2σ	100.0	100.0	99.2	100.0
IgG4 PAA	100.0	99.6	99.5	99.5
IgG4σ1	98.0	97.9	97.7	97.5
IgG4σ2	97.8	97.9	97.7	97.4

Representative data of SEC analysis. Abbreviations: IgG1 AA, IgG1 L234A/L235A; IgG1 FES, IgG1 L234F/L235E/P331S; IgG4 PAA, IgG4 S228P/F234A/L235A.

To obtain a more complete picture of the stability of the Fc variants, protein melting points and aggregation onset temperature were measured for each by a combination of intrinsic fluorescence and static light scattering. Table 5 summarizes the mid-point values for thermal transition, Tm1, and for onset of aggregation at Tagg = 266 nm. Tm values for the concentrated antibody variants are similar between the release lot and the week 4 samples. Tagg values of the 266 nm values are also similar for the release lot and week 4 samples. Thus, the variants have not changed their conformational or colloidal stabilities at high concentrations.

Table 5. Tm and temperature of aggregation (Tagg) of Fc variants at high concentration (40–50 mg/mL).

Antibody	Week	$Tm1$ (°C)	$Tagg_{266nm}$ (°C)
IgG1	R	69.3 ± 1.4	68.0 ± 0.1
IgG1	Week 0	68.2 ± 0.5	67.4 ± 0.6
IgG1	Week 4	68.3 ± 0.3	67.6 ± 0.8
IgG1σ	R	61.5 ± 1.0	68.9 ± 0.3
IgG1σ	Week 0	60.6 ± 0.2	67.8 ± 0.3
IgG1σ	Week 4	60.8 ± 0.1	67.9 ± 0.7
IgG1 AA	R	69.3 ± 0.5	68.3 ± 0.3
IgG1 AA	Week 0	68.2 ± 0.0	67.7 ± 0.2
IgG1 AA	Week 4	68.4 ± 0.1	68.1 ± 0.2
IgG1 FES	R	65.9 ± 0.8	67.7 ± 0.1
IgG1 FES	Week 0	64.9 ± 0.6	66.9 ± 0.3
IgG1 FES	Week 4	65.1 ± 0.1	67.5 ± 0.2

Table 5. *Cont.*

Antibody	Week	Tm1 (°C)	Tagg$_{266nm}$ (°C)
IgG2	R	67.5 ± 1.8	65.1 ± 0.8
IgG2	Week 0	66.5 ± 0.4	65.6 ± 0.1
IgG2	Week 4	66.7 ± 0.4	65.2 ± 1
IgG2σ	R	61.7 ± 0.8	61.9 ± 2.4
IgG2σ	Week 0	62.6 ± 0.2	66.2 ± 0.1
IgG2σ	Week 4	62.7 ± 0.2	66.8 ± 0.4
IgG4 PAA	R	66.4 ± 1.9	62.1 ± 0.5
IgG4 PAA	Week 0	66.2 ± 0.6	65.5 ± 1.3
IgG4 PAA	Week 4	66.3 ± 0.2	65.9 ± 0.1
IgG4σ1	R	61.5 ± 0.1	62.7 ± 0.9
IgG4σ1	Week 0	62.0 ± 0.2	65.5 ± 0.2
IgG4σ1	Week 4	62.0 ± 0.4	66.1 ± 0.0
IgG4σ2	R	61.6 ± 0.7	62.0 ± 1.5
IgG4σ2	Week 0	61.8 ± 0.1	65.6 ± 0.1
IgG4σ2	Week 4	62.1 ± 0.2	65.9 ± 0.1

Tm1 and Tagg$_{266nm}$ values = mean ± range; Ab samples are in PBS. R is the Release material; Week 0 is data after concentration of material; Week 4 is data after week 4 of incubation.

2.8.3. Stability of Low pH Treated Antibody Fc Variants

To assess whether the Fc variants are stable after low pH treatment for viral load reduction in manufacturing, antibody samples were treated at pH 3.5 for 6 h and dialyzed back to pH 7.4 in PBS. Dynamic light scattering analyses were performed to quantitate presence of soluble aggregates and sub-visible particles. Table 6 shows the average hydrodynamic radius (R_h), percent polydispersity (% Pd) values, and percent mass comparison between the release lot and the low pH treated samples. All samples yielded a peak with an average radius of 5–7 nm, average % Pd of less than 15%, and an average % Mass of greater than 99%. These features are indicative of a mono-dispersed species for the antibody variants.

Table 6. Dynamic light scattering data after low pH treatment.

Ab1	Avg R_h	Avg %Pd	Avg %Mass	Avg R_h	Avg %Pd	Avg %Mass
	Release Material			Low pH Material		
IgG1	5.4 ± 0.1	10.3 ± 2.1	100.0 ± 0.0	5.1 ± 0.2	12.2 ± 1.7	100.0 ± 0.0
IgG1σ	5.4 ± 0.0	9.6 ± 0.3	100.0 ± 0.0	5.0 ± 0.2	13.2 ± 0.8	100.0 ± 0.0
IgG1 AA	5.8 ± 0.3	16.7 ± 10.1	100.0 ± 0.1	5.5 ± 0.2	12.2 ± 7.1	100.0 ± 0.0
IgG1 FES	5.5 ± 0.0	7.7 ± 1.6	100 ± 0.0	5.2 ± 0.1	11.9 ± 1.3	99.7 ± 0.6
IgG2	5.8 ± 0.1	18.2 ± 5.2	99.6 ± 0.1	5.0 ± 0.1	13.2 ± 0.6	100.0 ± 0.0
IgG2σ	5.4 ± 0.0	13.0 ± 3.6	100.0 ± 0.0	5.1 ± 0.2	12.5 ± 1.5	99.6 ± 0.8
IgG4—PAA	5.3 ± 0.1	9.9 ± 0.6	100.0 ± 0.0	5.3 ± 0.0	10.4 ± 0.8	100.0 ± 0.0
IgG4σ1	5.4 ± 0.1	9.3 ± 0.5	100.0 ± 0.0	6.4 ± 0.2	15.5 ± 1.6	100.0 ± 0.0
IgG4σ2	5.4 ± 0.0	8.9 ± 0.4	100.0 ± 0.0	6.6 ± 0.1	14.8 ± 1.3	100.0 ± 0.0

Samples were tested at 1 mg/mL in triplicate and shown as mean ± standard deviation (STD). Avg—average; Percent polydispersity (% Pd); hydrodynamic radius (R_h).

2.9. Fc Crystal Structures

To assess the impact of the silencing mutations on the structure of the Fc, crystal structures of the Fc fragments of huIgG1σ, huIgG4σ1, and huIgG4σ2 were determined to resolutions of 1.90 Å, 1.90 Å, and 1.85 Å, respectively. Data collection and refinement statistics are summarized in Table 7. Crystals of all three Fc constructs belonged to space group P2$_1$2$_1$2$_1$, and had similar unit cell parameters. This isomorphism reflects a similarity in Fc packing within the crystal lattice where one intact Fc dimer is present in the asymmetric unit. In this crystal form, the A chain C$_H$2 domain is less ordered as evidenced by the reduced quality of the electron density map and elevated temperature factors of the atoms. The reason for this is that the C_H2 domains of the A and B chains have a difference in the crystal contacts. For the three Fc structures reported here and excluding the glycan contributions, the B chain has an average of 2082 ± 186 Å^2 of buried surface area through interaction with symmetry related Fc molecules compared with only 1477 ± 181 Å^2 of buried surface area for chain A. Consequently, for the three structures, the average B-factor calculated over residues 238–339 and glycan residues for chains B and A was 37.6 ± 4.5 Å^2 and 54.1 ± 1.8 Å^2, respectively.

The huIgG1σ Fc structure includes residues G236-S444 and A235-S444 for chains A and B, respectively, which are only those observed in the electron density map (Figure S1). The structure reveals conformational differences in ordered residues N-terminal to V240 and within the BC loop (residues S267 to V273) relative to wild-type huIgG1 Fc (e.g., PDB 3AVE [40]) and receptor-bound Fc structures (e.g., PDB 3SGK [47]). These regions can directly influence interactions with FcγRs. The lower hinge regions of wild-type huIgG1 structures, including those in complex with FcγRs, show extended conformations while that of huIgG1σ has a kinked conformation at S239 (Figures 12 and S1). The backbone psi angle value for S239 averaged over both chains in the huIgG1σ Fc structure is 10° compared to an average value of 145° for wild-type huIgG1 in structure 3AVE [40]. It should be noted that the altered lower hinge conformation is not a requirement of crystal packing (Figure S2). Interestingly, the kinked conformation is observed in structures of two other engineered Fcs in complex with FcγRIIb, a P238D mutant (PDB 3WJJ [48]) and an Fc having the P238D mutation in addition to six other mutations (PDB 3WJL [48]) (Figure 12).

Table 7. X-ray crystallographic data and refinement statistics.

	IgG1σ Fc	IgG4σ1 Fc	IgG4σ2 Fc
Crystal Data			
Space group	$P2_12_12_1$	$P2_12_12_1$	$P2_12_12_1$
Unit cell parameters			
a, b, c (Å)	73.36, 79.17, 101.44	74.74, 78.39, 97.39	74.52, 78.47, 97.51
α, β, γ (°)	90.0, 90.0, 90.0	90.0, 90.0, 90.0	90.0, 90.0, 90.0
Resolution (Å)	50.00–1.90 (1.95–1.90) [a]	50.00–1.90 (1.95–1.90)	50.0–1.85 (1.90–1.85)
Measured reflections	256,516 (19,366)	248,644 (18,842)	363,451 (27,057)
Unique reflections	47,175 (3427)	45,491 (3348)	49,481 (3648)
Completeness (%)	99.7 (99.7)	99.4 (99.9)	99.9 (100.0)
Redundancy	5.4 (5.7)	5.5 (5.6)	7.3 (7.4)
R_{merge} [b]	0.041 (0.539)	0.045 (0.588)	0.056 (0.670)
<I/σ> [c]	21.3 (2.9)	18.6 (2.8)	19.8 (3.0)
Refinement Statistics			
Resolution (Å)	33.28–1.90	32.70–1.90	47.26–1.85
Number of reflections	47,088	45,473	49,469
R_{work} (%) [d]	19.2	18.2	18.1
R_{free} (%) [d]	22.9	22.5	21.5
Number atoms	3772	3771	3897
Protein	3292	3320	3404
Carbohydrate	220	220	210
Solvent	260	231	283
Mean B-factor (Å^2)	41.9	45.9	40.0
Protein	40.4	43.7	37.9
Carbohydrate	58.5	75.7	66.2
Solvent	46.4	50.0	45.9
RMSD [e]			
Bond lengths (Å)	0.009	0.012	0.018

Table 7. *Cont.*

	IgG1σ Fc	IgG4σ1 Fc	IgG4σ2 Fc
Bond angles (°)	1.213	1.438	1.699
Ramachandran			
Favored (%)	98.1	98.8	98.1
Allowed (%)	1.9	1.2	1.9
Outliers (%)	0.0	0.0	0.0

[a] Values for high resolution shell are shown in parentheses. [b] $R_{merge} = \Sigma_{hkl}\Sigma_i (|<I_{hkl}> - I_{hkl,i}|)/\Sigma_{hkl}\Sigma_i \, I_{hkl,i}$, where hkl enumerates the unique reflections and i their symmetry-equivalent or multiply-measured contributors. I is the intensity for a given measurement. [c] $<I/\sigma>$ is the average x-ray reflection intensity (I) measurement divided by the standard deviation of that measurement (σ) for the whole X-ray data set. [d] $R = (\Sigma_{hkl} \, ||F_{obs}(hkl)| - |F_{calc}(hkl)||)/(\Sigma_{hkl} \, |F_{obs}(hkl)|)$, where $F_{obs}(hkl)$ and $F_{calc}(hkl)$ are the observed and calculated structure factors. R_{work} includes structure factors for the entire data set minus those data sequestered for computing R_{free}. R_{free} includes only the latter randomly chosen subset of the data. [e] RMSD (Root Mean Square Deviation) in the table describes how well the atomic bond distances and angles compare to idealized values determined from a set of reference structures.

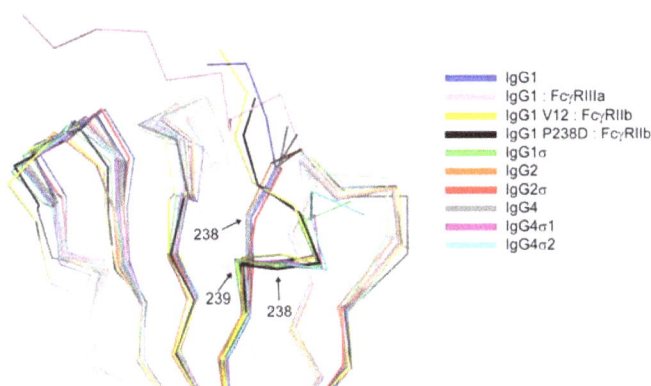

Figure 12. Structural consequence of mutation at position 238 reveals an altered conformation of the lower hinge. Backbone alignment of Fc C_H2 domains of huIgG1 (Protein Data Bank (PDB) 3AVE [40]), huIgG1 in complex with FcγRIIIa (chain B, PDB 3SGJ [47]), huIgG1 V12 variant in complex with FcγRIIb (PDB 3WJL [48]), huIgG1 P238D in complex with FcγRIIb (PDB 3WJJ [48]), huIgG1σ, huIgG2 (PDB 4HAF [49]), huIgG2σ (PDB 4L4J [29]), huIgG4 (PDB 4C54 [50]), huIgG4σ1, huIgG4σ2. The Cα positions of residues 238 and 239 are labeled.

The structure of huIgG4σ1 Fc includes residues A237-L443 and G236-S444 for chains A and B, respectively, that were modeled into the visible electron density. For the huIgG4σ2 Fc structure, residues S238-L443 and E233-S444 were modeled for the respective chains (Figure S1). The lower hinge residues in both chains of the huIgG4σ1 Fc structure and in chain B of the huIgG4σ2 Fc structure adopt a kinked conformation like that described above for huIgG1σ (Figures 12 and S1). In contrast, the same residues in chain A of the huIgG4σ2 structure adopt a wild-type-like extended conformation (Figure 12). All these Fcs having a kinked lower hinge also contain the mutation of P238, however, the structure of huIgG2σ (PDB 4L4J [29]) and chain A of huIgG4σ2 indicate that this mutation does not restrict the conformation to the kinked form.

Previous crystal structures of huIgG1, huIgG2 and huIgG4 have shown two distinct conformations for the BC loop (residues 267 to 273) and the FG loop (residues 322 to 332) that seem to correlate with the ability to bind to FcγRs. Denoting the two conformations as 'flipped-in' and 'flipped-out', both loops are in the flipped-in conformation in structures of wild-type huIgG1 Fc (PDB 3AVE [40]) and huIgG2 Fc (PDB 4HAF [49]), whereas they adopt the flipped-out conformation in structures of huIgG2σ (PDB 4L4J [29]), huIgG4 wild-type (PDB 4C54 [50]) and in huIgG4σ1 and huIgG4σ2 (Figure 13). Interestingly, the electron density map for huIgG1σ Fc reveals a deviation from this pattern such that the primary conformation of the BC loop is the flipped-out conformation while that of the FG loop is the wild-type flipped-in conformation. In chain A, for which the density is poorer, these are the only conformations evident. In chain B, density is also present for the flipped-in conformation of the BC loop, and this conformation refined to an occupancy of 40%. Density for the FG loop in chain B suggests an increased mobility for this loop, but was not well enough defined to model a second conformation. Whereas crystal packing precludes a flipped-out FG conformation for one of the two Fc chains in the huIgG1σ structure, a flipped-out conformation could be tolerated for the second (Figure S3). In the wild-type huIgG1 Fc structure (PDB 3AVE [40]), residue P271 in a flipped-out conformation of the BC loop would clash with the flipped-in FG loop. However, this clash is avoided in huIgG1σ by a deflection of the FG loop as a rigid body (Figure 13). In addition to the deviation in conformations of the BC and FG loops for the huIgG1σ Fc, the structural studies suggest an increased mobility for these two loops that may play a direct role in how this Fc interacts with the FcγRs.

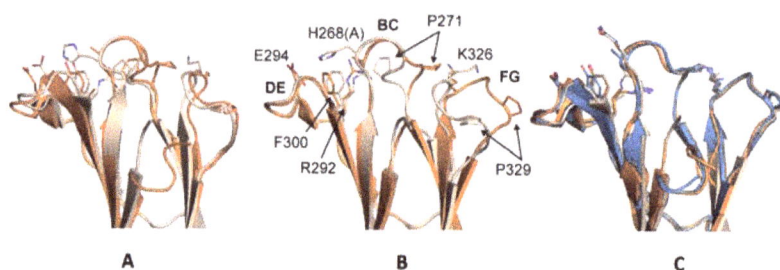

Figure 13. Conformational differences in BC and FG loops in WT and σ variants of huIgG. (**A**) Overlay of C$_{H}$2 domains (cartoon) from huIgG1 (wheat, PDB 3AVE chain A [40]) and huIgG1σ (orange, chain B); (**B**) huIgG2 (wheat, PDB 4HAF chain A [49]) and huIgG2σ (orange, PDB 4L4J chain A [29]); (**C**) huIgG4 (wheat, PDB 4C54 chain A [50]), huIgG4 σ1 (blue, chains A and B), and huIgG4 σ2 (orange, chains A and B). BC and FG loops are in the flipped-in conformation in huIgG1 (middle, wheat) and flipped-out conformation in huIgG2σ (middle, orange). Residues at positions 268, 271, 292, 294, 300, 326, and 329 are shown as sticks and are labeled in the middle panel with huIgG2 sequence (mutation at position 268 in huIgG2σ given in parentheses). N-terminal residues 234–236 are omitted from the cartoon of huIgG1 for clarity, and the alternate conformations modeled for the BC loop of huIgG1σ are shown.

2.10. Molecular Dynamics Simulations of the Fc FG and BC Loops

To further investigate the dynamic stability of the loops in wild-type huIgG1 and huIgG1σ that could potentially influence FcγR binding, explicit solvent molecular dynamics simulations were performed starting from the crystal structures of wild-type huIgG1 (PDB 3AVE [40]) and huIgG1σ. Both simulations were initiated from structures in which the FG loop was in the flipped-in conformation. For the wild-type simulation, the FG loop remained close (within 2Å C$_{\alpha}$ RMSD) to the flipped-in conformation for 57.1% of the simulation as compared to 44.4% in the case of huIgG1σ where the loop was found to be substantially more flexible (Figures 14 and 15). Notably, the flipped-out conformational state of the FG loop was not found to have significant population in either simulation. Figure S4 shows the spatial distribution of P329 in the FG loop, a key residue for binding to various FcγRs (e.g., [29,51]). The distribution of P329 is more tightly clustered in wild-type than in huIgG1σ suggesting that the FG loop in wild-type is preconfigured for receptor binding.

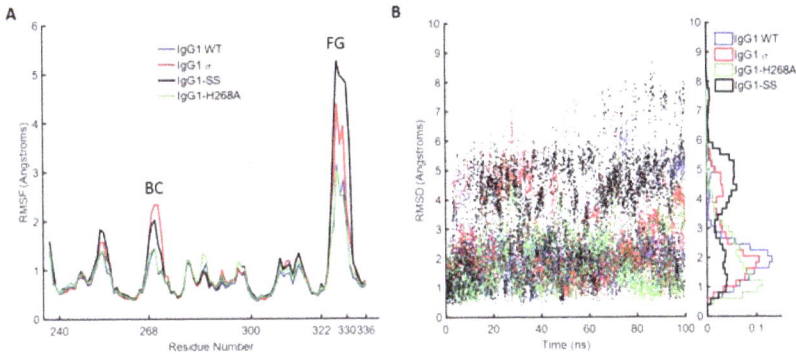

Figure 14. Molecular dynamics simulation of huIgG1 wild-type and mutants. (**A**) Root mean squared fluctuations (RMSF) of Cα atoms of chain B of C_H2 domain. The trajectories are aligned on the Cα atoms with respect to the non-BC and non-FG loop residues of chain B of C_H2 domain of the corresponding crystal structures. The peaks corresponding to the BC and FG loop are marked. The two loops are more flexible in huIgG1σ (red) as compared to wild-type (blue). The fluctuations in huIgG1-H268A (green, containing only the BC loop mutation) are closer to wild-type while those in huIgG1-SS (A330S P331S) (black, contains only the FG loop mutations) are closer to the silent variant. (**B**) Root mean squared deviation (RMSD) of the Cα atoms of the FG loop with respect to the flipped-in conformation. RMSD values are shown for all frames of the 100-ns trajectory. The right panel shows a normalized histogram of the RMSD values.

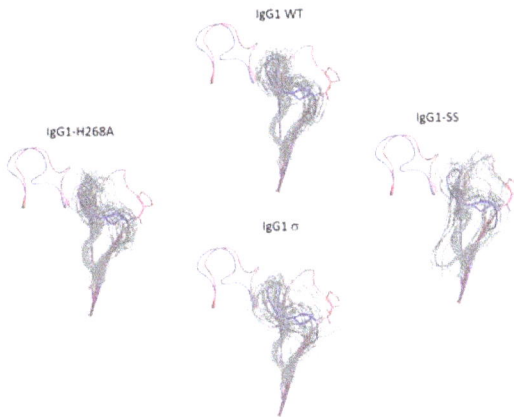

Figure 15. Conformational distribution of the FG loop of chain B of C_H2 domain from molecular dynamics simulations of huIgG1 wild-type and mutants. Each panel shows the flipped-in (blue cartoon) and flipped-out (red cartoon) conformations of the two loops. The flipped-in conformation is from the huIgG1 WT (PDB 3AVE [40]) while the flipped-out conformation is from huIgG4 WT (PDB 4C54 [50]). All simulations were started from the flipped-in conformation of the FG loop. The panels show fifty conformations (gray) sampled at an interval of 2 ns after aligning the molecular dynamics (MD) trajectory on the Cα atoms of the non-BC and non-FG residues of the C_H2 domain of chain B in the corresponding crystal structure. Figure S4 shows the conformation of P329 in the above loop conformations. The FG loop of huIgG1σ and huIgG1-SS is significantly more flexible than that of huIgG1 WT and huIgG1-H268A.

In addition to the FG loop, the simulations show that the BC loop is also more flexible in huIgG1σ than in the wild-type huIgG1 (Figures 14 and 16). In the wild-type simulation, the BC loop stays close to the initial flipped-in conformation throughout the simulation, but occupied both flipped-in and flipped-out states in the huIgG1σ simulation (Figure 16). This is clearly illustrated by the spatial distribution of residue P271 in the BC loop which shows a single cluster for wild-type but two distinct clusters for huIgG1σ (Figure S5) corresponding to the two conformational states. Interestingly, although the huIgG1σ simulation was initiated from the flipped-out conformation of the BC loop, the flipped-in conformation turns out to be more populated. The simulations may not be long enough to converge the populations of the two states, but strongly suggest that the BC loop of silent huIgG1σ is more flexible than that of the wild-type.

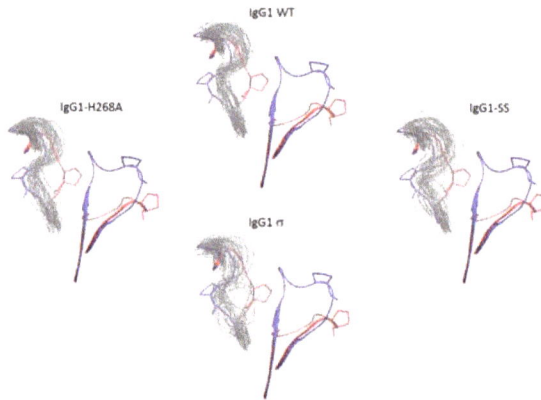

Figure 16. Conformational distribution of the BC loop of chain B of C_H2 domain from molecular dynamics simulations of huIgG1 wild-type and mutants. Each panel shows the flipped-in (blue cartoon) and flipped-out (red cartoon) conformations of the two loops. The flipped-in conformation is from the huIgG1 WT (PDB 3AVE [40]) while the flipped-out conformation is from huIgG4 WT (PDB 4C54 [50]). The huIgG1σ was started with the BC loop in the flipped-out conformation while the other simulations started from the flipped-in conformation. The panels show fifty conformations (gray) sampled at an interval of 2 ns after aligning the MD trajectory on the Cα atoms of the non-BC and non-FG residues of the C_H2 domain of chain B in the corresponding crystal structure. Figure S5 shows the conformation of P271 in the above loop conformations. In the wild-type and huIgG1-H268A simulations, the BC loop remained close to the initial flipped-in conformation of the loop. The huIgG1σ showed a broader distribution around the flipped-in state while huIgG1σ samples both flipped-in and flipped-out states though the flipped-in state was dominant.

HuIgG1 and huIgG1σ differ at position 268 (histidine in huIgG1, alanine in huIgG1σ) in the BC loop, and at positions 330 (alanine in huIgG1, serine in huIgG1σ) and 331 (proline in huIgG1, serine in huIgG1σ) in the FG loop. To investigate the impact of these mutations on the flexibility of the two loops, two additional simulations were performed after modelling the mutations in the huIgG1 wild-type crystal structure. One simulation (referred to as huIgG1-H268A) contained only the BC loop mutation, and the other (referred to as huIgG1-SS) contained only the two FG loop mutations. The flexibility of both loops in the huIgG1-H268A simulation was nearly identical to that in the wild-type simulation while the flexibility in huIgG1-SS simulation was closer to that in the huIgG1σ simulation (Figures 14, 15, S5 and S6). The mutant simulations suggest that the A330S P331S mutations in the FG loop are primarily responsible for the increased flexibility of the two loops in huIgG1σ; the H268A mutation in the BC loop has minimal impact.

3. Discussion

Many therapeutic antibodies and Fc fusion proteins employ Fc activity as part of their mechanisms of action. Fc engagement with FcγR can activate myeloid cell and NK cell activity as well as the generation of reactive species that induce apoptosis and release of inflammatory cytokines which are important for eliminating unwanted target cells (i.e., tumor cells) [52–55]. However, antibody targeting to cell surface receptors can pose potential safety risks since Fc activity could elicit ADCC, ADCP, CDC, and/or apoptosis which can cause tissue damage, depletion of target cells, and infusion reactions. Although numerous antibody engineering efforts to silence the Fc activity of huIgG1 and huIgG2 have been reported [3,16,24,25,33,34,56], we describe alternative novel mutations in huIgG1σ, huIgG4σ1, huIgG4σ2, which can provide design choices for future Fc silent huIgG antibodies or Fc fusions.

Several antibody panels (Ab1, Ab2, Ab3, Ab4, Ab5) with Fc silent mutations were constructed, purified, and tested in vitro for Fc binding and function. The huIgG1σ, huIgG4σ1, and huIgG4σ2 antibodies tested as immune complexes had minimal binding to FcγRI, FcγRIIa, FcγRIIb, and FcγRIIIa, compared to huIgG1 WT using concentration ranges that could be found in clinical dosing. Since the TNFα on target cells and FcγRIII on effector cells are both multivalent, the engagement of IgG1 molecules to both cells would involve avidity. Thus, ADCC activity and the induction of immune effector function which depend on avidity and could occur at lower concentrations compared with monovalent antigen (i.e., in plate assays) [57]. Therefore, the aforementioned cell based assays provide a more sensitive measure of the degree of silencing effector function and provide a meaningful biological readout. HuIgG1σ, huIgG4σ1 and huIgG4σ2 were more silent in ADCC activity when compared to huIgG1 AA, IgG1 FES, and IgG4 PAA. HuIgG4 WT was not included in the comparisons because such molecules can exchange to half-molecules in a dynamic process (Fab-arm exchange) [35]. Instead, subsequent studies were compared with HuIgG4 PAA, which has a stabilized hinge, reduced effector function, and has been used in therapeutic antibodies [25].

In vivo studies emphasize again the lack of immune functionality with the huIgG sigma variants. Results using FcγR-hu mice demonstrated significantly lower levels of T cell dependent activation (CD69 and CD25 upregulation) with the sigma variants compared to huIgG1 AA, IgG1 FES, and huIgG4 PAA. In vivo half-life in a transgenic mouse model of human FcRn and in cynomolgus monkeys indicated that the silent mutations did not alter PK properties compared to normal huIgG. In addition, potential immunogenicity, evaluated in silico (for protein sequence "hot spots" that favor immune response initiation) predicts minimal immunogenic risk for these silent Fc variants. Although ex vivo immune responses measured by T cell proliferation and IL-2 secretion, were not tested here, the huIgG2σ (reported previously) shows relatively low risk of clinical immunogenicity as determined by comparing the frequency and magnitude of ex vivo T cell responses [29]. The IgG1 and IgG4 sigma variants with very similar mutations and low predicted immunogenic risk are likely to have similar non-immunogenic profiles.

The impact of huIgG1σ, huIgG4σ1 and huIgG4σ2 mutations on biophysical properties was assessed for their effect on manufacturing. Thermal stress for all the sigma variants compared to their wild-type versions, suggests lower thermal stabilities in the C_H2 and hinge regions. However, there is little evidence that reduced thermal stability of these magnitudes translates into developability issues or in vivo stability issues. All Fc variants are stable for 4 weeks when concentrated (to 40–50 mg/ML), or subjected to low pH stress. Also, there are no additional post-translational modifications or significant changes in solution particle size associated with these Fc mutations.

In the analysis of Fc:FcγR structures, the packing of P329 in the C_H2 domain FG loop between two tryptophan side chains in the Fc receptor is a conserved feature of huIgG Fc interaction with FcγRI (e.g., 4W4O [51]), FcγRIIb (e.g., 3WJJ [48]), and FcγRIIIa (e.g., 3SGJ [47]). The receptor bound conformation of the FG loop is like its conformation in apo-structures of huIgG1 Fc and huIgG2 Fc suggesting that in these subtypes, the FG loop is preconfigured for receptor engagement (Figure S6). The structure of huIgG2σ, an engineered silent variant of huIgG2, first revealed a unique flipped conformation for the FG loop that was proposed to be responsible in part for the diminished receptor interaction of this variant. The same structure also revealed a unique, flipped conformation of the BC loop relative to wild-type structures of huIgG1 Fc and huIgG2 Fc. Herein, we denote the conformations of the BC and FG loops as observed in a prototypical structure of huIgG1 wild-type Fc (e.g., PDB 3AVE [40]) as flipped-in and the altered conformations of the same loops observed in the crystal structure of huIgG2σ (PDB 4L4J [29]) as flipped-out (Figure 13).

It has been proposed that, in the case of huIgG2σ, the H268A mutation in the BC loop abolished an electrostatic interaction with E294 in the DE loop resulting in the observed flipped conformation of the BC loop which in turn triggered the flip of the FG loop [29]. Indeed, alignment of the C_H2 domains from crystal structures of huIgG2 Fc and huIgG2σ Fc suggested that residue P271 in a flipped-out BC loop could clash with K326 in a flipped-in FG loop (Figure 4). A similar flipped-out conformation of the FG loop has since been observed in crystal structures involving huIgG4 Fc [50] (Figure 4). Davies et al. have suggested that sequence differences within the FG loop between huIgG1 and huIgG4 were primarily responsible for the FG loop flip in the latter, and not the absence of H268 [50]. Consistently, one C_H2 domain in PDB 4D2N [58], a structure of deglycosylated huIgG4 Fc, reveals an FG loop that, although partially disordered, appears to have a flipped-out conformation while the BC loop is maintained in a flipped-in conformation. The present crystal structure of the huIgG1σ Fc demonstrated the possibility of the coexistence of a flipped-out BC loop and a flipped-in FG loop. A clash between P271 and K326 was avoided by a rigid body displacement rather than conformational flip of the FG loop away from the flipped BC loop (Figure 4). Furthermore, MD simulations of a single C_H2 domain of huIgG1 wild-type and mutants were consistent with the hypothesis that positions 330 and 331 are of primary importance to the conformational stability of the FG loop. The simulations showed that the SS (A330S P331S) mutations dramatically increase the flexibility of the FG loop even in the absence of the H268A mutation. Also, just the H268A mutation, in the absence of the FG loop mutations, only marginally increases the flexibility of the two loops relative to wild-type.

The results of our in vitro and in vivo studies demonstrated that huIgG4σ1 and huIgG4σ2 are more silent than huIgG4 PAA as discussed above. Given that all the mutations for these variants were localized to the lower hinge region (Figure 2) and that structurally the dispositions of the BC and FG loops were identical to those observed in structures of wild-type huIgG4 Fc (Figure 4), these mutations could function either by directly disrupting hinge:receptor interactions or indirectly by altering lower hinge backbone conformation. Structures of huIgG1 in complex with FcγRI and FcγRIIIa have shown the importance for L235 in receptor engagement (Figure S6). This residue was mutated to alanine in both huIgG1σ as well as huIgG4σ1/2, and this mutation likely has a direct effect on receptor engagement. In contrast, an altered conformation of ordered lower hinge residues N-terminal to position 240 observed in structures of huIgG1σ, huIgG4σ1, and huIgG4σ2 relative to huIgG1 wild-type was like that observed in structures of huIgG1 P238D and huIgG1 C220S, E233D, G237D, P238D, H268D, P271G, A330R in complex with FcγRIIb (Figure 3). All Fc variants demonstrating this kinked conformation commonly share mutation of a conserved proline at position 238 and have an altered ability to engage Fc receptor. Thus, although P238 does make contacts with Fc receptor, it is likely that this residue also plays an important indirect role in maintaining receptor affinity, biasing the structure of the lower hinge toward a conformation competent for receptor engagement.

In summary, antibodies with huIgG1σ, huIgG4σ1, and huIgG4σ2 Fc regions have demonstrated minimal FcγR interactions by in vitro and in vivo methods. Immunogenicity, developability, and PK risks of these variants have been evaluated and determined to be comparable to that the huIgG1 WT (Table 8) and huIgG4 PAA. Thus, we propose that the huIgG1σ, huIgG4σ1 and huIgG4σ2 variants, which offer several IgG subtype choices, be considered along with the existing silent Fc structures for incorporation into antibody based biotherapeutic molecules.

Table 8. Summary of Silent Fc Designs.

	IgG1 WT	IgG1 AA	IgG1σ	IgG1 FES	IgG2σ	IgG4 PAA	IgG4σ1	IgG4σ2
Number of Mutations	0	2	7	3	7	3	5	6
Lack of Fc Immune Functions [1]	+	++/+	++++	++/+++	++++	+++	++++	++++
Developability [2]	++++	++++	++++	++++	+++	++++	++++	++++
PK [3]	++++	+++	++++	++++	+++	++++	++++	++++
Low Immunogenicity Risk [4] (*in silico*)	++++	++++	++++	ND	+++	++++	++++	+++
Low PBMC Immunogenicity [5]	++++	++++	ND	ND	++++	++++	ND	ND

ND indicates no data. [1] Lack of Fc Immune Functions was determined collectively by FcγR binding, ADCC, ADCP, CDC, in vivo T cell activation, and *in vivo* tumor cell inhibition studies. The least silent is denoted by + and ranges to ++++ as the most silent. [2] Developability was determined by thermal stability, stability of concentrated IgG samples and stability after low pH treatment. Good is +, intermediate is +++, and best is ++++. [3] PK was performed in human FcRn transgenic mice and in cynomolgus monkeys. Variants with longer half-life are denoted ++++, and ones with slightly shorter half-life are shown as +++. [4] Low Immunogenicity Risk was predicted using a T cell epitope analyses software. ++++ indicates low risk and +++ is medium to low risk. [5] Low PBMC immunogenicity data from a previous report are included because huIgG1σ and huIgG4σ variants have very similar sequences. ++++ indicates low risk.

4. Materials and Methods

4.1. Antibodies

Ab1 is a huIgG1 kappa antibody specific for tumor necrosis factor alpha (TNFα); Ab2 is a huIgG1 kappa monoclonal antibody specific for F glycoprotein of Respiratory Syncytial Virus (RSV); Ab3 is a huIgG1 bispecific antibody targeted against RSV and glycoprotein gp120 of the human immunodeficiency virus (HIV) envelope; Ab4 is an anti-mouse CD3ε-chain Ab, 145-2C11, and Ab5 is TA99, an anti-gp75 antibody which targets gp75 antigen on B16F10 melanoma cells [45,59]. H435A is a mutation which reduces binding to FcRn [60]. LoF refers to low fucosylated IgG which has increased ADCC/ADCP effector function in vitro and in vivo [42]. R10Z8E9 is a mouse anti-huIgG antibody that is specific for the C_H2 domain [61]. All antibodies were produced and purified at Sino Biologics by transient HEK cell transfection. Antibodies were purified using standard protein A chromatography and confirmed to be greater than 95% purity and low in endotoxin prior to experiments. The bispecific antibody was made using the DuoBody® technology (Genmab, Copenhagen, Denmark) [62], and confirmed to be greater than 95% purity.

4.2. Cell Lines

K2 cells are Sp2/0 mouse myeloma cells which express a mutant, transmembrane form of human TNFα [41,63]. K2 cells were cultured at 37 °C, 5% CO_2, in Iscove's Modified Dulbecco's Medium (IMDM) with GlutaMAX and 5% (*v/v*) heat-inactivated fetal bovine serum (FBS), 1× non-essential amino acids (NEAA), 1× sodium pyruvate, 0.5 μg/mL mycophenolic acid, 2.5 μg/mL hypoxanthine, and 50 μg/mL xanthine (MHX). Media components were purchased from Life Technologies (as 100×, Carlsbad, CA, USA) and the MHX components from Sigma (St. Louis, MO, USA).

B16F10 mouse melanoma cell line was obtained from the American Tissue Culture Collection. Cells were cultured in RPMI 1640 supplemented with 10% (*v/v*) FBS, 1× NEAA, and 1× sodium pyruvate.

4.3. Fc Gamma Receptor (FcγR) Binding

Binding of Ab1 variants to human FcγRs was assessed using an AlphaScreen (PerkinElmer, Boston, MA, USA) bead assay in a competition binding format. FcγRs were purchased from R&D Systems or Sino Biological Inc. (Beijing, China). Since the outcome for huIgG with a silent Fc in a FcγR binding assay may be a negative result, assay sensitivity was increased by introducing avidity to the test samples via cross-linking. This was achieved using a goat F(ab′)₂ anti-huIgG F(ab′)₂-specific fragment (Jackson ImmunoResearch, West Grove, PA, USA) in 1:1 molar ratio with the test huIgGs. Cross-linked test antibodies (Thermo Fisher Scientific, Waltham, USA) were added to Corning white half-well 96-well assay plates (Corning Inc., Corning, NY, USA) at the designated concentrations in competition with biotin-labeled huIgG Fc fragment (biot-Fc) at either 1 µg/mL (for FcγRI, -RIIa, and -RIIb assays) or 5 µg/mL (for FcγRIIIa assays). Biotinylated Fc fragment was used to prevent binding to the biot-Fc by the test article cross-linker described above. FcγRs as specified were added to a 200 ng/mL final concentration. Nickel chelate acceptor beads were added; followed by streptavidin donor beads. Plates were covered with foil adhesive plate sealers to protect from light, and placed on an orbital plate shaker with gentle shaking for 45 min at room temperature (RT). Subsequently, plates were read on the EnVision multi-label plate reader (Perkin-Elmer), and data plotted with GraphPad Prism v6.0 software (GraphPad, San Diego, CA, USA).

4.4. Competitive Binding to Recombinant Human FcRn

A competitive binding assay was used to assess relative affinities of different antibody samples to recombinant human FcRn (in-house expressed with transmembrane and cytoplasmic domains of FcRn replaced with a poly-histidine affinity tag). Ninety-six-well copper-coated plates (Thermo Scientific) were used to capture FcRn-His6 at 4 µg/mL in PBS, after which plates were washed with 0.15 M NaCl, 0.02% (*w/v*) Tween 20, and then incubated with blocking reagent (0.05 M MES, 0.025% (*w/v*) bovine serum albumin, 0.001% (*w/v*) Tween-20, pH 6.0, 10% (*v/v*) ChemiBlocker from Sigma-Aldrich, St. Louis, MO, USA. Plates were washed and serial dilutions of competitor test antibody in blocking reagent were added to plates in the presence of a fixed 4 µg/mL concentration of an indicator antibody (a biotinylated huIgG1). Plates were incubated at RT for 1 h, washed 3 times, and then incubated with a 1:10,000 dilution of HRP (Jackson ImmunoResearch Laboratories) at room temperature (RT) for 30 min to bind biotinylated antibody. Plates were washed and bound streptavidin-HRP was detected by adding TMB peroxidase substrate (Fitzgerald, Acton, MA, USA). Color development was stopped by addition of 0.5 M HCl. Optical densities were determined with a SpectraMax Plus384 plate reader (Molecular Devices, Sunnyvale, CA, USA) at 450 nm wavelength, and data plotted with GraphPad Prism v6.0 software.

4.5. Antibody-Dependent Cell-Mediated Cytotoxicity (ADCC)

Peripheral mononuclear cells (PBMCs) were isolated from heparinized blood from in-house donors. Eighty (80) mL of blood from 2 donors were typically used per experiment. Blood was diluted 2-fold with PBS and 30 mL was layered over 15 mL of Ficoll-Paque (Perkin-Elmer, Waltham, MA, USA) in a 50 mL conical centrifuge tube. Tubes were centrifuged at $400\times g$ at RT for 30 min. The upper plasma supernatant was removed and the interface white cell layer was collected and washed twice with PBS to remove Ficoll and majority of platelets. Cells were resuspended in IMDM-5% heat inactivated FBS with $1\times$ sodium pyruvate, $1\times$ NEAA, and $1\times$ penicillin-streptomycin ($100\times$ from Life Technologies, Carlsbad, CA, USA) for culturing overnight at 37 °C, 5% CO_2.

K2 cells were used as target cells at a ratio of 50 effector cells per 1 target cell. Target cells were pre-labeled with BATDA (bis (acetoxymethyl) 2,2′:6′,2″-terpyridine-6,″-dicarboxylate, DELFIA® EuTDA, PerkinElmer) for 25 min at 37 °C, washed 3 times in culture medium (IMDM with Glutamax, 10% (*v/v*) heat-inactivated Fetal Bovine serum (FBS), $1\times$ non-essential amino acids (NEAA), $1\times$ sodium pyruvate, $1\times$, penicillin-streptomycin; all from Life Technologies and resuspended in culture medium. Target cells (2×10^5 cells/mL, 50 µL) were added to test antibody (100 µL) in 96-well U-bottom plates, then effector cells (1×10^7 cells/mL, 50 µL) were added. Plates were centrifuged at $200\times g$ for 3 min, incubated at 37 °C for 2 h, and then centrifuged again at $200\times g$ for 3 min. A total of 20 µL of supernatant was removed per well, and cell lysis was measured by the addition of 200 µL of the DELFIA Europium-based reagent (PerkinElmer). Fluorescence was measured using an Envision 2101 Reader (PerkinElmer). Data were normalized to maximal cytotoxicity with 0.7% (*w/v*) Triton X-100 (Sigma-Aldrich, St. Louis, MO, USA) or 10% (*v/v*) Lysis Buffer (DELFIA®

PerkinElmer) and minimal lysis using target cells in the absence of any Ab. Samples were tested in duplicate. Percent specific lysis was calculated to be (sample lysis − minimal lysis) divided by (maximal lysis-minimal lysis) × 100. Data were fit to a sigmoidal dose-response model using GraphPad Prism v6.0 software.

4.6. Antibody-Dependent Cellular Phagocytosis (ADCP)

Monocytes were isolated from human PBMCs using a Monocyte Isolation Kit (Miltenyi, Auburn, AL, USA) and differentiated into macrophages for 1 week by culturing with 10 ng/mL recombinant human granulocyte-macrophage colony-stimulating factor (GM-CSF) and 10 ng/mL recombinant human IL-4 (both from R & D Systems, Minneapolis, MN, USA) in IMDM with Glutamax, 10% heat-inactivated FBS, 1× NEAA, 1× sodium pyruvate, 1× penicillin-streptomycin. Macrophages were labeled with PKH26 (fluorescent dye for cell membrane, Sigma) and K2 cells were labeled with PKH67 (Sigma). Labeled cells in IMDM-10% (*v/v*) heat-inactivated FBS media without phenol red were incubated for 5 h with a macrophage to K2 cell ratio of 1 effector to 1 target cell in the presence of test antibody. Two-color flow cytometry analyses were performed with a MACSQuant Flow Cytometer (Miltenyi) using optimal compensation in the B1 (PKH67) and B2 (PKH26) channels and gating on single cells. Dual-labeled cells (PKH26+/PKH67+) were considered to represent phagocytosis of K2 target cells by macrophages. Percent ADCP or phagocytosis of target cells was calculated to be 100× number of dual-labeled cells (macrophage + target) divided by the total number of target cells in the population (phagocytosed + non-phagocytosed) after >50,000 cell counts. The percent specific ADCP was obtained by subtracting from each sample the background value (macrophage + target incubated without Ab) [5].

4.7. Complement-Dependent Cytotoxicity (CDC)

K2 cells were used as target cells for CDC assays. A total of 50 μL of cells was added to wells of a 96-well plates for a final concentration of 8×10^4 cells per well in IMDM with Glutamax, 10% (*w/v*) heat-inactivated PBS, 1× NEAA, 1× sodium pyruvate, 1x penicillin-streptomycin. An additional 50 μL was added to the wells with or without test Abs and plates were incubated at 37 °C for 2 h. A total of 50 μL of 10% (*w/v*) rabbit complement (Invitrogen, Carlsbad, CA, USA) was added to the wells, and plates incubated for 20 min at 37 °C. All samples were performed in triplicate. The plates were centrifuged at 200× *g* for 3 min, 50 μL of supernatant was removed to separate plates, and CDC was measured with a LDH cytotoxicity detection kit (Roche, Indianapolis, IN, USA). Absorbance was measured using a Spectra Max Plus 384 (PerkinElmer). Data were fit to a sigmoidal dose-response model using GraphPad Prism v6.0 software. Maximal cytotoxicity was obtained with Triton X-100 (Sigma-Aldrich) and spontaneous release with cells and complement alone. Specific cell lysis was calculated as follows: Cytotoxicity (%) = 100 × (optical density (OD) of sample − OD of spontaneous release)/(OD of maximal lysis − OD of spontaneous release).

4.8. Animals

The FcγR-humanized (FcγR-hu) mice used in the T-cell activations studies express the different human FcγRs: CD16a (FcγRIIIa), CD16b (FcγRIIIb), CD32a (FcγRIIa), CD32b (FcγRIIb) and CD64 (FcγRI) and their endogenous mouse FcγRs have been inactivated [43]. Three strains of these C57BL/6 mice (8–10 weeks old) were used: FcγRα null females, FcγR-hu hemizygous (hemi) females, and FcγR-hu homozygous (homo) females.

Human FcRn transgenic animals (8–10 weeks old) used in PK studies were derived from C57BL/6 mice [64]. Tg32 mice (B6.Cg-Fcgrt[tmLDcr]Tg(FCGRT)32Dcr from The Jackson Laboratory) have their endogenous mouse FcRn α gene knocked out and are transgenic with the human FcRn α gene under the control of the native human gene promoter [65,66]. The FcRn transgenic strain show clinical chemical parameters like those found in wild-type mice except for endogenous huIgG levels, which are greatly reduced in these mice [67]. Tg32 hemi referred to mice hemizygous for the FcRn transgene, the latter derived by mating homozygous transgenic mice with FcRn α knockout mice (transgene copy number reduced by half).

Naive cynomolgus monkeys used in the PK study at WuXi AppTec., (Suzhong, China) were approximately 2 to 3.5 years old and weighed between 2.5 and 3.0 kg.

4.9. T Cell Activation

FcγR-hu mice were used in the in vivo binding and T cell activation studies. The preliminary study was done with three strains of mice (8–10 weeks old): FcRα null females, FcγR-hu (hemi) females, and FcγR-hu homozygous (homo) females. Test antibody was injected into the intraperitoneal cavity of the mice at 0.5 mg/kg, 10 mL/kg. Approximately 24 h later, mice were euthanized by CO_2 asphyxiation and their spleens removed and placed into tubes containing cold RPMI-1640, 5% (*v/v*) heat-inactivated FBS, 1% (*w/v*) L-glutamine.

Mouse splenocytes were prepared from 3 to 6 mice as single-cell suspensions from each individual spleen on the day of harvest. They were washed with media, followed by anucleated red blood cell depletion using hypotonic RBC lysis solution (eBiosciences). Splenocytes were analyzed by flow cytometry for cell surface expression of CD25 and CD69 T cell activation markers. Cells were resuspended in staining buffer consisting of PBS for viability staining (IR Live Dead, Invitrogen), washed, and then incubated with anti-CD16/32 (2.4G2, BD Biosciences, San Jose, CA, USA) to block nonspecific binding. Immunostaining was done in the presence of APC-CD25, FITC-CD8a, PE-CD4, PerCP-CD69, PE-Vio770-CD3ε (BD Biosciences, Biolegend or Miltenyi) at 4 °C for 30 min protected from light, and followed by two washes.

Cells were analyzed on the MACSQuant Analyzer (Miltenyi). Analyses of the multivariate data were performed using FlowJo v10 software (FlowJo, Ashland, OR, USA). The percent of CD8+ and CD3+ cells that were also positive for CD25 (or CD69) expression were based on data collected with greater than 50,000 cells from each sample.

4.10. B16F10 Syngeneic Tumor Model

FcγR-hu mice (8–10 weeks old female) were used for the syngeneic mouse lung tumor model. B16F10 tumor cells were intravenously injected into tail vein at 2×10^5 cells/mouse, 0.2 mL of 1×10^6 cells/mL in PBS [44,45]. At 30–60 min post IV injection, 0.2 mL of the test antibody sample was injected into the peritoneum. Antibody doses were given on day 0, 2, 4, 7, 9 and 11. On day 21, the mice were sacrificed and the lungs were weighed and scored for the number of metastases. The lung tumor index was determined by lung weight and tumor grade [68].

Tumor index = lung weight × grade for animal tumor

Grading:

1. Less than 10 tumor foci
2. 10–100 tumor foci
2.5. More than 100 foci, but countable
3. One lobe of the lung is full of tumor
4. Both lobes are full of tumor
5. Lungs are full of tumor and tumor growing out into cavities

4.11. Pharmacokinetics (PK)

For mouse antibody PK studies, female Tg32 hemi mice were injected with test antibody intravenously via tail vein at a dose of 2 mg/kg into 3 or 4 animals per group. Serial retro-orbital bleeds were obtained from CO_2-anesthesized mice at indicated time points and terminal bleed was taken by cardiac puncture. After 30 min at RT, blood samples were centrifuged at 2500 rpm for 15 min and serum collected for analyses. All PK studies were approved by the Institutional Animal Care and Use Committee at Janssen Research & Development, LLC (Spring House, PA, USA).

An electrochemiluminescent immunoassay was used to measure human antibody concentration in mouse sera. Briefly, Streptavidin Gold multiarray 96-well plates 96-well plates (Meso Scale Discovery, Rockville, MD, USA) were coated with 50 μL/well of 1 μg/mL biotinylated F(ab')$_2$ goat anti-human IgG (H + L, Jackson Immunochemical) in Starting Block T20 (Thermo Scientific) overnight at 4 °C and washed with Tris-buffered saline with 0.05% (*w/v*) Tween 20 (TBST from Sigma). Standards and serum samples were prepared in sample buffer (1% (*w/v*) bovine serum albumin in TBST and 20 mM EDTA) added to plates and incubated for 2 h at RT on a shaker. Plates were washed and incubated for 1 h with 1 μg/mL MSD-Sulfo (ruthenium)-labeled with a pan huIgG1 Ab, R10Z8E9. Plates were washed, 1× Read buffer T was added and plates were read on the MSD Sector Imager 6000 (Meso Scale Discovery).

Terminal half-life (t$_{1/2}$) calculations of the elimination phase for PK studies were determined using a 1-phase exponential decay model fitted by linear regression of natural log concentration versus time using Prism version 6.0 software. The least squares nonlinear decay model was weighted by 1/fitted concentration. Half-life calculations of the elimination phase were determined using the formula t$_{1/2}$= −ln2/β, where β is the slope of the line fitted by the least square regression analysis starting after the first dose.

Monkey PK studies were performed at WuXi Apptec in China. Three animals were IV injected with test antibody at 1.5 mg/kg, and 1 mL of blood was collected via a cephalic vein at pre-dose, and at day 1, day 3, day 5, day 8, day 15 and day 22 post-dose. Serum samples were prepared and PK analyses were performed at Frontage (Shanghai) using a similar MSD format. Protocols were reviewed and approved by WuXi AppTec Institutional Animal Care and Use Committee (IACUC) prior to procedures (MGMT-011; TECH-030).

4.12. ImmunoFilter^{TM} Analyses

The amino acid sequences for the different variants were analyzed in silico using ImmunoFilter^{TM}, an HLA class II-peptide binding prediction tool to predict comparative immunogenicity in humans (v2.7, Xencor, Inc., Monrovia, CA, USA, examples [69–71]). This prediction tool uses an immunochemical data set of peptide agretope binding to class II major histocompatibility complex (MHC) which can assess potential immunogenicity for more than 95% of U.S. population, based on empirical binding data. Output provided includes raw binding scores for peptides across a sequence of interest, standardized binding scores, binding probabilities to each allelic combination, and summary *IScores*, which are weighted, population-relevant values. For analyses, peptides from the wild-type and variant sequences with 100% identity to each other were excluded by application of a tolerance threshold, and only peptides spanning the sequences of interest were included. Resulting *IScores* were averaged across all loci, and plotted. Higher *IScores* indicate a higher predicted immunogenic risk (PIR).

4.13. Developability

To assess whether antibodies with silent Fc regions would have good manufacturing properties, biophysical analytical tests were performed for thermal stability and solubility.

4.13.1. Differential Scanning Calorimetry (DSC)

DSC experiments were performed using a MicroCal Auto VP-capillary DSC system (Malvern Instruments Ltd., Malvern, UK) in which temperature differences between the reference and sample cell were continuously measured and converted to power units. Samples were heated from 25 °C to 110 °C at a rate of 1 °C/min. A pre-scan time of 10 min and a filtering period of 10 s were used for each run. DSC measurements were made at sample concentrations of approximately 0.5 mg/mL in 1× PBS buffer in duplicate. Analysis of the resulting data was performed using MicroCal Origin 7 software (MicroCal, Northampton, MA, USA).

4.13.2. Concentration Assessment

Antibody samples were concentrated by centrifugation at $2250 \times g$ using Amicon Ultra-15 centrifugal filter units with Ultracel-30 membranes (Sigma-Aldrich). Samples were inspected for signs of precipitation until volumes were reached for a concentration of 40–50 mg/mL. The sequence-predicted absorbance constants (A280/mg/mL) for each antibody were used to calculate sample concentrations at absorbance 280 nm.

4.13.3. Dynamic Light Scattering (DLS)

Particle size and size distributions were determined using DLS on a DynaPro Plate Reader (Wyatt Technologies Corporation, Santa Barbara, CA, USA) at 23 °C. For each analysis, 30 μL of each sample at 1 mg/mL was placed in a 384-well black polystyrene plate with a clear flat bottom (Corning, CLS3540). Triplicate measurements were performed for each sample with each measurement consisting of 20 runs.

4.13.4. Static Light Scattering (SLS), Thermal Melting (Tm) and Thermal Aggregation (Tagg) Analyses

Tm was determined for each sample using intrinsic fluorescence with the Uncle instrument (Unchained Labs, Pleasanton, CA, USA). Tagg was assessed by Static Light Scattering (SLS) to monitor protein aggregation using the same instrument. For the combined Tm and Tagg method, antibody was loaded and run with a thermal ramp from 15 to 95 °C; at a ramp rate of 0.3 °C/min.

4.13.5. Size Exclusion Chromatography (SEC)

Samples were separated over a TOSOH TSKgel BioAssist G3SWxL column (7.8 mm × 30 cm, 5 μm, TOSOH) that had been equilibrated with PBS supplemented with 500 mM NaCl, at a flow rate of 0.5 mL/min using an Agilent 1100-series HPLC (Agilent Technologies, Santa Clara, CA, USA). A target of 100–200 μg of total protein was injected per run. Peaks were monitored using absorbance at 280 nm. Data analysis of species found in each sample was performed using ChemStation software (Agilent Technologies).

4.13.6. Low pH Treatment

Exposure to low pH was performed for accelerated stability testing. Protein samples were prepared at a concentration of 1 mg/mL. Samples were dialyzed for 6 h into 0.05 M sodium acetate buffer, pH 3.5; then dialyzed for 16 h in 0.1 M PBS, pH 7.4, and stored at 4 °C prior to analyses.

4.14. Crystallography

Recombinant huIgG1σ Fc and huIgG4σ2 Fc were transiently expressed in HEK 293 cells and purified by Protein A affinity chromatography at Sino Biological Inc. (China). Recombinant huIgG4σ1 Fc was transiently expressed in 293 Expi cells and purified in two steps using Protein A affinity chromatography and size exclusion chromatography (Superdex 200 PG) at Aldevron (Fargo, ND, USA). Proteins were delivered in 20 mM Tris, 50 mM NaCl, pH 7.5 at concentrations ranging from 2 to 4 mg/mL. Fc molecules were further concentrated to 10–13 mg/mL prior to crystallization. Crystallization experiments employed the vapor-diffusion method. Crystallization drops were set up using a Mosquito liquid handling robot (TTP Labtech, Melbourn, UK) in 96-well Corning 3550 (huIgG1σ and huIgG4σ2) or MRC 2 well (huIgG4σ1) crystallization plates. Diffraction quality crystal were obtained for all three Fc regions in 9–10% (w/v) PEG 20,000, 0.1 M sodium acetate, pH 5.5 (reservoir condition for huIgG4σ2 additionally contained 5% (w/v) MPD). Prior to data collection, crystals were cryo-protected in reservoir supplemented with 20% (w/v) glycerol and flash frozen in liquid N_2. X-ray diffraction data for huIgG1σ and huIgG4σ1 were collected on the IMCA-CAT beam line (17-ID) at the Advanced Photon Source (APS) at Argonne National Laboratory equipped with a DECTRIS Pilatus 6M pixel array detector. Diffraction data for huIgG4σ2 were collected by Shamrock

Structures, LLC on the SER-CAT beam line (22-ID) at APS equipped with a Rayonix 300HS CCD detector. All data were processed with the program XDS [72].

Initial phases were determined by the method of molecular replacement with the program Phaser [73] as implemented in the CCP4 suite of programs [74]. Individual C_H2 and C_H3 domains isolated from chain A of PDB 3AVE [40] (huIgG1 Fc) were provided as search models for huIgG1σ; and for huIgG4σ1 and huIgG4σ1, individual C_H2 and C_H3 domains were used from a previously refined internal structure of wild-type huIgG4 Fc. Phaser positioned the equivalent of one Fc dimer in the asymmetric unit in space group $P2_12_12_1$. The structures underwent rounds of rebuilding and refinement with the programs Crystallographic Object-Oriented Toolkit (COOT) [75] and Phenix [76,77] respectively. Data collection and refinement statistics are provided in Table 2. The atomic coordinates and structure factors are archived in the Protein Data Bank under the accession numbers 5W5L, 5W5M, and 5W5N corresponding to huIgG1σ Fc, huIgG4 σ1 Fc, and huIgG4 σ2 Fc, respectively.

4.15. Molecular Dynamics Simulations

Explicit solvent MD simulations were performed on intact Fc to investigate the flexibility of BC and FG loops. Two simulations were initiated from the available crystal structures: IgG1 WT using the crystal structure (PDB 3AVE) of IgG1 wild-type Fc and IgG1σ using the crystal structure described here. Additional two simulations were initiated after modelling mutations in the wild-type crystal structure: IgG1-H268A containing H268A mutation and IgG1-SS containing A330S and P331S mutations. The simulations included both chains of C_H2 and C_H3 domains, and the glycans present in the corresponding crystal structures. The simulations were set up in the Maestro graphical user interface and run using the Desmond program (multisim version 3.8.5.19 and mmshare version 3.5) [78], both part of the Schrodinger 2016-3 suite [79]. The systems were protonated at neutral pH and centered in an orthorhombic box such that the minimum distance from any protein atom to the box wall was 10 Å. The box was solvated using SPC [80] water molecules and counter ions were added to neutralize the system. OPLS3 force field [81] was used as the potential energy function for the protein. Replica Exchange Solute Tempering [82] (REST) MD simulations were performed at 300 K and 1 bar using 16 replicas with the BC (residue numbers 267–273) and FG (residue numbers 322–333) residues of only chain B specified as "hot". REST MD simulations were designed to enhance sampling of the hot subset of the full simulation system. Note that sampling is enhanced only in chain B since chain A loops are cold. To get an idea of the degree by which sampling is enhanced by REST, we compared the backbone fluctuations of the two chains in the C_H2 domain. Figure S7 shows that it was advantageous to use the REST approach for all systems except IgG1σ. Default implementation of REST in Desmond was used for initial equilibration, setting up energy function of each replica, specifying exchange protocol and recording trajectory data. Simulations were performed on AWS (Amazon Web Services) cloud computing platform with each simulation employing 8 NVIDIA Tesla K80 GPU cards (Nvidia, Santa Clara, CA, USA). The production run was 100 nanoseconds long and the final trajectory from each replica contained 4166 conformations saved at an interval of 24 ps. Number of atoms in the simulations was approximately 53,000 and a single simulation took approximately 74 hours. A time step of 2 femtoseconds was used and exchanges were attempted at an interval of 1.2 picoseconds. In all simulations, the acceptance ratio of exchanges between the adjacent replicas was observed to be between 0.2 and 0.4. The results presented here correspond to the trajectory from the physical replica for which the energy function is unperturbed.

5. Conclusions

A requirement for truly silent human Fc designs is evident as more antibodies are being developed for immunotherapy. The Fc engineered human mutations described here as IgG1σ, IgG4σ1 and IgG4σ2 demonstrate equal or lower immune functionality than the corresponding IgG2σ design. Lack of immune effector function is demonstrated in vitro with FcγR binding, cytotoxicity assays, and in vivo with T cell activation and tumor inhibition studies. Crystal structures and simulations of these Fc variants reveal altered conformational preferences within the lower hinge and BC and FG loops relative to wild-type IgG, providing a structural rationalization for diminished Fc receptor engagement. Immunogenicity predictions, pharmacokinetic studies in mice and monkeys, and biophysical analyses also support these novel mutations as optimized silent Fc choices for the development of therapeutic antibodies.

Supplementary Materials: The following are available online at www.mdpi.com/2073-4468/6/3/12/s1; Table S1: Cynomolgus monkey PK parameters for Ab2. Figure S1: Electron density map about lower hinge residues, Figure S2: Altered lower hinge conformation of huIgG1σ is not a requirement of crystal packing, Figure S3: Crystal packing does not prevent huIgG1σ FG loop from adopting a flipped-out conformation, Figure S4: Conformation distribution of P329, Figure S5: Conformation distribution of P271, Figure S6: Interaction of lower hinge and FG loop with FcγR, Figure S7: REST MD enhances loop sampling.

Acknowledgments: The authors thank protein expression and purification groups and the Structural Biology Group in Antibody Drug Discovery in Biologics Research at Janssen R & D, LLC, and Sino Biologics, China, for the generation and purification of huIgG reagents used in this study. Additional thanks go to members of Antitope and WuXi AppTec for their efforts in T-cell epitope analysis and PK, respectively, and Lingjie Xu of Janssen Research & Development, China, for performing the monkey PK data analyses. Special thanks goes to Chichi Huang for continual encouragement, Elena Catterton and Ekaterina Shatalova for diligent patent literature and summary; and to Joani Wendel for formatting this manuscript. Use of the IMCA-CAT beamline 17-ID at the Advanced Photon Source was supported by the companies of the Industrial Macromolecular Crystallography Association through a contract with Hauptman-Woodward Medical Research Institute. This research used resources of the Advanced Photon Source, a U.S. Department of Energy (DOE) Office of Science User Facility operated for the DOE Office of Science by Argonne National Laboratory under Contract No. DE-AC02-06CH11357.

Author Contributions: S.H.T. and S.G.M. designed and performed research, collected, analyzed and interpreted data, performed statistical analysis, and wrote the manuscript; A.A.A. designed and conducted crystallography studies and contributed to manuscript writing; S.S. performed the molecular simulations and contributed to manuscript writing; R.E. and D.S. designed and performed the DSC experiments and data analyses; X.L. performed the high concentration stability and SEC experiments and data analyses; A.G. performed the thermal stability, colloidal stability and DLS experiments and data analyses; S.-J.W. conceived and designed the biophysical analysis experiments and contributed to biophysical part of manuscript; R.D. performed the animal PK studies; G.L.G. advised on crystallography, modeling strategy, and edited manuscript; J.L. advised on crystallography and modeling strategy; M.L.C. interpreted data, contributed to writing and editing the manuscript; B.J.S. designed the strategy for antibody mutations, interpreted data and contributed to writing of manuscript.

Conflicts of Interest: Janssen Research & Development, LLC provided funding for the research. All authors are employees of Janssen Research & Development, LLC.

References

1. Gergely, J.; Sarmay, G.; Rajnavolgyi, E. Regulation of antibody production mediated by Fc gamma receptors, IgG binding factors, and IgG Fc-binding autoantibodies. *Crit. Rev. Biochem. Mol. Biol.* **1992**, *27*, 191–225. [CrossRef] [PubMed]
2. Roopenian, D.C.; Akilesh, S. FcRn: The neonatal Fc receptor comes of age. *Nat. Rev. Immunol.* **2007**, *7*, 715–725. [CrossRef] [PubMed]
3. Strohl, W.R. Optimization of Fc-mediated effector functions of monoclonal antibodies. *Curr. Opin. Biotechnol.* **2009**, *20*, 685–691. [CrossRef] [PubMed]
4. Shields, R.L.; Namenuk, A.K.; Hong, K.; Meng, Y.G.; Rae, J.; Briggs, J.; Xie, D.; Lai, J.; Stadlen, A.; Li, B.; et al. High resolution mapping of the binding site on human IgG1 for Fc gamma RI, Fc gamma RII, Fc gamma RIII, and FcRn and design of IgG1 variants with improved binding to the Fc gamma R. *J. Biol. Chem.* **2001**, *276*, 6591–6604. [CrossRef] [PubMed]
5. Lazar, G.A.; Dang, W.; Karki, S.; Vafa, O.; Peng, J.S.; Hyun, L.; Chan, C.; Chung, H.S.; Eivazi, A.; Yoder, S.C.; et al. Engineered antibody Fc variants with enhanced effector function. *Proc. Natl. Acad. Sci. USA* **2006**, *103*, 4005–4010. [CrossRef] [PubMed]
6. Stavenhagen, J.B.; Gorlatov, S.; Tuaillon, N.; Rankin, C.T.; Li, H.; Burke, S.; Huang, L.; Vijh, S.; Johnson, S.; Bonvini, E.; et al. Fc optimization of therapeutic antibodies enhances their ability to kill tumor cells in vitro and controls tumor expansion in vivo via low-affinity activating Fcgamma receptors. *Cancer Res.* **2007**, *67*, 8882–8890. [CrossRef] [PubMed]
7. Czajkowsky, D.M.; Hu, J.; Shao, Z.; Pless, R.J. Fc-fusion proteins: New developments and future perspectives. *EMBO Mol. Med.* **2012**, *4*, 1015–1028. [CrossRef] [PubMed]

8. Pollreisz, A.; Assinger, A.; Hacker, S.; Hoetzenecker, K.; Schmid, W.; Lang, G.; Wolfsberger, M.; Steinlechner, B.; Bielek, E.; Lalla, E.; et al. Intravenous immunoglobulins induce CD32-mediated platelet aggregation in vitro. *Br. J. Dermatol.* **2008**, *159*, 578–584. [CrossRef] [PubMed]
9. Kontermann, R.E.; Brinkmann, U. Bispecific antibodies. *Drug Discov. Today* **2015**, *20*, 838–847. [CrossRef] [PubMed]
10. Daeron, M. Fc receptor biology. *Ann. Rev. Immunol.* **1997**, *15*, 203–234. [CrossRef] [PubMed]
11. Labrijn, A.F.; Aalberse, R.C.; Schuurman, J. When binding is enough: Nonactivating antibody formats. *Curr. Opin. Immunol.* **2008**, *20*, 479–485. [CrossRef] [PubMed]
12. Strohl, W.R. Fusion proteins for half-life extension of biologics as a strategy to make biobetters. *BioDrugs* **2015**, *29*, 215–239. [CrossRef] [PubMed]
13. Tao, M.H.; Morrison, S.L. Studies of aglycosylated chimeric mouse-human IgG. Role of carbohydrate in the structure and effector functions mediated by the human IgG constant region. *J. Immunol.* **1989**, *143*, 2595–2601. [PubMed]
14. Sazinsky, S.L.; Ott, R.G.; Silver, N.W.; Tidor, B.; Ravetch, J.V.; Wittrup, K.D. Aglycosylated immunoglobulin G1 variants productively engage activating Fc receptors. *Proc. Natl. Acad. Sci. USA* **2008**, *105*, 20167–20172. [CrossRef] [PubMed]
15. Hristodorov, D.; Fischer, R.; Linden, L. With or without sugar? (A)glycosylation of therapeutic antibodies. *Mol. Biotechnol.* **2013**, *54*, 1056–1068. [CrossRef] [PubMed]
16. Jacobsen, F.W.; Stevenson, R.; Li, C.; Salimi-Moosavi, H.; Liu, L.; Wen, J.; Luo, Q.; Daris, K.; Buck, L.; Miller, S.; et al. Engineering an IgG scaffold lacking effector function with optimized developability. *J. Biol. Chem.* **2017**, *292*, 1865–1875. [CrossRef] [PubMed]
17. Chappel, M.S.; Isenman, D.E.; Everett, M.; Xu, Y.Y.; Dorrington, K.J.; Klein, M.H. Identification of the Fc gamma receptor class I binding site in human IgG through the use of recombinant IgG1/IgG2 hybrid and point-mutated antibodies. *Proc. Natl. Acad. Sci. USA* **1991**, *88*, 9036–9040. [CrossRef] [PubMed]
18. Smith, S.L. Ten years of orthoclone OKT3 (muromonab-CD3): A review. *J. Transpl. Coord. Off. Publ. N. Am. Transpl. Coord. Organ.* **1996**, *6*, 109–119, quiz 120–101.
19. Alegre, M.L.; Peterson, L.J.; Xu, D.; Sattar, H.A.; Jeyarajah, D.R.; Kowalkowski, K.; Thistlethwaite, J.R.; Zivin, R.A.; Jolliffe, L.; Bluestone, J.A. A non-activating "humanized" anti-CD3 monoclonal antibody retains immunosuppressive properties in vivo. *Transplantation* **1994**, *57*, 1537–1543. [CrossRef] [PubMed]
20. Bruhns, P.; Iannascoli, B.; England, P.; Mancardi, D.A.; Fernandez, N.; Jorieux, S.; Daeron, M. Specificity and affinity of human Fcgamma receptors and their polymorphic variants for human IgG subclasses. *Blood* **2009**, *113*, 3716–3725. [CrossRef] [PubMed]
21. Ravetch, J.V.; Kinet, J.P. Fc receptors. *Ann. Rev. Immunol.* **1991**, *9*, 457–492. [CrossRef] [PubMed]
22. Nimmerjahn, F.; Ravetch, J.V. Divergent immunoglobulin G subclass activity through selective Fc receptor binding. *Science* **2005**, *310*, 1510–1512. [CrossRef] [PubMed]
23. Brezski, R.J.; Georgiou, G. Immunoglobulin isotype knowledge and application to Fc engineering. *Curr. Opin. Immunol.* **2016**, *40*, 62–69. [CrossRef] [PubMed]
24. Alegre, M.L.; Collins, A.M.; Pulito, V.L.; Brosius, R.A.; Olson, W.C.; Zivin, R.A.; Knowles, R.; Thistlethwaite, J.R.; Jolliffe, L.K.; Bluestone, J.A. Effect of a single amino acid mutation on the activating and immunosuppressive properties of a "humanized" OKT3 monoclonal antibody. *J. Immunol.* **1992**, *148*, 3461–3468. [PubMed]
25. Xu, D.; Alegre, M.L.; Varga, S.S.; Rothermel, A.L.; Collins, A.M.; Pulito, V.L.; Hanna, L.S.; Dolan, K.P.; Parren, P.W.; Bluestone, J.A.; et al. In vitro characterization of five humanized OKT3 effector function variant antibodies. *Cell. Immunol.* **2000**, *200*, 16–26. [CrossRef] [PubMed]
26. Mueller, J.P.; Giannoni, M.A.; Hartman, S.L.; Elliott, E.A.; Squinto, S.P.; Matis, L.A.; Evans, M.J. Humanized porcine VCAM-specific monoclonal antibodies with chimeric IgG2/G4 constant regions block human leukocyte binding to porcine endothelial cells. *Mol. Immunol.* **1997**, *34*, 441–452. [CrossRef]
27. An, Z.; Forrest, G.; Moore, R.; Cukan, M.; Haytko, P.; Huang, L.; Vitelli, S.; Zhao, J.Z.; Lu, P.; Hua, J.; et al. IgG2m4, an engineered antibody isotype with reduced Fc function. *MAbs* **2009**, *1*, 572–579. [CrossRef] [PubMed]
28. Oganesyan, V.; Gao, C.; Shirinian, L.; Wu, H.; Dall'Acqua, W.F. Structural characterization of a human Fc fragment engineered for lack of effector functions. *Acta Crystallogr. Sect. D Biol. Crystallogr.* **2008**, *64*, 700–704. [CrossRef] [PubMed]

29. Vafa, O.; Gilliland, G.L.; Brezski, R.J.; Strake, B.; Wilkinson, T.; Lacy, E.R.; Scallon, B.; Teplyakov, A.; Malia, T.J.; Strohl, W.R. An engineered fc variant of an IgG eliminates all immune effector functions via structural perturbations. *Methods* **2014**, *65*, 114–126. [CrossRef] [PubMed]
30. Dillon, T.M.; Ricci, M.S.; Vezina, C.; Flynn, G.C.; Liu, Y.D.; Rehder, D.S.; Plant, M.; Henkle, B.; Li, Y.; Deechongkit, S.; et al. Structural and functional characterization of disulfide isoforms of the human IgG2 subclass. *J. Biol. Chem.* **2008**, *283*, 16206–16215. [CrossRef] [PubMed]
31. Thakkar, S.V.; Sahni, N.; Joshi, S.B.; Kerwin, B.A.; He, F.; Volkin, D.B.; Middaugh, C.R. Understanding the relevance of local conformational stability and dynamics to the aggregation propensity of an IgG1 and IgG2 monoclonal antibodies. *Protein Sci. Publ. Protein Soc.* **2013**, *22*, 1295–1305. [CrossRef] [PubMed]
32. Wypych, J.; Li, M.; Guo, A.; Zhang, Z.; Martinez, T.; Allen, M.J.; Fodor, S.; Kelner, D.N.; Flynn, G.C.; Liu, Y.D.; et al. Human IgG2 antibodies display disulfide-mediated structural isoforms. *J. Biol. Chem.* **2008**, *283*, 16194–16205. [CrossRef] [PubMed]
33. Schlothauer, T.; Herter, S.; Koller, C.F.; Grau-Richards, S.; Steinhart, V.; Spick, C.; Kubbies, M.; Klein, C.; Umana, P.; Mossner, E. Novel human IgG1 and IgG4 Fc-engineered antibodies with completely abolished immune effector functions. *Protein Eng. Des. Sel. PEDS* **2016**, *29*, 457–466. [CrossRef] [PubMed]
34. Escobar-Cabrera, E.; Lario, P.; Baardsnes, J.; Schrag, J.; Durocher, Y.; Dixit, S. Asymmetric Fc engineering for bispecific antibodies with reduced effector function. *Antibodies* **2017**, *6*, 7. [CrossRef]
35. Angal, S.; King, D.J.; Bodmer, M.W.; Turner, A.; Lawson, A.D.; Roberts, G.; Pedley, B.; Adair, J.R. A single amino acid substitution abolishes the heterogeneity of chimeric mouse/human (IgG4) antibody. *Mol. Immunol.* **1993**, *30*, 105–108. [CrossRef]
36. Goldberg, A.; Geppert, T.; Schiopu, E.; Frech, T.; Hsu, V.; Simms, R.W.; Peng, S.L.; Yao, Y.; Elgeioushi, N.; Chang, L.; et al. Dose-escalation of human anti-interferon-alpha receptor monoclonal antibody medi-546 in subjects with systemic sclerosis: A phase 1, multicenter, open label study. *Arthritis Res. Ther.* **2014**, *16*, R57. [CrossRef] [PubMed]
37. Edelman, G.M.; Cunningham, B.A.; Gall, W.E.; Gottlieb, P.D.; Rutishauser, U.; Waxdal, M.J. The covalent structure of an entire gammaG immunoglobulin molecule. *Proc. Natl. Acad. Sci. USA* **1969**, *63*, 78–85. [CrossRef] [PubMed]
38. Kabat, E.A.; National Institutes of Health (U.S.) Office of the Director. *Sequences of Proteins of Immunological Interest*, 5th ed.; DIANE Publishing: Collingdale, PA, USA, 1991.
39. Lefranc, M.P.; Giudicelli, V.; Ginestoux, C.; Bodmer, J.; Muller, W.; Bontrop, R.; Lemaitre, M.; Malik, A.; Barbie, V.; Chaume, D. Imgt, the international immunogenetics database. *Nucleic Acids Res.* **1999**, *27*, 209–212. [CrossRef] [PubMed]
40. Matsumiya, S.; Yamaguchi, Y.; Saito, J.; Nagano, M.; Sasakawa, H.; Otaki, S.; Satoh, M.; Shitara, K.; Kato, K. Structural comparison of fucosylated and nonfucosylated Fc fragments of human immunoglobulin G1. *J. Mol. Biol.* **2007**, *368*, 767–779. [CrossRef] [PubMed]
41. Scallon, B.J.; Moore, M.A.; Trinh, H.; Knight, D.M.; Ghrayeb, J. Chimeric anti-TNF-α monoclonal antibody cA2 binds recombinant transmembrane TNF-α and activates immune effector functions. *Cytokine* **1995**, *7*, 251–259. [CrossRef] [PubMed]
42. Scallon, B.; McCarthy, S.; Radewonuk, J.; Cai, A.; Naso, M.; Raju, T.S.; Capocasale, R. Quantitative in vivo comparisons of the Fc gamma receptor-dependent agonist activities of different fucosylation variants of an immunoglobulin G antibody. *Int. Immunopharmacol.* **2007**, *7*, 761–772. [CrossRef] [PubMed]
43. Smith, P.; DiLillo, D.J.; Bournazos, S.; Li, F.; Ravetch, J.V. Mouse model recapitulating human Fcgamma receptor structural and functional diversity. *Proc. Natl. Acad. Sci. USA* **2012**, *109*, 6181–6186. [CrossRef] [PubMed]
44. Bevaart, L.; Jansen, M.J.; van Vugt, M.J.; Verbeek, J.S.; van de Winkel, J.G.; Leusen, J.H. The high-affinity IgG receptor, FcgammaRI, plays a central role in antibody therapy of experimental melanoma. *Cancer Res.* **2006**, *66*, 1261–1264. [CrossRef] [PubMed]
45. Boross, P.; Jansen, J.H.M.; van Tetering, G.; Nederend, M.; Brandsma, A.; Meyer, S.; Torfs, E.; van den Ham, H.-J.; Meulenbroek, L.; de Haij, S.; et al. Anti-tumor activity of human IgG1 anti-gp75 TA99 mab against B16F10 melanoma in human FcgammaRI transgenic mice. *Immunol. Lett.* **2014**, *160*, 151–157. [CrossRef] [PubMed]
46. Garber, E.; Demarest, S.J. A broad range of Fab stabilities within a host of therapeutic IgGs. *Biochem. Biophys. Res. Commun.* **2007**, *355*, 751–757. [CrossRef] [PubMed]

47. Ferrara, C.; Grau, S.; Jager, C.; Sondermann, P.; Brunker, P.; Waldhauer, I.; Hennig, M.; Ruf, A.; Rufer, A.C.; Stihle, M.; et al. Unique carbohydrate-carbohydrate interactions are required for high affinity binding between FcgammaRIII and antibodies lacking core fucose. *Proc. Natl. Acad. Sci. USA* **2011**, *108*, 12669–12674. [CrossRef] [PubMed]

48. Mimoto, F.; Katada, H.; Kadono, S.; Igawa, T.; Kuramochi, T.; Muraoka, M.; Wada, Y.; Haraya, K.; Miyazaki, T.; Hattori, K. Engineered antibody Fc variant with selectively enhanced FcγRIIb binding over both FcγRIIa(R131) and FcγRIIa(H131). *Protein Eng. Des. Sel.* **2013**, *26*, 589–598. [CrossRef] [PubMed]

49. Teplyakov, A.; Zhao, Y.; Malia, T.J.; Obmolova, G.; Gilliland, G.L. IgG2 Fc structure and the dynamic features of the IgG CH2-CH3 interface. *Mol. Immunol.* **2013**, *56*, 131–139. [CrossRef] [PubMed]

50. Davies, A.M.; Rispens, T.; Ooijevaar-de Heer, P.; Gould, H.J.; Jefferis, R.; Aalberse, R.C.; Sutton, B.J. Structural determinants of unique properties of human IgG4-Fc. *J. Mol. Biol.* **2014**, *426*, 630–644. [CrossRef] [PubMed]

51. Kiyoshi, M.; Caaveiro, J.M.; Kawai, T.; Tashiro, S.; Ide, T.; Asaoka, Y.; Hatayama, K.; Tsumoto, K. Structural basis for binding of human IgG1 to its high-affinity human receptor FcgammaRI. *Nat. Commun.* **2015**, *6*, 6866. [CrossRef] [PubMed]

52. Desjarlais, J.R.; Lazar, G.A.; Zhukovsky, E.A.; Chu, S.Y. Optimizing engagement of the immune system by anti-tumor antibodies: An engineer's perspective. *Drug Discov. Today* **2007**, *12*, 898–910. [CrossRef] [PubMed]

53. Nimmerjahn, F.; Ravetch, J.V. Fcγ receptors: Old friends and new family members. *Immunity* **2006**, *24*, 19–28. [CrossRef] [PubMed]

54. Nimmerjahn, F.; Ravetch, J.V. The antiinflammatory activity of IgG: The intravenous IgG paradox. *J. Exp. Med.* **2007**, *204*, 11–15. [CrossRef] [PubMed]

55. Nimmerjahn, F.; Ravetch, J.V. Fcγ receptors as regulators of immune responses. *Nat. Rev. Immunol.* **2008**, *8*, 34–47. [CrossRef] [PubMed]

56. Cole, M.S.; Stellrecht, K.E.; Shi, J.D.; Homola, M.; Hsu, D.H.; Anasetti, C.; Vasquez, M.; Tso, J.Y. Hum291, a humanized anti-CD3 antibody, is immunosuppressive to T cells while exhibiting reduced mitogenicity in vitro. *Transplantation* **1999**, *68*, 563–571. [CrossRef] [PubMed]

57. Nimmerjahn, F.; Ravetch, J.V. Analyzing antibody-Fc receptor interactions. *Methods Mol. Biol.* **2008**, *415*, 151–162. [PubMed]

58. Davies, A.M.; Jefferis, R.; Sutton, B.J. Crystal structure of deglycosylated human IgG4-Fc. *Mol. Immunol.* **2014**, *62*, 46–53. [CrossRef] [PubMed]

59. Hara, I.; Takechi, Y.; Houghton, A.N. Implicating a role for immune recognition of self in tumor rejection: Passive immunization against the brown locus protein. *J. Exp. Med.* **1995**, *182*, 1609–1614. [CrossRef] [PubMed]

60. Firan, M.; Bawdon, R.; Radu, C.; Ober, R.J.; Eaken, D.; Antohe, F.; Ghetie, V.; Ward, E.S. The MHC class I-related receptor, FcRn, plays an essential role in the maternofetal transfer of γ-globulin in humans. *Int. Immunol.* **2001**, *13*, 993–1002. [CrossRef] [PubMed]

61. Stubenrauch, K.; Wessels, U.; Lenz, H. Evaluation of an immunoassay for human-specific quantitation of therapeutic antibodies in serum samples from non-human primates. *J. Pharm. Biomed. Anal.* **2009**, *49*, 1003–1008. [CrossRef] [PubMed]

62. Labrijn, A.F.; Meesters, J.I.; de Goeij, B.E.; van den Bremer, E.T.; Neijssen, J.; van Kampen, M.D.; Strumane, K.; Verploegen, S.; Kundu, A.; Gramer, M.J.; et al. Efficient generation of stable bispecific IgG1 by controlled Fab-arm exchange. *Proc. Natl. Acad. Sci. USA* **2013**, *110*, 5145–5150. [CrossRef] [PubMed]

63. Scallon, B.; Cai, A.; Solowski, N.; Rosenberg, A.; Song, X.Y.; Shealy, D.; Wagner, C. Binding and functional comparisons of two types of tumor necrosis factor antagonists. *J. Pharm. Exp. Ther.* **2002**, *301*, 418–426. [CrossRef]

64. Roopenian, D.C.; Christianson, G.J.; Sproule, T.J. Human FcRn transgenic mice for pharmacokinetic evaluation of therapeutic antibodies. *Methods Mol. Biol.* **2010**, *602*, 93–104. [PubMed]

65. Proetzel, G.; Wiles, M.V.; Roopenian, D.C. Genetically engineered humanized mouse models for preclinical antibody studies. *BioDrugs Clin. Immunother. Biopharm. Gene Ther.* **2014**, *28*, 171–180. [CrossRef] [PubMed]

66. Proetzel, G.; Roopenian, D.C. Humanized FcRn mouse models for evaluating pharmacokinetics of human IgG antibodies. *Methods* **2014**, *65*, 148–153. [CrossRef] [PubMed]

67. Stein, C.; Kling, L.; Proetzel, G.; Roopenian, D.C.; de Angelis, M.H.; Wolf, E.; Rathkolb, B. Clinical chemistry of human FcRn transgenic mice. *Mamm. Genome Off. J. Int. Mamm. Genome Soc.* **2012**, *23*, 259–269. [CrossRef] [PubMed]

68. Gautam, A.; Waldrep, J.C.; Densmore, C.L.; Koshkina, N.; Melton, S.; Roberts, L.; Gilbert, B.; Knight, V. Growth inhibition of established B16-F10 lung metastases by sequential aerosol delivery of p53 gene and 9-nitrocamptothecin. *Gene Ther.* **2002**, *9*, 353–357. [CrossRef] [PubMed]

69. Bernett, M.J.; Dahiyat, B.; Desjarlais, J.; Lazar, G.A.; Moore, G.L. Antibodies with Modified Isoelectric Points and Immunofiltering. US 14/853,622, 10 March 2016.

70. Jacobs, S. Stabilized Fibronectin Domain Compositions, Methods and Uses. EP20110775616, 22 June 2016.

71. Leung, D.D.M.; Swanson, B.A.; Tang, Y.; Luan, P.; Witcher, D.R. Anti-Hepcidin Antibodies and Uses Thereof. EP20120172307, 28 November 2012.

72. Kabsch, W. Xds. *Acta Crystallogr. Sect. D Biol. Crystallogr.* **2010**, *66*, 125–132. [CrossRef] [PubMed]

73. McCoy, A.J.; Grosse-Kunstleve, R.W.; Adams, P.D.; Winn, M.D.; Storoni, L.C.; Read, R.J. Phaser crystallographic software. *J. Appl. Crystallogr.* **2007**, *40*, 658–674. [CrossRef] [PubMed]

74. Winn, M.D.; Ballard, C.C.; Cowtan, K.D.; Dodson, E.J.; Emsley, P.; Evans, P.R.; Keegan, R.M.; Krissinel, E.B.; Leslie, A.G.; McCoy, A.; et al. Overview of the CCP4 suite and current developments. *Acta Crystallogr. Sect. D Biol. Crystallogr.* **2011**, *67*, 235–242. [CrossRef] [PubMed]

75. Emsley, P.; Cowtan, K. Coot: Model-building tools for molecular graphics. *Acta Crystallogr. Sect. D Biol. Crystallogr.* **2004**, *60*, 2126–2132. [CrossRef] [PubMed]

76. Afonine, P.V.; Grosse-Kunstleve, R.W.; Chen, V.B.; Headd, J.J.; Moriarty, N.W.; Richardson, J.S.; Richardson, D.C.; Urzhumtsev, A.; Zwart, P.H.; Adams, P.D. Phenix.Model_vs_data: A high-level tool for the calculation of crystallographic model and data statistics. *J. Appl. Crystallogr.* **2010**, *43*, 669–676. [CrossRef] [PubMed]

77. Adams, P.D.; Afonine, P.V.; Bunkoczi, G.; Chen, V.B.; Davis, I.W.; Echols, N.; Headd, J.J.; Hung, L.W.; Kapral, G.J.; Grosse-Kunstleve, R.W.; et al. Phenix: A comprehensive python-based system for macromolecular structure solution. *Acta Crystallogr. Sect. D Biol. Crystallogr.* **2010**, *66*, 213–221. [CrossRef] [PubMed]

78. Bowers, K.J.; Chow, E.; Xu, H.; Dror, R.O.; Eastwood, M.P.; Gregersen, B.A.; Klepeis, J.L.; Kolossvary, I.; Moraes, M.A.; Sacerdoti, F.D.; et al. Scalable algorithms for molecular dynamics simulations on commodity clusters. In Proceedings of the 2006 ACM/IEEE Conference on Supercomputing ACM, Tampa, FL, USA, 11–17 November 2006.

79. Schrödinger. Available online: http://www.schrodinger.com (accessed on 1 March 2016).

80. Robinson, G.W.Z.; Zhu, S.B.; Singh, S.; Evans, M.W. *Water in Biology, Chemistry, and Physics: Experimental Overviews and Computational Methodologies*; World Scientific: Singapore, 1996; p. 9.

81. Harder, E.; Damm, W.; Maple, J.; Wu, C.; Reboul, M.; Xiang, J.Y.; Wang, L.; Lupyan, D.; Dahlgren, M.K.; Knight, J.L.; et al. Opls3: A force field providing broad coverage of drug-like small molecules and proteins. *J. Chem. Theory Comput.* **2016**, *12*, 281–296. [CrossRef] [PubMed]

82. Liu, P.; Kim, B.; Friesner, R.A.; Berne, B.J. Replica exchange with solute tempering: A method for sampling biological systems in explicit water. *Proc. Natl. Acad. Sci. USA* **2005**, *102*, 13749–13754. [CrossRef] [PubMed]

antibodies

MDPI

Article

Host Cell Proteins in Biologics Manufacturing: The Good, the Bad, and the Ugly

Martin Kornecki [1], Fabian Mestmäcker [1], Steffen Zobel-Roos [1], Laura Heikaus de Figueiredo [2], Hartmut Schlüter [2] and Jochen Strube [1,*]

[1] Institute for Separation and Process Technology, Clausthal University of Technology, Leibnizstr. 15, 38678 Clausthal-Zellerfeld, Germany; kornecki@itv.tu-clausthal.de (M.K.); mestmaecker@itv.tu-clausthal.de (F.M.); zobel-roos@itv.tu-clausthal.de (S.Z.-R.)

[2] Institute of Clinical Chemistry, Department for Mass Spectrometric Proteomics, University Medical Center Hamburg-Eppendorf, Martinistr. 52, 20246 Hamburg, Germany; l.heikaus@uke.de (L.H.d.F.); hschluet@uke.de (H.S.)

* Correspondence: strube@itv.tu-clausthal.de

Received: 17 August 2017; Accepted: 10 September 2017; Published: 16 September 2017

Abstract: Significant progress in the manufacturing of biopharmaceuticals has been made by increasing the overall titers in the USP (upstream processing) titers without raising the cost of the USP. In addition, the development of platform processes led to a higher process robustness. Despite or even due to those achievements, novel challenges are in sight. The higher upstream titers created more complex impurity profiles, both in mass and composition, demanding higher separation capacities and selectivity in downstream processing (DSP). This creates a major shift of costs from USP to DSP. In order to solve this issue, USP and DSP integration approaches can be developed and used for overall process optimization. This study focuses on the characterization and classification of host cell proteins (HCPs) in each unit operation of the DSP (i.e., aqueous two-phase extraction, integrated countercurrent chromatography). The results create a data-driven feedback to the USP, which will serve for media and process optimizations in order to reduce, or even eliminate nascent critical HCPs. This will improve separation efficiency and may lead to a quantitative process understanding. Different HCP species were classified by stringent criteria with regard to DSP separation parameters into "The Good, the Bad, and the Ugly" in terms of pI and MW using 2D-PAGE analysis depending on their positions on the gels. Those spots were identified using LC-MS/MS analysis. HCPs, which are especially difficult to remove and persistent throughout the DSP (i.e., "Bad" or "Ugly"), have to be evaluated by their ability to be separated. In this approach, HCPs, considered "Ugly," represent proteins with a MW larger than 15 kDa and a pI between 7.30 and 9.30. "Bad" HCPs can likewise be classified using MW (>15 kDa) and pI (4.75–7.30 and 9.30–10.00). HCPs with a MW smaller than 15 kDa and a pI lower than 4.75 and higher than 10.00 are classified as "Good" since their physicochemical properties differ significantly from the product. In order to evaluate this classification scheme, it is of utmost importance to use orthogonal analytical methods such as IEX, HIC, and SEC.

Keywords: upstream; downstream; host cell protein; CHO; ATPE; iCCC

1. Introduction

The amounts of biotechnology products produced worldwide, prescription as well as over-the-counter drugs, are estimated to account for around 50% of the most successful pharmaceutical products by the year 2020 [1]. Oncology constitutes the biggest therapeutic sector, with an annual growth rate of around 12.5% and sales of approximately $83.2 billion in 2015. Among the five top-selling oncological products, three will be monoclonal antibodies by the year 2020 [2]. The manufacturing process of biopharmaceuticals such as monoclonal antibodies (e.g., IgG, immunoglobulin G) is divided into upstream (USP) and downstream processing (DSP) [3–8]. The production of the monoclonal antibody in bioreactors (BR) using mammalian cells as an expression host and the separation of the liquid phase from the cells using centrifuges or filters is defined as USP [9]. The subsequent DSP is designed to separate side components like host cell proteins (HCP) or host cell DNA (hDNA) from the main component [6,7,10]. The most common unit operations used in the DSP are typically chromatography and filtration.

The commercial success of monoclonal antibodies of course led to significantly increased demand in their production scale [11]. Coping with these demands without significantly changing the approved manufacturing facilities almost forced companies to follow the route of increasing titers within the existing facilities.

Hence, compared to earlier yields of a couple of grams per liter, today antibody concentrations of up to 25 g/L can be achieved using a modified perfusion process [12,13]. Routinely, antibody concentrations of between three and five grams per liter can be generated in fed-batch processes [3,14,15]. However, increasing product titers at constant volumes due to higher cell concentrations will lead to capacity limitations in the DSP, which has to be compensated for by longer process times, higher material consumption, and corresponding costs [4]. This will significantly shift the cost of goods from the USP to the DSP [5]. Therefore, DSP technologies are required that circumvent this upcoming "downstream bottleneck," handling high titer volumes [4,16,17].

Optimizations in the USP concepts have led to increasing product titers. Along with this, raised impurity profiles have been observed [8,9]. Various compositions of the cultivation broth present challenges in the DSP of biotechnologically produced proteins. Considering the generic platform production process for antibodies, unit operations like centrifugation, micro- and ultrafiltration, protein A affinity chromatography, two orthogonal virus inactivation steps, ion-exchange (IEX), and hydrophobic interaction chromatography (HIC) are being used [5]. For the characterization of protein purification stages, key performance parameters can be used. These are typically resolution, speed, recovery, and capacity, as seen in Figure 1.

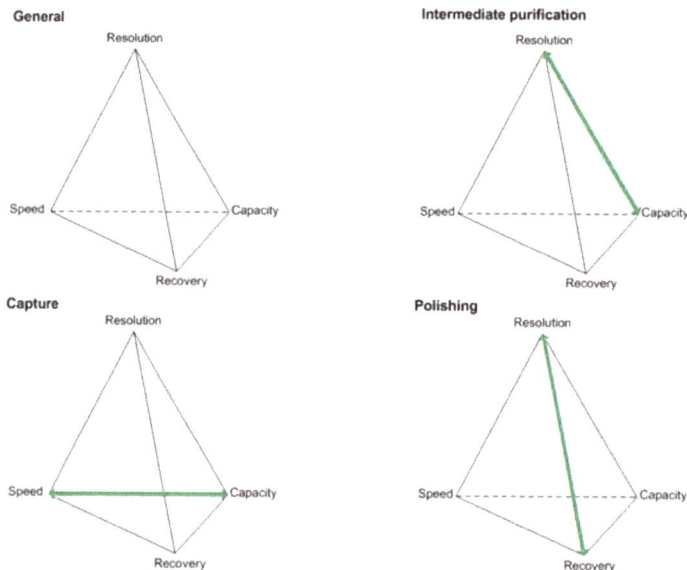

Figure 1. Key performance parameters of the capture, intermediate purification, and polishing step for protein purifications according to [18].

The objectives vary depending on the purification stage in focus, and therefore generate different challenges that have to be addressed during process optimization [18]. For example, the protein A affinity chromatography, used as a capture step, will reach its capacity limitation due to increasing product titers. This can be problematic since this criterion characterizes the capture step and is one of its two objectives. Moving downstream, selectivity challenges occurring during the intermediate purification and polishing step will prevent each step from reaching one of its objectives. Selectivity challenges are going to affect the resolution in IEX and HIC separation operations when HCP resemble the product in terms of pI and hydrophobicity, respectively [12,18].

Impurities like HCPs, which resemble the desired product only in one characteristic (e.g., pI), may challenge the IEX but can probably be easily separated from the product by an additional chromatographic step (e.g., HIC). Impurities similar to the product in more than one characteristic (e.g., pI and hydrophobicity) will be troublesome during purification and polishing. Therefore, the pI and hydrophobic distribution of the impurity spectrum can negatively affect IEX and HIC separations, respectively.

Critical performance parameters regarding the separation efficiency during the capture of monoclonal antibodies using affinity chromatography are capacity limitations as well as (un-)specific HCP co-elution [19]. Consequently, new approaches and technology are needed in order to circumvent future bottlenecks and separation challenges [4].

Furthermore, the existing challenges in process engineering have worsened since regulatory agencies demand higher product quality, an advanced understanding of the process and product, as well as batch-independent product quality [20–22]. Bioprocess engineering will probably focus in regulated industries on quality by design and process analytical technology mechanisms, in order to design, analyze, and control manufacturing processes [23]. This shall lead to improved process control by knowledge-based and statistical methods, which ultimately guarantees the process' robustness.

For example, monoclonal antibodies and fragments represent an interesting group of biopharmaceuticals due to their broad field of application (e.g., analysis or diagnostic). Those glycoproteins are structurally complex and differ in various formats, as can be seen in Figure 2. IgG is the most common format as a biopharmaceutical drug [2].

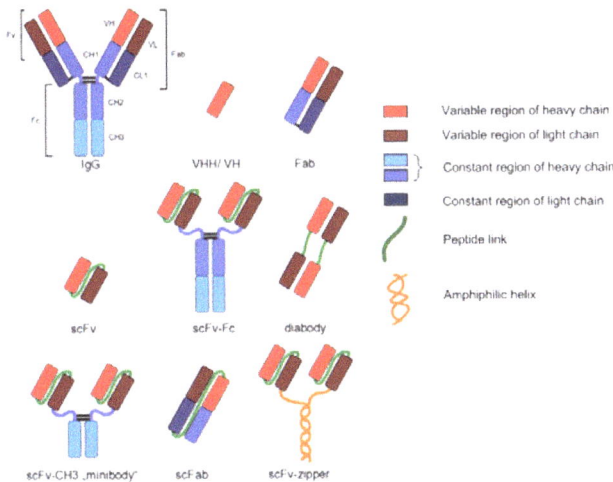

Figure 2. Various formats of recombinant antibodies [24].

The post-translational modifications, especially glycosylations, of these proteins are of utmost importance for their correct function [20]. The immense diversity of glycosylation patterns impacts the functionality, immunogenicity, and pharmacokinetics of the antibody [24,25]. Due to this, posttranslational modifications should be considered critical quality attributes (CQA) and verified throughout the manufacturing of monoclonal antibodies [26–30]. Antibody N-glycans can be quantitatively determined by normal phase chromatography after N-glycosidase digestion and glycan labeling, for example [31]. The most prevalent N-linked glycosylation patterns at the Cγ2 domain of the heavy chain (Fc) of an immunoglobulin G (IgG) are depicted in Figure 3, where the most common glycosylation of IgG is shown in section D.

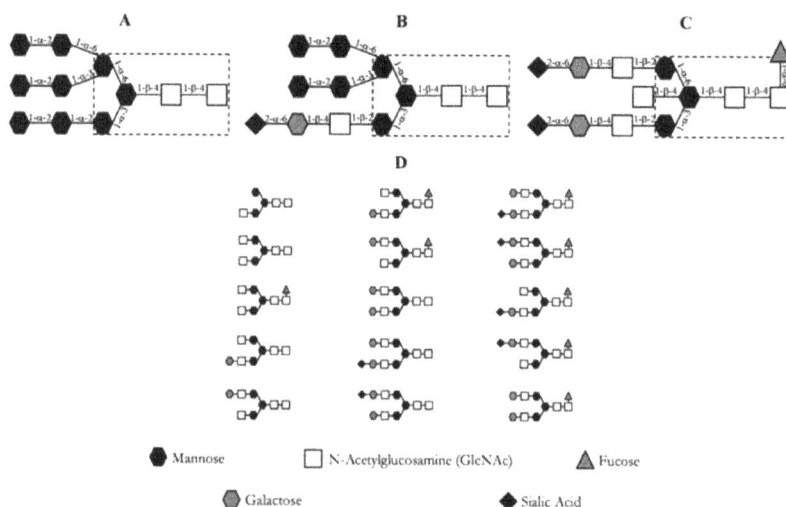

Figure 3. Common glycosylation patterns of an IgG. (**A**) high mannose content; (**B**) hybrid; (**C**) complex biantennary oligosaccharide with core fucosylation; (**D**) most prevalent oligosaccharide structures of IgG [20].

Besides critical process parameters (CPP) like pH, pO_2 and pCO_2, more often impurities play an important role in affecting CQA of the biopharmaceutical product. For example, extracellular proteases and glycosidases accumulating during the cultivation negatively influence the CQA of monoclonal antibodies [14,32–35]. The impurity spectrum consists of a multiplicity of different substances (HCP, hDNA, virus, cells, and cell debris). In this integration approach, HCPs are considered as the primary impurity based on their broad composition and range of isoelectric point (pI), molecular weight (MW), and hydrophobicity, as can be seen in Table 1 [36–40]. They exhibit no constant level, composition, or property distribution. HCPs caused by secretion or cell lysis can range in pI (2–11), MW (10–200 kDa), and variable hydrophobicity, and are therefore difficult to separate if their physicochemical properties resemble the product of interest.

Table 1. Physicochemical properties of the main impurities during the production of biopharmaceuticals, according to [38].

Class	pI	MW (kDa)	Hydrophobicity	Origin	Cause
HCP	2–11	10–200	Variable	Host cells	Secretion, lysis
hDNA	2–3	90–1000	Low	Host cells	Lysis
Insulin	5.3–5.5	5.8	Low	Media	Supplement
Virus	4–7.5	200–7200	Variable	Host cells, media	Contamination
Endotoxins	1–4	3–40	Variable	Media, contamination	Contamination

Primary recovery and purification steps for a biopharmaceutical DSP are based on physicochemical properties in order to efficiently purify the product. However, especially in the case of increasing product titers, a sub-population of impurities (i.e., HCP), which negatively affect the product quality, may remain with the desired protein and represent a certain risk [40]. Therefore, it is of critical importance to validate qualitatively and quantitatively the separation efficiency of each unit operation in the DSP.

This assessment will lead to an expanded understanding of each unit operation by classifying the impurities into "The Good, the Bad, and the Ugly":

- Impurities, which can be separated easily from the main component, are considered "the Good." They possess physicochemical properties significantly different from the protein of interest (i.e., pI, MW, hydrophobicity). As a result, they may be separated by only one unit operation in an efficient way (ion exchange in terms of charge differences).
- Side components showing more similarity to the product are more difficult to separate or are persistent throughout (i.e., not separable from the product) and thus are considered as "the Bad" or "the Ugly."

By characterizing the HCP criteria for an efficient DSP, it is possible to gain a deeper understanding of the process and preserve the quality of the product. This categorization can be used for an USP DSP integration approach towards an efficient production process by circumventing the generation or accumulation of "Bad" and "Ugly" impurities (Figure 4).

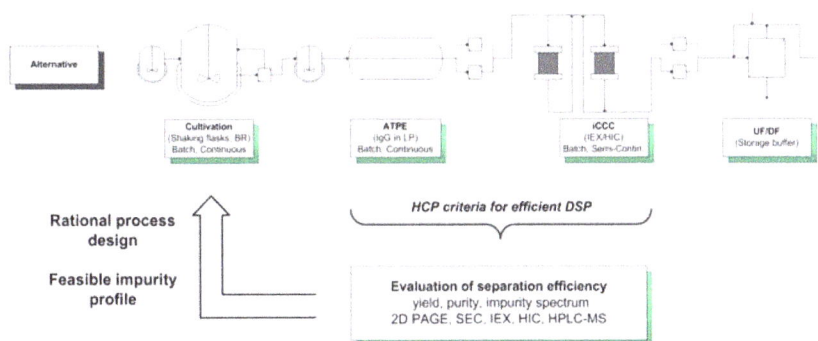

Figure 4. USP DSP integration approach for a systematic development of a bioprocess.

The considered process for the production of monoclonal antibodies utilizes mammalian cell cultivations. Afterwards, the aqueous two-phase extraction (ATPE) is used as a cell harvesting or capture step, depending on the system composition used [41–45]. Following the ATPE, the integrated counter current chromatography (iCCC), which is a combination of an IEX and HIC, is employed as a purification and polishing step. This combination of chromatographic columns leads to a highly purified product [46].

The integration approach begins with a data-driven characterization of HCP occurring in the broth and in each unit operation. The separation efficiency is determined by analytical methods (i.e., 2D SDS-PAGE, SEC, IEX, HIC, and HPLC-MS/MS). The SEC chromatograms qualitatively describe the impurity spectrum and can be used for a determination of impurities in the molecular weight range of the considered product (150 kDa). The IEX and HIC are used for characterizing the charge and hydrophobicity of the HCPs. 2D SDS-PAGE analysis, combined with HPLC-MS/MS measurements, is used for the identification and, of utmost importance, classification of "The Good, the Bad, and the Ugly" HCPs. This classification is done by evaluating the molecular weight, isoelectric point, and hydrophobicity of the HCPs, as seen in Table 2.

Afterwards, these findings are used in rational process design in order to minimize or even eliminate "Ugly" HCPs, which cannot be easily separated from the product (Figure 4).

Table 2. Analytical methods used for the characterization of HCP.

Characteristic	Method	Orthogonal Method
Isoelectric point	2D-SDS PAGE	IEX; HPLC-MS/MS
Molecular weight	SEC	2D-SDS PAGE; HPLC-MS/MS
Hydrophobicity	HIC	-

One possible process design optimization procedure is the improvement of media components. Media optimization is capable of changing the broth's HCP composition towards a population that is easier to separate or at least exhibits a lower HCP concentration. In addition, an optimized medium not only shifts the HCP profile but also improves the cell growth and product titer, which is depicted in Table 3 [47].

Table 3. Improved parameters by using an optimized medium according to [47].

Parameter	Optimized medium
Titer increase	Factor 2.5
Cell growth	Factor 2–2.3
IgG/HCP	65%
HCP profile	Shift

The shifted HCP profile can be seen in the 2D-SDS PAGE comparison in Figure 5.

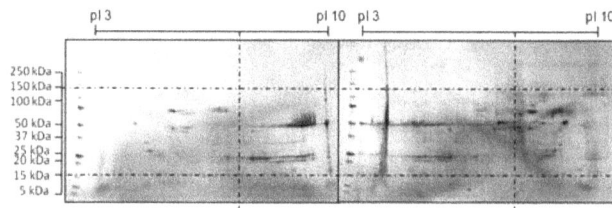

Figure 5. Comparison of 2D-SDS PAGE of a reference (**right**) and optimized medium (**left**) during a CHO cultivation according to [47]. Media was improved by a three-level DoE design.

In this work, the results of the characterization of the HCP profile from a mAb production process are presented. Process-related data as well as analysis-related data are used for the characterization of the process and for the classification of HCPs. The results of each analytical method are critically evaluated in order to determine a process flow being suitable for USP DSP integration and process optimization. Analytical methods such as SEC, 2D-PAGE, IEX, HIC as well as HPLC-MS/MS were used in order to identify critical HCP in the cell-free broth and during each unit operation (i.e., ATPE, IEX, and HIC).

2. Results and Discussion

A schematic overview of the considered alternative process as well as process- and analysis-related data are shown in Figure 6.

The HCP criteria for an efficient DSP have to be evaluated for each unit operation, according to Figure 4. Here, the classification of HCP focuses on the broth, the broth after diafiltration and on a side component fraction after HIC separation. Process-related data such as titer, yield, and purity of each unit operation are shown in Table 4.

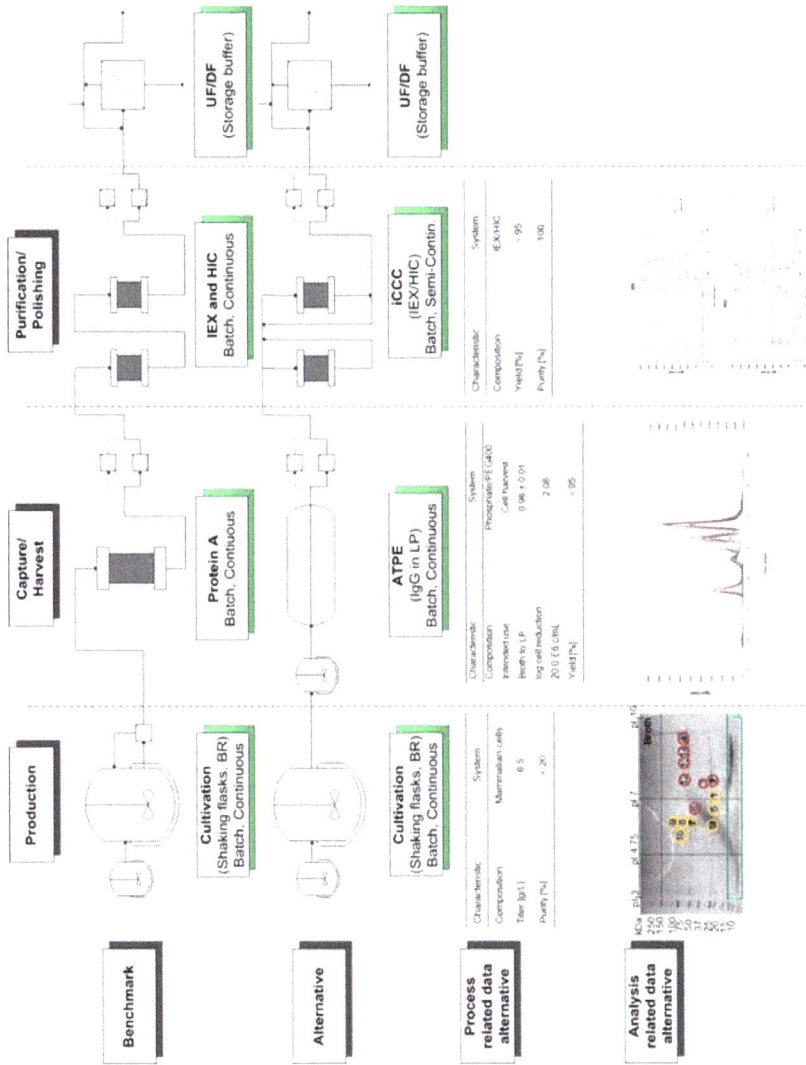

Figure 6. Schematic overview of the considered alternative process in comparison of the benchmark manufacturing route. In addition, process and analysis related data are shown and discussed in the text.

This analytical procedure focuses on the classification and characterization of HCPs. Therefore, each unit operation of the DSP has to be evaluated by its separation efficiency, using analytical methods such as 2D-SDS PAGE, IEX, HIC, and SEC to determine HCP criteria for an efficient DSP, as seen in Figure 3. Protein A and size-exclusion chromatography are used to determine yield and purity, respectively. Each unit operation was loaded with the native broth in order to determine their separation efficiency.

Table 4. Process related data of the cultivation, ATPE, and iCCC. Yield and purity were determined using protein A chromatography and SEC, respectively.

	Cultivation	ATPE	iCCC
System	Mammalian cells	PEG400/40 wt% PO$_4$	IEX/HIC combination
Titer/yield	6.5 g/L	>95%	>95%
Log cell reduction 20.0 E6 cells/mL	-	2.08	-
Purity	<20%	up to 80% *	100%

* Protein-based according to SEC.

The fraction number five occurring on the HIC was chosen due to the high side component content near the target product, as seen in Figure 7. In the following, the classification of the HCPs will be performed by 2D-PAGE gels, as depicted in Figure 8.

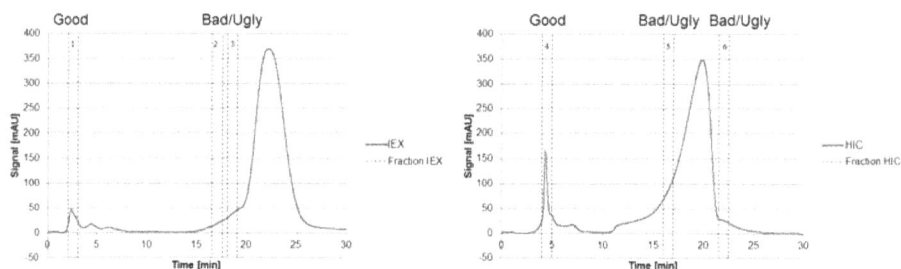

Figure 7. Chromatograms of an analytical IEX (**left**) and HIC (**right**) measurement of the diafiltrated cell-free CHO supernatant. The vertical sections represent the number of fractions taken, representing "Good", "Bad," and "Ugly" impurities.

The classification criterion of the considered HCP was selected by comparing their pI and molecular weight to the target product, as seen in Table 5.

Table 5. Classification of the "Good, Bad, and Ugly" HCP in comparison to the physicochemical properties of the monoclonal antibody (mAb). MW, molecular weight; pI, isoelectric point.

Characteristic	mAb	Good	Bad	Ugly
MW [kDa]	144.2	<15	>15	>15
pI [−]	8.30	<4.75 >10.00	4.75–7.30 9.30–10.00	7.30–9.30

While considering the 2D-PAGE gels, proteins with a MW lower than 15 kDa can be considered "Good" since they can be separated by using diafiltration subsequent to an ATPE with a suitable MW cutoff. Therefore, this filtration step is coupled to a buffer change, which is necessary for the use of the iCCC, since the specific light phase contains PEG400, resulting in a more viscous solution, which would make the chromatographic steps more difficult to handle.

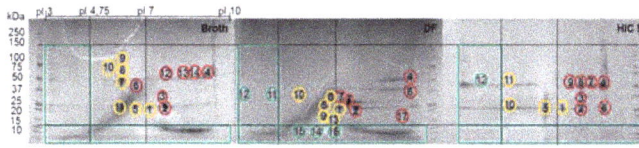

Figure 8. 2D SDS-PAGE of the broth, diafiltrated broth (DF) and HIC fraction. Green circles represent "Good", yellow circles "Bad," and red circles "Ugly" HCP.

Proteins larger than 15 kDa have to be separated by another unit operation, which is based on other physicochemical properties (i.e., pI, hydrophobicity). Therefore, "Bad" and "Ugly" proteins possess a MW larger than 15 kDa. The horizontal line at 150 kDa represents the target protein in its functional condition. The vertical lines depict the isoelectric point at 4.75 and 7.0. Impurities with a pI of 4.75 can be subjected to a possible precipitation step using hydrochloric acid, which significantly reduces their concentration [42]. Those impurities exhibit a different pI than the target protein and can efficiently be separated by an IEX and are therefore classified as "Good." Experimental IEX data show a distinctly different interaction with the stationary phase due to their surface charge distribution, as seen in Figure 9, which resemble the "Good" HCP. They elute near the void volume and can be easily separated. A similar train of thought can be conducted while characterizing the HIC chromatogram. As can be seen in Figure 9, the target product gets concentrated by each cycle in the iCCC mode.

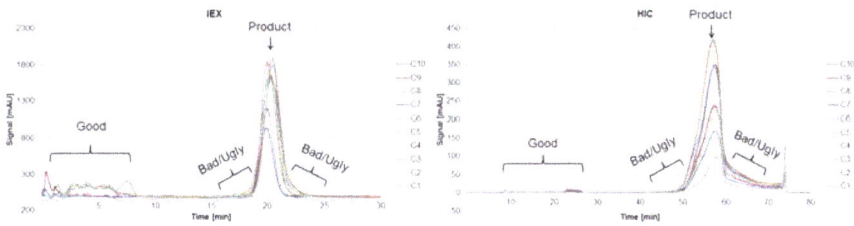

Figure 9. Chromatograms of the IEX (**left**) and HIC (**right**) after various cycles in the iCCC mode.

Impurities with a pI range close to the target protein (i.e., 7.30–9.30) are more difficult to separate via an IEX and are therefore considered "Ugly." However, since the separation efficiency of the IEX will depend on the column, buffer solution, and process parameters, this range can vary depending on the system used. Impurities with characteristics in between those of "Good" and "Ugly" are defined as "Bad." They possess a pI of 4.75–7.30 and 9.30–10.00 and can be difficult to separate when other physicochemical properties (i.e., hydrophobicity) resemble the target product. This dependency can of course also occur with "Ugly" HCP but since they are already classified as difficult to separate, they will not be characterized differently. Regarding the "Good" HCP, this dependency will not occur even if other physicochemical properties show close similarity to the product, since at least one physicochemical attribute is significantly different from the target product. The pI is restricted to 10 due to the pH gradient used in the IEF prior to 2D gel electrophoresis.

The 2D-PAGE analysis seen in Figure 8 is suitable for the visualization of the side component spectrum. However, the sample preparation requires reducing agents such as DTT, which destroys the protein's structure by reducing the disulfide bonds. This preparation procedure results in spots on the gel, which do not resemble their native structure in the supernatant. Following the aforementioned classification and separation system, proteins with a MW lower than 15 kDa but with an isoelectric point near the target product will sometimes be classified as "Good" since they can be separated by filtration. Hence, the native protein can be "Ugly" even if it appears as "Good" in the gel (assuming no change in surface charge). Thus, it is of the utmost importance to use orthogonal analytical methods to validate the classification. In terms of MW, SEC analysis can be conducted in order to determine the size distribution of side components, as seen in Figure 10. The advantage of using SEC analysis is the determination of the side component's native MW distribution as well as their qualitative mass in proportion to the product's signal. The disadvantage is the less sensitive detection of low mass content side components as well as proteins resulting in a signal overlap with the mAb.

In order to identify proteins present in the 2D-PAGE gel spots, their tryptic peptides were identified via analysis with liquid chromatography (LC) coupled to tandem mass spectrometry (MS/MS) and a database search. The numbers in the gels in Figure 8 indicate the spots that were analyzed using LC-MS/MS. The first five spots occurred in every gel. The subsequent spots were unique in each gel. The identified peptides and their corresponding proteins of each spot in these gels are presented in the appendix (Tables A1–A3). The identified proteins of the first recurrent spots are listed alongside with their MW and pI in Table 6.

As can be seen in Table 6, the MW and pI of the spots analyzed with 2D-PAGE do not correspond to the value in the protein database. This is a result of proteins existing in different species due to posttranslational modifications and proteolytic processing (proteolytic degradation and sample preparation, respectively). In contrast, the theoretically calculated pI values obtained by the ExPASy computation tool (http://web.expasy.org/compute_pi/) represent the unmodified full length amino acid sequence of a defined protein. SEC analysis, for example, is a non-invasive analytical method for the determination of MW distribution of side components, if the salt concentration used in aqueous eluents allows for separation based on molecular size exclusion alone due to the hydrodynamic radius [48]. Nevertheless, for a systematic integration approach, the classification of HCPs based on their physicochemical properties can lead to an enhanced process understanding, especially in the DSP.

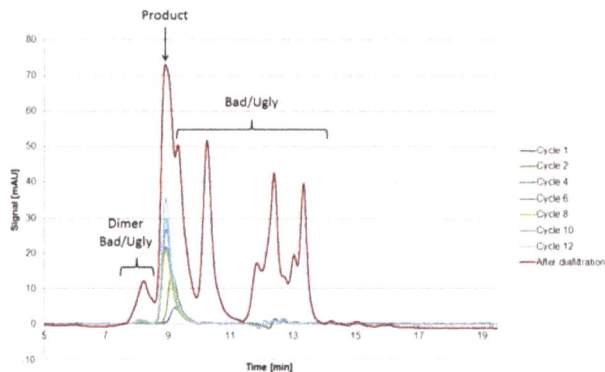

Figure 10. SEC chromatograms after various cycles as well as the broth after diafiltration.

Table 6. Classification of CHO proteins identified via LC-MS/MS analysis and characterized by 2D-PAGE gel. Comparison of theoretical (UniProt; pI calculated according to the amino acid sequences) and observed MW and pI with respect to the spot position on the 2D-PAGE gel.

Spot Gel	MW Gel	pI Gel	Class Gel	MW [1]	pI [2]	Class MS	Protein	UniProt Accession Number
1	25	7.0	Bad	81.56	5.69	Bad	Glutathione S-transferase Mu7-like protein	A0A061IN16
2	25	7.5	Ugly	102.7	6.02	Bad	Actin, cytoplasmic 1	A0A069C7Y3
3	30	7.6	Ugly	38.03	6.08	Bad	Purine nucleoside phosphorylase-like protein	A0A061ILE8
4	50	9.4	Bad	72.13	7.23	Ugly	Pyruvate kinase	A0A098KXF7
5	25	6.3	Bad	38.03	6.08	Bad	Purine nucleoside phosphorylase-like protein	A0A061ILE8

[1] Theoretical values according to the unmodified full length protein according to UniProt; [2] Theoretical values calculated using the ExPASy computation tool (http://web.expasy.org/compute_pi/).

3. Materials and Methods

Chinese hamster ovary cells (CHO DG44) were used for the production of a monoclonal antibody. The culture conditions were 37 °C, 5% carbon dioxide, and 130 rpm. The cultivations were carried out in shake flasks in a serum-free medium.

The ATP system applied consisted of 44.5% broth, 15.5% PEG400 (Merck KGaA, Darmstadt, Germany), and 40% of a 40 wt% phosphate buffer. All the components were weighed. The extraction was carried out at pH 6.0 in 50-mL beakers at room temperature. The system was mixed for 15 min at 140 rpm in an incubator shaker. Phase separation took place within 30 min in a separatory funnel.

The broth was diafiltrated using a SARTOFLOW® Slice 200 Benchtop system from Sartorius Stedim (Germany). A 10 kDa Hydrosart® (Sartorius Stedim, Göttingen, Germany) was utilized as a membrane module.

The iCCC (integrated counter-current chromatography) is run by using Fractogel® EMD SO_3^- (s) and Fractogel® EMD Phenyl(s) (Merck KGaA, Darmstadt, Germany). The buffers consisted of a 20 mM sodium phosphate buffer (Na_2HPO_4, NaH_2PO_4) as well as a 20 mM sodium phosphate buffer with 1 M Na_2SO_4.

The product was quantified by Protein A chromatography (PA ID Sensor Cartridge, Applied Biosystems, Bedford, MA, USA). Dulbecco's PBS buffer (Sigma-Aldrich, St. Louis, MO, USA) was used as a loading buffer at pH 7.4 and as an elution buffer at pH 2.6. The absorbance was monitored at 280 nm.

The size exclusion chromatography was done by using a Yarra™ 3 μm SEC-3000 column (Phenomenex Ltd., Aschaffenburg, Germany) with 0.1 M Na_2SO_4, 0.1 M Na_2HPO_4, and 0.1 M NaH_2PO_4 (Merck KGaA, Germany) as a buffer system.

Isoelectric focusing was carried out using IPG strips (ReadyStripTM IPG Strips, linear, pH 3–10, BIO-RAD, Hercules, CA, USA) and an isoelectric focusing unit of Hoefer (Hoefer Inc., Holliston, MA, USA). A subsequent SDS PAGE was carried out using gels (Criterion TGX Precast Gel, 4–15% Bis-Tris, BIO-RAD), buffers, and an electrophoresis chamber from BIO-RAD. The resulting gels were colored by Coomassie Brilliant Blue G-250 (VWR International, Radnor, PA, USA).

For the identification of proteins, selected 2D GE spots were cut out and reduced into 1-mm² pieces. After reduction of the disulfide bonds with 10 mM DL-dithiothreitol (Sigma-Aldrich) and alkylation with 50 mM iodoacetamide (Sigma-Aldrich), an in-gel proteolytic digestion was performed with 8 ng/μL trypsin (Promega, Madison, WI, USA) at 37 °C overnight. The peptides were extracted from the gel with 65% acetonitrile and 5% acetic acid in water and the solvent was evaporated to complete dryness. The peptides were re-suspended in 20 μL 0.1% formic acid (Fluka) and subjected to LC-MS/MS analysis with a nano-flow ultra-performance liquid chromatography (nano-UPLC) system (nanoACQUITY, Waters, Manchester, UK) coupled via an electrospray-ionization (ESI) source to a tandem mass spectrometer (MS/MS) consisting of a quadrupole and a orbitrap mass analyzer (Orbitrap QExcactive, Thermo Scientific, Bremen, Germany). Four microliters of each sample were loaded onto a reversed-phase (RP) trapping column (Symetry C18 Trap Column; 100 Å, 5 μm, 180 μm × 20 mm) and washed with 1% buffer B for 5 min. The peptides were eluted onto a RP capillary column (nanoAcquity Peptide BEH analytical column; 130 Å, 1.7 μm, 75 μm × 200 mm) and separated by a gradient from 3 to 35% buffer B in 35 min (250 nL/min). Eluting peptides were ionized and desorbed by ESI in the positive mode using a fused-silica emitter (I.D. 10 μm, New Objective, Woburn, MA, USA) at a capillary voltage of 1800 V. Data-dependent acquisition mode was used with the following parameters: MS level over a m/z range from 400 to 1500, with a resolution of 70,000 FWHM at m/z 200. Maximum injection time was set to 120 ms for an AGC target of 1E6. For MS/MS analysis the top 12 signals were isolated in a 2 m/z window and fragmented with a normalized HCD collision energy of 25. Fragment spectra were recorded with a resolution of 17,500 FWHM at m/z 200. Maximum injection time was set to 60 ms for an AGC target 5E5.

LC–MS raw data were processed with MaxQuant (Max Planck Institute of Biochemistry, Planegg, Germany) algorithms (version 1.5.8.3). Protein identification was carried out with Andromeda against a hamster (*Cricetulus griseus*) (www.uniprot.org, downloaded on 31 January 2017) and a contaminant database. The searches were performed using a precursor mass tolerance set to 10 ppm and fragment mass tolerance set to 20 ppm. For peptide identification, two missed cleavages were allowed, a carbamidomethylation on the cysteine as a fixed modification and oxidation of the methionine as a variable modification. A maximum of five modifications per peptide were allowed.

4. Conclusions

The presented approach of integrating USP and DSP is based on the classification and characterization of impurities generated during USP. This will lead to a deeper quantitative process understanding and identification of issues in the DSP early on. Here, the HCPs were categorized into "The Good, the Bad, and the Ugly" by evaluating their physicochemical properties compared to the monoclonal antibody. In this approach "Good" impurities possess a MW lower than 15 kDa and a pI lower than 4.75. "Ugly" impurities on the other hand exhibit a pI of 7.3–9.3, whereas "Bad" impurities feature a pI between 4.75 and 7.3 as well as between 9.3 and 10.0. In order to evaluate the classification system for the generated HCPs, orthogonal analytical methods are of utmost importance. IEX and SEC analysis were conducted for the identification of impurities. Theoretical pI and MW calculated based on the amino acid sequence differ from the experimental values obtained in 2D gel electrophoresis. This is due to not considering posttranslational modifications, as well as in vivo and ex vivo proteolytic processing.

Nevertheless, it is possible to characterize HCP based on pI and MW properties. In order to fully categorize the separation efficiency of each unit operation in the DSP as well as of their combinations, the HCP profile has to be determined with the aforementioned analytical methods in future approaches. This portfolio can of course be extended by adding supplementary methods like NMR technologies, preferably online [49].

Considering the significant amount of work in terms of characterization, monitoring, and removal of impurities and contaminations created by the USP step, as well as the time and cost associated with their removal, it may be worthwhile to reflect in more detail how these impurities and product variations are generated in the first place. Work to this end already started some time ago. Initial results and corresponding concepts for a more balanced integrated process design will be presented in the near future.

Acknowledgments: The authors would especially like to acknowledge Petra Gronemeyer (now Boehringer/Biberach) for her outstanding contribution to this topic during her PhD studies, for which she has been highly esteemed at conferences.

Author Contributions: Martin Kornecki conceived and designed the experiment as well as wrote the paper. Martin Kornecki performed the SDS-PAGE experiments. Fabian Mestmäcker performed the chromatographic (iCCC, IEX, HIC) experiments. Laura Heikaus de Figueiredo performed the LC-MS/MS experiments and analyzed the data. All mentioned authors interpreted the data. Hartmut Schlüter as well as Jochen Strube substantively revised the work and contributed the materials and analysis tools. Jochen Strube is responsible for conception and supervision.

Conflicts of Interest: The authors declare no conflict of interest.

Appendix A

Table A1. Peptides and their corresponding proteins identified via LC-MS/MS in spots of 2D gel of the broth. Molecular weight (MW) and isoelectric point (pI) of the unmodified full length protein according to UniProt.

#	Gel MW	Gel pI	MW (UniProt)	pI (UniProt)	Primary Accession Number (UniProt)	Number of Unique Peptides	Protein
1	25	7.2	25.76	6.45	A0A061HUZ2	9	Platelet-activating factor
			81.56	5.69	A0A061IN16	11	Glutathione S-transferase Mu 7-like protein
			25.88	6.43	A0A061HYZ1	9	Peroxiredoxin-6-like protein
2	25	7.9	30.28	7.22	A0A061IFC9	4	Carbonic anhydrase
			89.52	6.23	A0A061IJC4	4	Glutathione S-transferase Mu 1-like protein
			81.56	5.69	A0A061IN16	4	Glutathione S-transferase Mu 7-like protein
3	30	7.6	72.13	7.23	A0A098KXF7	8	Pyruvate kinase
			38.03	6.08	A0A061ILE8	6	Purine nucleoside phosphorylase-like protein
			32.23	9.11	A0A061IAK4	5	l-lactate dehydrogenase A chain
4	50	9.6	45.28	8.48	A0A061IB69	7	Fructose-bisphosphate aldolase
			72.13	7.23	A0A098KXF7	15	Pyruvate kinase
			102.7	6.02	A0A069CTY3	5	Actin, cytoplasmic 1
5	25	6.6	27.39	6.34	A0A061I2E1	8	Proteasome subunit
			89.52	6.23	A0A061IJC4	8	Glutathione S-transferase Mu 1-like protein
			81.56	5.69	A0A061IN16	8	Glutathione S-transferase Mu 7-like protein
6	45	6.7	50.57	5.93	C3GR73	11	Rab GDP diss. inhib.
			52.79	6	A0A098KXB1	10	Cytosol aminopeptidase-like protein
			44.67	7.54	A0A061IJI8	9	Alpha-enolase
7	50	6.1	52.79	6	A0A098KXB1	20	Aminopeptidase
			72.13	7.23	A0A098KXF7	29	Pyruvate kinase
			145.1	8.37	A0A061HU29	15	Glucose-6-phosphate 1-dehydrogenase
8	57	6	73.86	5.56	A0A061I5D1	22	Heat shock protein
			74.72	5.29	A0A061HWC7	9	Plastin-3
			69.64	5.57	A0A061I5U1	9	Heat shock-related protein 2
9	80	6.1	72.13	7.23	A0A098KXF7	11	Pyruvate kinase
			117.7	5.42	C3IBG3	8	Ubiquitin activating enzyme E1
			73.86	5.56	A0A061I5D1	7	Heat shock protein
10	70	5.6	73.86	5.56	A0A061I5D1	9	Heat shock protein
			68.43	5.55	A0A061IIQ2	5	Vitamin K-dependent protein S
			85.71	5.2	A0A061IAX6	5	Dipeptidyl peptidase 3
11	25	6	25.88	6.43	A0A061HYZ1	9	Peroxiredoxin
			89.52	6.23	A0A061IJC4	19	Glutathione S-transferase Mu 1-like protein
			81.56	5.69	A0A061IN16	14	Glutathione S-transferase Mu 7-like protein
12	55	7.8	72.13	7.23	A0A098KXF7	37	Pyruvate kinase
			52.79	6	A0A098KXB1	10	Cytosol aminopeptidase-like protein
			73.86	5.56	A0A061I5D1	4	Heat shock protein
13	50	8.6	52.79	6	A0A098KXB1	19	Cytosol aminopeptidase-like protein
			72.13	7.23	A0A098KXF7	2	Pyruvate kinase
			145.1	8.37	A0A061HU29	20	Glucose-6-phosphate 1-dehydrogenase
14	50	9.2	72.13	7.23	A0A098KXF7	3	Pyruvate kinase
			44.67	7.54	A0A061IJI8	3	Alpha-enolase
			42.69	8.78	A0A061HV36	3	Eukaryotic translation initiation factor 2 subunit 3-like protein

Table A2. Peptides and their corresponding proteins identified via LC-MS/MS in spots of 2D gel of the diafiltrated broth. Molecular weight (MW) and isoelectric point (pI) of the unmodified full length protein according to UniProt.

#	Gel MW	Gel pI	MW (UniProt)	pI (UniProt)	Primary accession number (UniProt)	Number of Unique Peptides	Protein
1	25	7	26.96	5.38	A0A06I6A0	5	Glutathione S-transferase A4-like protein
			81.56	5.69	A0A06I1N16	8	Glutathione S-transferase Mu 7-like protein
2	25	7.5	38.03	6.08	A0A06I1LE8	8	Purine nucleoside phosphorylase-like protein
			38.03	6.08	A0A06I1LE8	6	Purine nucleoside phosphorylase
			102.7	6.02	A0A069C7Y3	2	Actin, cytoplasmic 1
			43.35	6.48	A0A06I1JG8	2	Prostaglandin reductase 1-like protein
3	30	7.4	45.28	8.48	A0A06I1B69	1	Fructose-bisphosphate aldolase
4	50	9.4	89.52	6.23	A0A06I1JC4	2	Glutathione S-transferase
5	25	6.6	38.03	6.08	A0A06I1LE8	7	Purine nucleoside phosphorylase
6	37	9.4	89.52	6.23	A0A06I1JC4	5	Glutathione S-transferase
			45.28	8.48	A0A06I1B69	4	Fructose-bisphosphate aldolase
			43.35	6.48	A0A06I1JG8	2	Prostaglandin reductase 1-like protein
			361.89	4.81	A0A06I1H02	2	Desmoglein-4-like protein
7	30	7	38.03	6.08	A0A06I1LE8	4	Purine nucleoside phosphorylase
			128.68	6.78	A0A06I1K77	3	Exosome component 10 isoform 1
			11.37	11.36	C3H2T6	2	Histone H4
8	30	6.8	27.79	4.7	A0A06I1CS6	4	Protein sigma
			102.7	6.02	A0A069C7Y3	5	Actin, cytoplasmic 1
			361.89	4.81	A0A06I1H02	4	Desmoglein-4-like protein
9	17	6.6	11.37	11.36	C3H2T6	4	Histone H4
			14.99	10.2	A0A06I1P52	4	Histone H2B
10	30	5.6	52.25	5.35	A0A06I1ML2	13	Annexin
			268.7	5.69	A0A06I1P39	10	Filamin-B isoform 4
			50.99	6.94	A0A06I1BI4	4	Cathepsin F
11	30	4.5	52.25	5.35	A0A06I1ML2	2	Annexin
			14.73	9.87	A0A06I1QB8	3	Ubiquitin-60S
12	30	2.8	38.03	6.08	A0A06I1LE8	8	Purine nucleoside phosphorylase
			89.52	6.23	A0A06I1JC4	8	Glutathione S-transferase
			101.51	5.12	A0A06I1RD9	5	AP complex subunit beta
13	15	6.7	38.03	6.08	A0A06I1LE8	4	Purine nucleoside phosphorylase
			89.52	6.23	A0A06I1JC4	3	Glutathione S-transferase
14	15	6.1	38.31	5.33	A0A06I1EW1	3	Nuclear migration protein nudC-like protein
			38.31	5.33	A0A06I1EW1	5	Nuclear migration protein nudC-like protein
			89.52	6.23	A0A06I1JC4	4	Glutathione S-transferase
			54.11	5.01	A0A06I1DB2	3	Prelamin-A/C-like isoform 1
15	12	5.6	17.16	7.8	A0A06I1OI3	4	SH3 binding protein
			17.19	5.94	C3HBD4	3	Nucleoside diphosphate kinase
16	17	6.6	23.42	5.1	C3CXB0	3	Rho GDP
			89.52	6.23	A0A06I1JC4	8	Glutathione S-transferase
			102.7	6.02	A0A069C7Y3	2	Actin, cytoplasmic 1
17	16	9.2	14.73	9.87	A0A06I1QB8	3	Ubiquitin-60S
			102.7	6.02	A0A069C7Y3	3	Actin, cytoplasmic 1
			23.42	5.1	C3CXB0	3	Rho GDP

Table A3. Peptides and their corresponding proteins identified via LC-MS/MS in spots of 2D gel of the HIC fraction. Molecular weight (MW) and isoelectric point (pI) of the unmodified full length protein according to UniProt.

#	Gel MW	Gel pI	MW (UniProt)	pI (UniProt)	Primary Accession Number (UniProt)	Number of Unique Peptides	Protein
1	25	6.9	-	-	G3H2T6	-	-
2	25	7.6	11.37	11.36	A0A069C7Y3	2	Histone H4
3	30	7.6	102.7	6.02	A0A061ILE8	2	Actin, cytoplasmic 1
			38.03	6.08	A0A069C7Y3	2	Purine nucleoside
4	50	8.4	102.7	6.02	A0A061IJI8	1	Actin, cytoplasmic 1
			44.67	7.54	A0A098KXF7	8	Alpha-enolase
			72.13	7.23	A0A061ILE8	8	Pyruvate kinase
5	25	6.3	38.03	6.08	A0A061ILE8	2	Purine nucleoside
6	25	8.4	38.03	6.08	A0A061IJI8	6	Purine nucleoside
7	50	8	44.67	7.54	A0A061IJI8	4	Alpha-enolase
8	47	7.6	102.7	6.02	A0A061ILE8	1	Actin, cytoplasmic 1
			72.13	7.23	A0A098KXF7	5	Pyruvate kinase
			44.67	7.54	A0A061ICE4	3	Alpha-enolase
9	50	7.2	102.7	6.02	A0A061QB8	2	Actin, cytoplasmic 1
			59.76	9.22	A0A069C7Y3	4	ATP synthase subunit
			14.73	9.87	A0A061QB8	2	Ubiquitin-60S ribosomal protein L40-like isoform 2
			102.7	6.02	A0A069C7Y3	2	Actin, cytoplasmic 1
10	25	5.1	-	-	-	-	-
11	50	4.8	211.66	5.42	A0A0G1I4N6	1	CAP-Gly domain-containing linker protein 1
12	52	4	-	-	-	-	-

(-) Spots, which were not able to be identified.

References

1. EvaluatePharma. *World Preview 2016, Outlook to 2022*; EvaluatePharma: London, UK, 2016; pp. 1–39.
2. EvaluatePharma. *World Preview 2015, Outlook to 2020*; EvaluatePharma: London, UK, 2015; pp. 1–39.
3. Li, F.; Vijayasankaran, N.; Shen, A.; Kiss, R.; Amanullah, A. Cell culture processes for monoclonal antibody production. *mAbs* **2010**, *2*, 466–479. [CrossRef] [PubMed]
4. Gronemeyer, P.; Ditz, R.; Strube, J. Trends in Upstream and Downstream Process Development for Antibody Manufacturing. *Bioengineering* **2014**, *1*, 188–212. [CrossRef]
5. Sommerfeld, S.; Strube, J. Challenges in biotechnology production—Generic processes and process optimization for monoclonal antibodies. *Chem. Eng. Process. Process Intensif.* **2005**, *44*, 1123–1137. [CrossRef]
6. Birch, J.R.; Racher, A.J. Antibody production. *Adv. Drug Deliv. Rev.* **2006**, *58*, 671–685. [CrossRef] [PubMed]
7. Liu, H.F.; Ma, J.; Winter, C.; Bayer, R. Recovery and purification process development for monoclonal antibody production. *mAbs* **2010**, *2*, 480–499. [CrossRef] [PubMed]
8. Shukla, A.A.; Thömmes, J. Recent advances in large-scale production of monoclonal antibodies and related proteins. *Trends Biotechnol.* **2010**, *28*, 253–261. [CrossRef] [PubMed]
9. Jain, E.; Kumar, A. Upstream processes in antibody production: Evaluation of critical parameters. *Biotechnol. Adv.* **2008**, *26*, 46–72. [CrossRef] [PubMed]
10. Strube, J.; Grote, F.; Josch, J.P.; Ditz, R. Process development and design of downstream processes. *Chemie-Ingenieur-Technik* **2011**, *83*, 1044–1065. [CrossRef]
11. Gagnon, P. Technology trends in antibody purification. *J. Chromatogr. A* **2012**, *1221*, 57–70. [CrossRef] [PubMed]
12. Kelley, B. Industrialization of mAb production technology: The bioprocessing industry at a crossroads. *mAbs* **2009**, *1*, 440–449. [CrossRef]
13. Chon, J.H.; Zarbis-Papastoitsis, G. Advances in the production and downstream processing of antibodies. *New Biotechnol.* **2011**, *28*, 458–463. [CrossRef] [PubMed]
14. Park, J.H.; Jin, J.H.; Lim, M.S.; An, H.J.; Kim, J.W.; Lee, G.M. Proteomic Analysis of Host Cell Protein Dynamics in the Culture Supernatants of Antibody-Producing CHO Cells. *Sci. Rep.* **2017**, *7*, 44246. [CrossRef] [PubMed]
15. Reinhart, D.; Damjanovic, L.; Kaisermayer, C.; Kunert, R. Benchmarking of commercially available CHO cell culture media for antibody production. *Appl. Microbiol. Biotechnol.* **2015**, 4645–4657. [CrossRef] [PubMed]
16. Strube, J.; Sommerfeld, S.; Lohrmann, M. Process Development and Optimization for Biotechnology Production—Monoclonal Antibodies. In *Bioseparation and Bioprocessing*, 2nd ed.; Subramanian, G., Ed.; Wiley-VCH: Weinheim, Germany, 2007.
17. Strube, J.; Grote, F.; Ditz, R. Bioprocess Design and Production Technology for the Future. In *Biopharmaceutical Production Technology*; Subramanian, G., Ed.; Wiley-VCH: Weinheim, Germany, 2012.
18. GE Healthcare. *Strategies for Protein Purification. Handbook*; GE Healthcare: Little Chalfont, UK, 2010.
19. Levy, N.E.; Valente, K.N.; Choe, L.H.; Lee, K.H.; Lenhoff, A.M. Identification and Characterization of Host Cell Protein Product-Associated Impurities in Monoclonal Antibody Bioprocessing. *Biotechnol. Bioeng.* **2014**, *111*, 904–912. [CrossRef] [PubMed]
20. Del Val, I.J.; Kontoravdi, C.; Nagy, J.M. Towards the implementation of quality by design to the production of therapeutic monoclonal antibodies with desired glycosylation patterns. *Biotechnol. Prog.* **2010**, *26*, 1505–1527. [CrossRef] [PubMed]
21. Hinz, D.C. Process analytical technologies in the pharmaceutical industry: The FDA's PAT initiative. *Anal. Bioanal. Chem.* **2006**, *384*, 1036–1042. [CrossRef] [PubMed]
22. Mercier, S.M.; Rouel, P.M.; Lebrun, P.; Diepenbroek, B.; Wijffels, R.H.; Streefland, M. Process analytical technology tools for perfusion cell culture. *Eng. Life Sci.* **2016**, *16*, 25–35. [CrossRef]
23. Hakemeyer, C.; McNight, N.; St. John, R.; Meier, S.; Trexler-Schmidt, M.; Kelley, B.; Zettl, F.; Puskeiler, R.; Kleinjans, A.; Lim, F.; et al. Process characterization and Design Space definition. *Biologicals* **2016**, *44*, 306–318. [CrossRef] [PubMed]
24. Frenzel, A.; Hust, M.; Schirrmann, T. Expression of recombinant antibodies. *Front. Immunol.* **2013**, *4*, 1–20. [CrossRef] [PubMed]

25. Shields, R.L.; Lai, J.; Keck, R.; O'Connell, L.Y.; Hong, K.; Gloria Meng, Y.; Weikert, S.H.A.; Presta, L.G. Lack of fucose on human IgG1 N-linked oligosaccharide improves binding to human Fc\gammaRIII and antibody-dependent cellular toxicity. *J. Biol. Chem.* **2002**, *277*, 26733–26740. [CrossRef] [PubMed]

26. Richter, V.; Kwiatkowski, M.; Omidi, M.; Omidi, A.; Robertson, W.D.; Schlüter, H. Mass spectrometric analysis of protein species of biologics. *Pharm. Bioprocess.* **2013**, *1*, 381–404. [CrossRef]

27. Hakemeyer, C.; Pech, M.; Lipok, G.; Herrmann, A. Characterization of the influence of cultivation parameters on extracellular modifications of antibodies during fermentation. *BMC Proc.* **2013**, *7*, P85. [CrossRef]

28. Kunert, R.; Reinhart, D. Advances in recombinant antibody manufacturing. *Appl. Microbiol. Biotechnol.* **2016**, *100*, 3451–3461. [CrossRef] [PubMed]

29. Jefferis, R. Glycosylation as a strategy to improve antibody-based therapeutics. Nature reviews. *Drug Discovery* **2009**, *8*, 226–234. [CrossRef] [PubMed]

30. Alt, N.; Zhang, T.Y.; Motchnik, P.; Taticek, R.; Quarmby, V.; Schlothauer, T.; Beck, H.; Emrich, T.; Harris, R.J. Determination of critical quality attributes for monoclonal antibodies using quality by design principles. *Biologicals* **2016**, *44*, 1–15. [CrossRef] [PubMed]

31. Brunner, M.; Fricke, J.; Kroll, P.; Herwig, C. Investigation of the interactions of critical scale-up parameters (pH, pO$_2$ and pCO$_2$) on CHO batch performance and critical quality attributes. *Bioprocess Biosyst. Eng.* **2016**, 1–13. [CrossRef] [PubMed]

32. Chee Furng Wong, D.; Tin Kam Wong, K.; Tang Goh, L.; Kiat Heng, C.; Gek Sim Yap, M. Impact of dynamic online fed-batch strategies on metabolism, productivity and N-glycosylation quality in CHO cell cultures. *Biotechnol. Bioeng.* **2005**, *89*, 164–177. [CrossRef] [PubMed]

33. Gao, S.X.; Zhang, Y.; Stansberry-Perkins, K.; Buko, A.; Bai, S.; Nguyen, V.; Brader, M.L. Fragmentation of a highly purified monoclonal antibody attributed to residual CHO cell protease activity. *Biotechnol. Bioeng.* **2011**, *108*, 977–982. [CrossRef] [PubMed]

34. Gramer, M.J.; Goochee, C.F. Glycosidase activities of the 293 and NS0 cell lines, and of an antibody-producing hybridoma cell line. *Biotechnol. Bioeng.* **1994**, *43*, 423–428. [CrossRef] [PubMed]

35. Robert, F.; Bierau, H.; Rossi, M.; Agugiaro, D.; Soranzo, T.; Broly, H.; Mitchell-Logean, C. Degradation of an Fc-fusion recombinant protein by host cell proteases: Identification of a CHO cathepsin D protease. *Biotechnol. Bioeng.* **2009**, *104*, 1132–1141. [CrossRef] [PubMed]

36. Tait, A.S.; Hogwood, C.E.M.; Smales, C.M.; Bracewell, D.G. Host cell protein dynamics in the supernatant of a mAb producing CHO cell line. *Biotechnol. Bioeng.* **2012**, *109*, 971–982. [CrossRef] [PubMed]

37. Hogwood, C.E.M.; Tait, A.S.; Koloteva-Levine, N.; Bracewell, D.G.; Smales, C.M. The dynamics of the CHO host cell protein profile during clarification and protein A capture in a platform antibody purification process. *Biotechnol. Bioeng.* **2013**, *110*, 240–251. [CrossRef] [PubMed]

38. Singh, N.; Arunkumar, A.; Chollangi, S.; Tan, Z.G.; Borys, M.; Li, Z.J. Clarification technologies for monoclonal antibody manufacturing processes: Current state and future perspectives. *Biotechnol. Bioeng.* **2016**, *113*, 698–716. [CrossRef] [PubMed]

39. Valente, K.N.; Lenhoff, A.M.; Lee, K.H. Expression of difficult-to-remove host cell protein impurities during extended Chinese hamster ovary cell culture and their impact on continuous bioprocessing. *Biotechnol. Bioeng.* **2015**, *112*, 1232–1242. [CrossRef] [PubMed]

40. Wang, X.; Hunter, A.K.; Mozier, N.M. Host cell proteins in biologics development: Identification, quantitation and risk assessment. *Biotechnol. Bioeng.* **2009**, *103*, 446–458. [CrossRef] [PubMed]

41. Eggersgluess, J.K.; Richter, M.; Dieterle, M.; Strube, J. Multi-stage aqueous two-phase extraction for the purification of monoclonal antibodies. *Chem. Eng. Technol.* **2014**, *37*, 675–682. [CrossRef]

42. Gronemeyer, P.; Ditz, R.; Strube, J. Implementation of aqueous two-phase extraction combined with precipitation in a monoclonal antibody manufacturing process. *Chimica Oggi/Chem. Today* **2016**, *34*, 66–70.

43. Eggersgluess, J.A.N.K.; Both, S.; Strube, J. Process Development for the Extraction of Biomolecules. *Chimica Oggi/Chem. Today* **2012**, *30*, 4.

44. Asenjo, J.A.; Andrews, B.A. Aqueous two-phase systems for protein separation: A perspective. *J. Chromatogr. A* **2011**, *1218*, 8826–8835. [CrossRef] [PubMed]

45. Azevedo, A.M.; Gomes, A.G.; Rosa, P.A.J.; Ferreira, I.F.; Pisco, A.M.M.O.; Aires-Barros, M.R. Partitioning of human antibodies in polyethylene glycol-sodium citrate aqueous two-phase systems. *Sep. Purif. Technol.* **2009**, *65*, 14–21. [CrossRef]

46. Zobel, S.; Helling, C.; Ditz, R.; Strube, J. Design and operation of continuous countercurrent chromatography in biotechnological production. *Ind. Eng. Chem. Res.* **2014**, *53*, 9169–9185. [CrossRef]

47. Gronemeyer, P.; Ditz, R.; Strube, J. DoE based integration approach of upstream and downstream processing regarding HCP and ATPE as harvest operation. *Biochem. Eng. J.* **2016**, *113*, 158–166. [CrossRef]

48. Ahmed, U.; Saunders, G. The Effect of NaCl Concentration on Protein Size Exclusion Chromatography. Application Note. Available online: http://cn.agilent.com/cs/library/applications/SI-02416.pdf (accessed on 17 July 2017).

49. Roch, P.; Mandenius, C.-F. On-line monitoring of downstream bioprocesses. *Curr. Opin. Chem. Eng.* **2016**, *14*, 112–120. [CrossRef]

antibodies

MDPI

Article

Epitope Sequences in Dengue Virus NS1 Protein Identified by Monoclonal Antibodies

Leticia Barboza Rocha [1,†], Rubens Prince dos Santos Alves [2,†], Bruna Alves Caetano [1],
Lennon Ramos Pereira [2], Thais Mitsunari [1], Jaime Henrique Amorim [2,‡],
Juliana Moutinho Polatto [1], Viviane Fongaro Botosso [3], Neuza Maria Frazatti Gallina [4],
Ricardo Palacios [5], Alexander Roberto Precioso [5], Celso Francisco Hernandes Granato [6],
Danielle Bruna Leal Oliveira [7] , Vanessa Barbosa da Silveira [7] , Daniela Luz [1],
Luís Carlos de Souza Ferreira [2] and Roxane Maria Fontes Piazza [1,*]

[1]   Laboratório de Bacteriologia, Instituto Butantan, São Paulo, 05503-900 SP, Brazil;
      leticia.rocha@butantan.gov.br (L.B.R.); bruna.caetano@butantan.gov.br (B.A.C.);
      thais.mitsunari@butantan.gov.br (T.M.); juliana.yassuda@butantan.gov.br (J.M.P.);
      daniedaluz@yahoo.com.br (D.L.)
[2]   Laboratório de Desenvolvimento de Vacinas, Instituto de Ciências Biomédicas, Universidade de São Paulo,
      São Paulo, 05508-000 SP, Brazil; rubens.bmc@gmail.com (R.P.d.S.A.); lennon_rp@usp.br (L.R.P.);
      jh.biomedico@gmail.com (J.H.A.); lcsf@usp.br (L.C.d.S.F.)
[3]   Laboratório de Virologia, Instituto Butantan, São Paulo, 05503-900 SP, Brazil;
      viviane.botosso@butantan.gov.br
[4]   Divisão de Desenvolvimento Tecnológico e Produção; Instituto Butantan, São Paulo, 05503-900 SP, Brazil;
      neuza.gallina@butantan.gov.br
[5]   Divisão de Ensaios Clínicos e Farmacovigilância, Instituto Butantan, São Paulo, 05503-900 SP, Brazil;
      ricardo.palacios@butantan.gov.br (R.P.); alexander.precioso@butantan.gov.br (A.R.P.)
[6]   Departamento de Medicina, Disciplina de Doenças Infecciosas e Parasitárias,
      Universidade Federal de São Paulo, São Paulo, 04023-062 SP, Brazil; celso.granato@grupofleury.com.br
[7]   Laboratório de Virologia Molecular e Clínica, Departamento de Microbiologia, Instituto de Ciências
      Biomédicas, Universidade de São Paulo, São Paulo, 05508-000 SP, Brazil; danibruna@gmail.com (D.B.L.O.);
      vanessa.silveirabio@gmail.com (V.B.d.S.)
*   Correspondence: roxane@butantan.gov.br; Tel.: +55-11-2627-9724
†   These authors contributed equally to the present work.
‡   Present address: Laboratório de Microbiologia, Centro das Ciências Biológicas e da Saúde, Universidade
      Federal do Oeste da Bahia, Barreiras, 47805-100. Bahia, Brazil.

Received: 8 August 2017; Accepted: 22 September 2017; Published: 15 October 2017

Abstract: Dengue nonstructural protein 1 (NS1) is a multi-functional glycoprotein with essential functions both in viral replication and modulation of host innate immune responses. NS1 has been established as a good surrogate marker for infection. In the present study, we generated four anti-NS1 monoclonal antibodies against recombinant NS1 protein from dengue virus serotype 2 (DENV2), which were used to map three NS1 epitopes. The sequence [193]AVHADMGYWIESALNDT[209] was recognized by monoclonal antibodies 2H5 and 4H1BC, which also cross-reacted with Zika virus (ZIKV) protein. On the other hand, the sequence [25]VHTWTEQYKFQPES[38] was recognized by mAb 4F6 that did not cross react with ZIKV. Lastly, a previously unidentified DENV2 NS1-specific epitope, represented by the sequence [127]ELHNQTFLIDGPETAEC[143], is described in the present study after reaction with mAb 4H2, which also did not cross react with ZIKV. The selection and characterization of the epitope, specificity of anti-NS1 mAbs, may contribute to the development of diagnostic tools able to differentiate DENV and ZIKV infections.

Keywords: dengue virus; NS1; Zika virus; mAbs; antibody recognition; amino acid sequences

---

1. Introduction

Dengue fever is an important mosquito-borne and the most prevalent and costly arbovirus affecting humans, caused by one of the four serotypes of dengue virus (DENV 1–4) [1]. In the last decade, a large number of dengue epidemics have occurred, which resulted in enormous economic and human loss in parts of Asia and South America [2,3]. Considering Brazil only, more than three million cases of confirmed dengue infections occurred between 2015 and 2017, with 70 cases per 100,000 inhabitants [4].

The DENV genome is composed of a single positive-sense RNA that encodes a single viral polyprotein that is further processed by viral and host proteases into three structural proteins (C, prM/M, and E) and seven nonstructural proteins (NS1, NS2A, NS2B, NS3, NS4A, NS4B, and NS5). NS1 is the first nonstructural protein to be translated and is essential to virus replication [5]. It is a conserved N-linked glycoprotein with a variable molecular mass of 46–55 kDa, which depends on its glycosylation status [6]. The NS1 protein can be found as a dimer associated with vesicular compartments within the cell, where it plays an important role as an essential cofactor in the virus replication process [7]. Alternatively, NS1 can be secreted into the extracellular space as a hexameric lipoprotein particle [8] that interacts with several plasma proteins [9,10].

The recent introduction of the Zika virus (ZIKV) to the American continent represented a regional and worldwide public health challenge [11]. The close evolutionary relationship between DENV and ZIKV is reflected by the high sequence conservation of both structural and non-structural proteins [12]. In this aspect, the identification of monoclonal antibodies (mAbs) able to react specifically with DENV or cross-react with ZIKV proteins is a relevant feature for the validation of the diagnostic tools based on the NS1 protein.

In pioneering work by Falconar et al. [8], the immunogenic regions of DENV2 NS1 employing mAbs were extensively studied. Recently, certain studies have been using new methods to predict the binding epitopes of proteins to specific antibodies [13,14]. This approach was also applied to identify binding epitopes of DENV NS1 protein serotypes [15–17]. Also, the crystal structure of the DENV2 NS1 protein (PDB code: 4O6B) has been solved in both dimeric and hexameric configurations [6], which provides a useful guide for the selection of potential epitopes for therapy and vaccine strategies.

In the present study, recombinant DENV2 NS1 was used to immunize mice and generate murine mAbs. Four mAbs were isolated, purified, characterized and tested for reactivity with native NS1 produced by all DENV serotypes in Vero-infected cells and also for cross-reactivity with ZIKV NS1.

2. Results

2.1. Isolation and Characterization of NS1-Specific DENV mAbs

Fusion of popliteal lymph node cells, from mice immunized with DENV2 rNS1, with a non-Ig-secreting or synthesizing line derived from a cell line created by fusing a BALB/c mouse spleen cell and the mouse myeloma P3X63Ag8 (SP2/O-Ag14) mouse myeloma cells, generated 25 secretory hybridomas. Among them, four hybridomas were selected by enzyme-linked immunosorbent assay (ELISA) and sub cloned by limiting dilution and named as 4F6, 4H2, 4H1BC, and 2H5. The clones were expanded, supernatants collected and mAbs purified for further characterization. Accordingly, mAbs 4F6 and 4H2 were characterized as IgG2a (immunoglobulin G), and 2H5 and 4H1BC as IgG1. The affinity constants were similar (10^{-8} M) as well as their reactivity with and limits of detection of NS1 (Table 1).

Table 1. Characteristics of the monoclonal antibodies (mAbs) against dengue virus (DENV) nonstructural protein 1 (NS1).

Name	4F6	4H2	2H5	4H1BC
IgG Subtype [a]	IgG2a	IgG2a	IgG1	IgG1
DENV2 NS1 reactivity [b]	Yes	Yes	Yes	Yes
Dissociation Constant (KD) [c]	1.1×10^{-8} M	6.2×10^{-8} M	7.3×10^{-8} M	8.4×10^{-8} M
Detection limit [d]	16 ng/mL	32 ng/mL	32 ng/mL	32 ng/mL
Epitope sequence [e]	25VHTWTEQYKFQPES38	127ELHNQTFLIDGPETAEC143	193AVHADMGYWIESALNDT209	193AVHADMGYWIESALNDT209
DENV (1–4) reactivity [f]	No	Yes	No	No
ZIKV reactivity [g]	No	No	Yes	Yes

[a] The Ig isotype and IgG subtypes were performed by enzyme-linked immunosorbent assay (ELISA) using anti-IgA, anti-IgM, anti-IgG1, anti-IgG2a, anti-IgG2b and anti-IgG3 coated onto microplates; [b] The Dengue virus serotype 2 (DENV2) NS1 reactivity was evaluated by indirect ELISA and immunoblotting using rNS1; [c] Dissociation constant was performed by ELISA using different concentrations of rNS1; [d] Detection limit was evaluated by ELISA using different concentrations of rNS1; [e] The conservancy of DENV2 NS1 epitopes recognized by specific mAbs in a peptide array was analyzed among the four serotypes of DENV, using three samples of NS1 amino acid sequences as representative of each DENV serotype; [f,g] DENV (1–4) and Zika virus (ZIKV) reactivity was evaluated by immunofluorescence in Vero cells infected with the specific virus strains.

The recognition pattern of the four NS1 mAbs was evaluated by ELISA using either intact or heat-denatured rNS1. All NS1 mAbs recognized the intact rNS1 protein, and although mAb 4F6 reacted similarly with the intact and the heated-treated rNS1 (Figure 1A), the other three mAbs (4H2, 2H5 and 4H1BC) reacted more efficiently with the intact protein (Figure 1B–D, respectively), which indicated that the recognized epitopes were, at least, partially represented by conformational structures. All four MAbs also recognized rNS1 in an immunoblot assay (Figure S1).

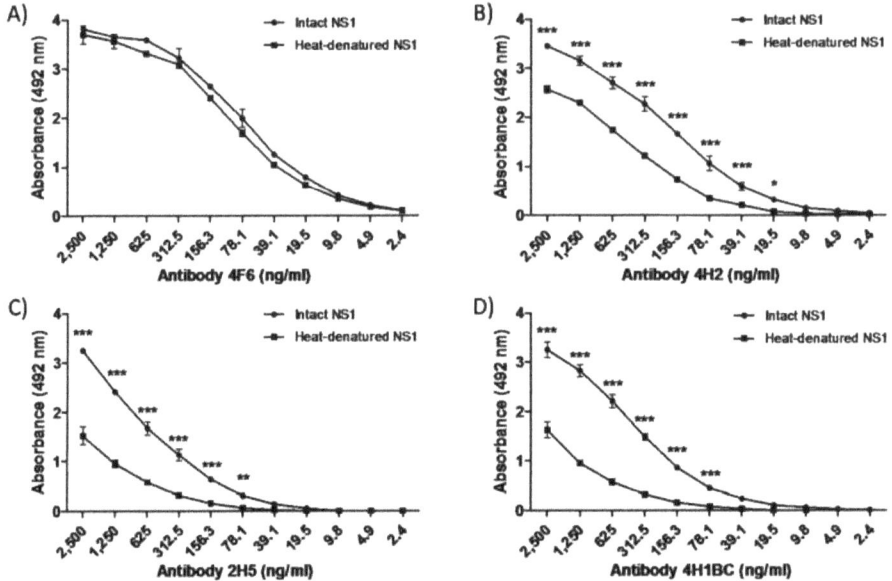

Figure 1. Characterization of nonstructural protein 1-specific (NS1) monoclonal antibodies (mAbs) reactivity by enzyme-linked immunosorbent assay (ELISA). Reactivity of mAbs to heated-treated or intact rNS1, as solid phase-bound antigens. The mAbs 4F6 (**A**), 4H2 (**B**), 2H5 (**C**) and 4H1BC (**D**) were serially diluted (log2) from an initial concentration of 2.5 µg/mL. Each well was adsorbed with 400 ng of rNS1. Heat denaturation was performed at 100 °C for 10 min. Statistical analyses were performed by two-way variance analysis followed by Bonferroni's post-test. (*** $p < 0.01$; ** $p < 0.1$; * $p < 0.5$).

2.2. Detection of Native DENV2 NS1 and Epitope Mapping

After selection, mAbs were tested by immunofluorescence assays using fixed DENV2-infected Vero cells. All four mAbs recognized the native viral NS1 expressed in infected cells, as shown in Figure 2. To localize the specific mAbs binding sites/epitopes, peptide mapping array experiments were performed (Figure S2). The results showed that mAb 4F6 reacted with the peptide corresponding to the sequence [25]VHTWTEQYKFQPES[38] of NS1 (Table 1), which is located in an external loop of the protein 3D structure (Figure 3). The 4H2 mAb recognized the peptide corresponding to the sequence [127]ELHNQTFLIDGPETAEC[143] of NS1 (Table 1), which is located in beta-sheets in an external region of the protein 3D structure (Figure 4). The other two mAbs (2H5 and 4H1BC) showed the same binding specificity and recognized the peptide [193]AVHADMGYWIESALNDT[209] (Table 1). This sequence was also located in a beta-sheet structure, located in an internal region of the protein (Figure 5). The analysis of epitope conservancy in several strains of DENV serotypes as well as Zika strains is detailed in Table S1.

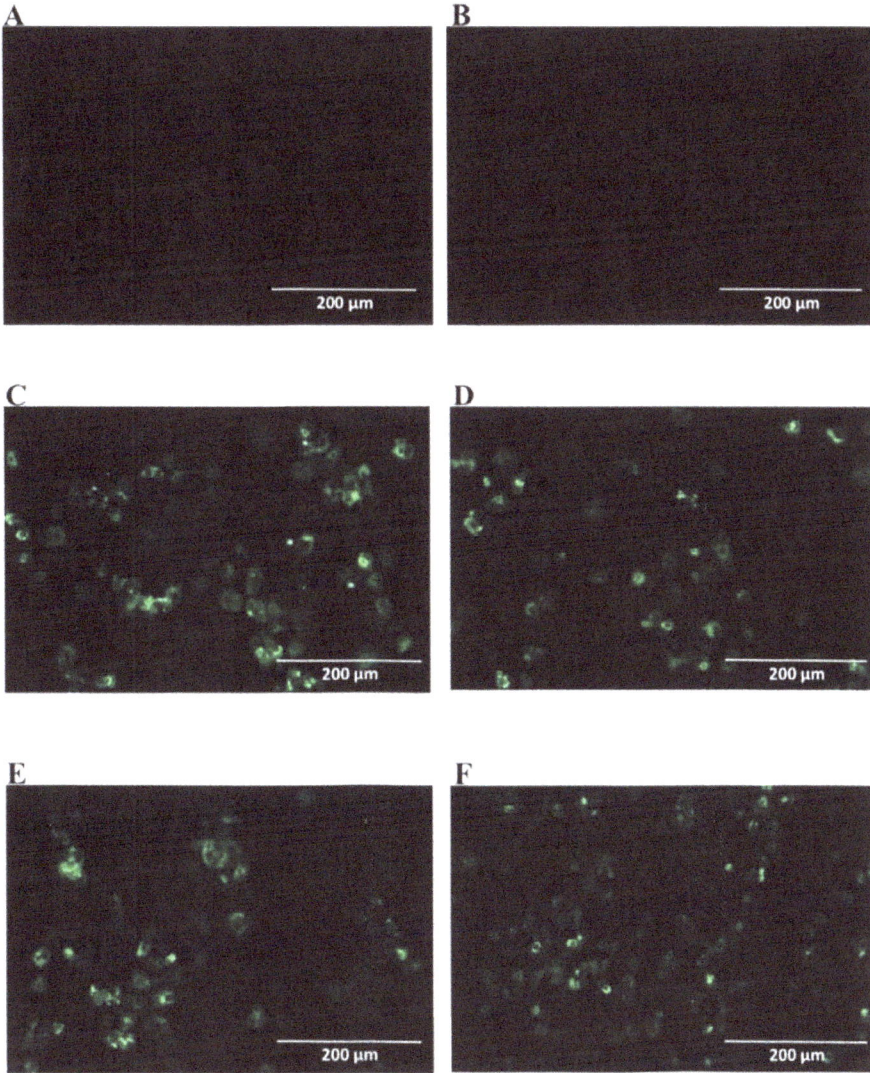

Figure 2. Reactivity of NS1-specific mAbs to dengue-serotype 2-infected Vero cells. Cells were infected with a multiplicity of infection (MOI) of 0.5, fixed, permeabilized and treated with each of the tested mAbs 48 h post infection. Then, cells were labeled with Alexa fluor® conjugated goat-anti mouse IgG. The negative controls: Mock-infected cells treated with a pool of mAbs anti-NS1 (**A**) and DENV2-infected cells labeled only with secondary antibody (**B**); Tested mAbs: (**C**) 4F6; (**D**) 4H2; (**E**) 2H5 and (**F**) 4H1BC. Magnification of 200×.

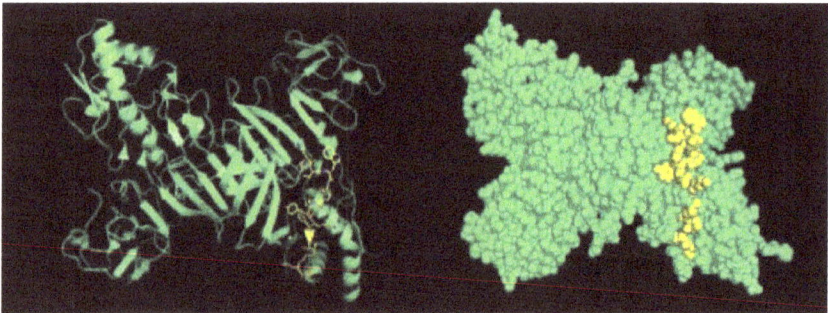

Figure 3. Three-dimensional structural model of a NS1 dimer and regions corresponding to epitopes recognized by 4F6 mAb. The NS1 3D model was generated by the program Python Molecular (PyMOL) in green. The sequence ^{25}VHTWTEQYKFQPES^{38} is highlighted in yellow.

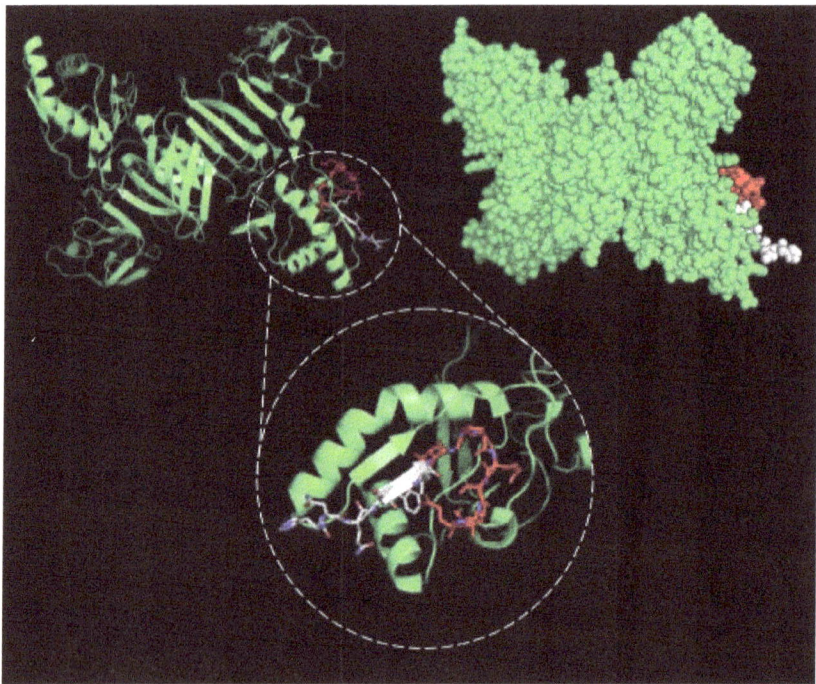

Figure 4. Three-dimensional structural model of a NS1 dimer and regions corresponding to epitopes recognized by 4H2 mAb. The NS1 3D model was generated by the program PyMOL in green. The sequence ^{127}ELHNQTFLIDGPETAEC^{143} is highlighted in red and white. In the detail the red structure represents the novel nine-amino acid sequence described herein.

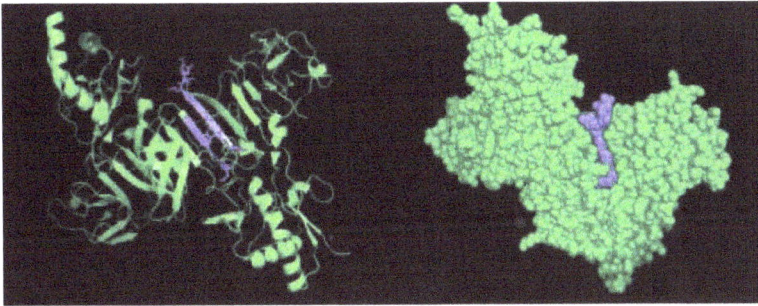

Figure 5. Three-dimensional structural model of a NS1 dimer and regions corresponding to epitopes recognized by 2H5 and 4H1BC mAbs. The NS1 3D model was generated by the program PyMOL in green. The sequence [193]AVHADMGYWIESALNDT[209] is highlighted in blue.

2.3. Analyses of mAbs' Cross-Reactivity with Different DENV Serotype and ZIKV

Since NS1 shares a high homology with amino acid sequences found among different flavivirus, the selected mAbs were tested for recognition of native ZIKV NS1 by immunofluorescence assay, using fixed ZIKV-infected Vero cells. In Figure 6, two mAbs are observed to cross-react with native ZIKV-NS1 in this test (2H5 and 4H1BC) (Figure 6C,D, Table 1). The other two mAbs (4F6 and 4H2) were specific for DENV NS1 (Figure 6A,B, Table 1). We also tested in vitro the reactivity of 4F6 and 4H2 mAbs with DENV serotypes, other than DENV2, and only the 4H2 mAb reacted with all four DENV serotypes (Figure 7, Table 1).

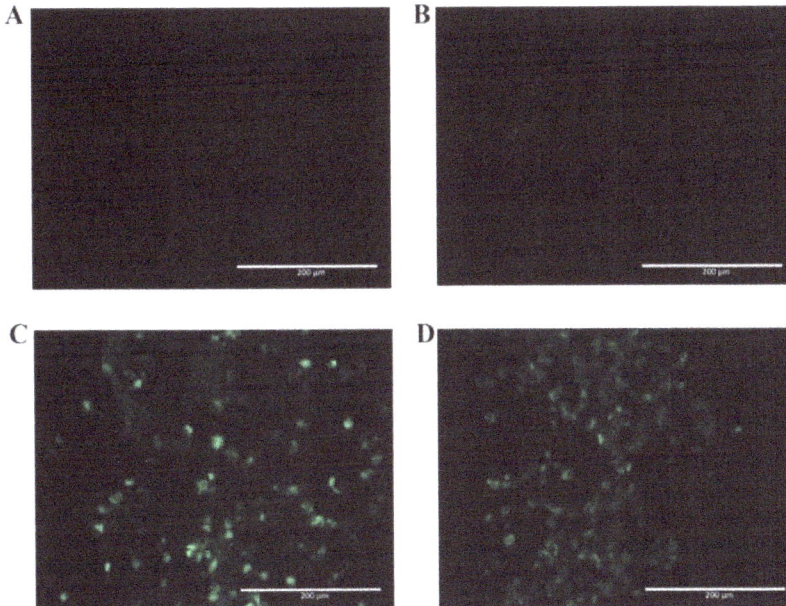

Figure 6. Reactivity of NS1-specific mAbs to zika virus-infected Vero cells. Cells were infected with a MOI of 0.05. 72 h post infection, cells were fixed, permeabilized and treated with each of the tested mAbs. Then, cells were labeled with FITC-conjugated goat-anti mouse IgG. Tested mAbs: (**A**) 4F6; (**B**) 4H2; (**C**) 2H5 and (**D**) 4H1BC. Magnification of 200×.

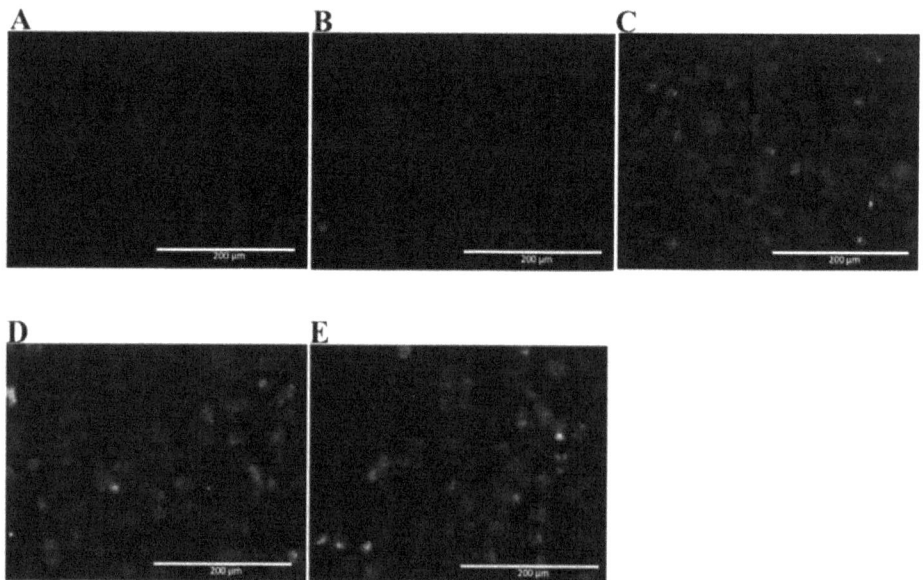

Figure 7. Reactivity of 4H2 mAb to Vero cells infected with DENV of different serotypes. Vero cells were infected with a MOI of 0.5, fixed, permeabilized and treated with mAb 4H2 48 h post infection. Then, cells were labeled with Alexa fluor® conjugated goat-anti mouse IgG. The negative controls: (**A**) Mock infected cells treated with a pool of mAbs anti-NS1 and (**B**) DENV-infected cells labeled only with secondary antibody; Tested DENV serotypes: (**C**) DENV1; (**D**) DENV3 and (**E**) DENV4. Magnification of 200×.

3. Discussion

The DENV NS1 has been used as a target antigen against dengue infection either for vaccines, antiviral drug design or diagnostic methods. Indeed, this protein is secreted by infected cells during the acute phase and circulates in the blood at high concentrations [19]. Nevertheless, the NS1 shares parts of its amino acid sequence among flavivirus. In the present study, we generated four mAbs against DENV2 recombinant NS1 and analyzed their reactivity with the dimeric-NS1 form. The mAbs were also reactive with the native NS1 produced in infected cells but showed different features. The epitope sequences recognized by different mAbs have been recently described and been considered the strategic point for understanding these interactions [15–17].

The mAbs 2H5 and 4H1BC showed a similar recognition pattern and share the same epitope binding, [193]AVHADMGYWIESALNDT[209]. This epitope has been reported as one of the immunodominant B cell epitopes in DENV2 NS1 [20]. This sequence was described in silico and is buried in a beta-sheet structure [15,17]. The recognition of the heat-denatured rNS1 was lower for these mAbs when compared with the non-denatured rNS1, suggesting that these mAbs recognize mainly a conformational epitope. Indeed, by immunofluorescence, both 2H5 and 4H1BC mAbs reacted with dengue virus serotype 2 infected Vero cells. However, they cross-reacted with native ZIKV NS1 in Vero infected cells. The in silico analyses of the similarity of this peptide sequence between different flaviviruses showed that this epitope is highly conserved in these virus but not Yellow fever, Japanese encephalitis and West Nile viruses (Table S2).

The mAb 4F6 recognizes the DENV2 complex-conserved LD2 epitope [25]VHTWTEQYKFQPES[38], located on the surface of NS1 fusion loop [8,15,17]. A previous study showed a mAb that binds in this motif is able to recognize the purified NS1 hexamer from all four DENV serotypes [21]. However,

4F6 mAb was able to detect only DENV2 native monomeric NS1 by immunofluorescence but no other serotypes. The divergent results may be accounted to methodological issues, since immunofluorescence is less sensitive and aims to detect infected cells expressing mainly monomeric intracellular NS1, while the purified hexamers were used in an ELISA-based detection system. Hence, this epitope may be exposed depending on the NS1 oligomeric level and the DENV serotype.

The fourth mAb obtained, 4H2, recognized the amino acid sequence [127]ELHNQTFLIDGPET AEC[143]. A preceding work described a shorter sequence, [125]STESHNQTFL[134] exposed in the same loop of DENV2 NS1 [15]. Interestingly, the sequence herein described has nine additional amino acids not previously reported as a B cell epitope and shifting the exposed region to a beta-sheet structure. It recognizes the native protein assessed by immunofluorescence of the four DENV serotypes infected Vero cells, but it did not cross-react with native ZIKV NS1 in Vero-infected cells.

Differentiation of DENV and ZIKV infections is a challenge for current serological tests, particularly in areas where both viruses circulate and co-infection can occur. Thus, mAbs, like 4H2, may be particularly useful for the development of an immunofluorescence based-assay that minimizes the risks associated with false positive results among ZIKV-infected subjects.

4. Materials and Methods

4.1. Viral Strains and Viral Antigen

The obtention of purified DENV2 NS1 dimers was achieved after denaturation/refolding steps of the protein expressed in *E. coli* followed by affinity chromatography, as previously reported [22]. This recombinant protein was utilized as an antigen for monoclonal antibody development and characterization. Four dengue serotypes and one Zika virus strain were used for further characterization of the mAbs obtained: a dengue virus serotype 2 JHA1 strain [23,24], a rDEN1Δ30 vaccine strain obtained by Δ30 deletion in 3′UTR of DENV1 Western Pacific strain [25], a rDEN3Δ30/31-7164 vaccine strain obtained by Δ30 and Δ31 deletions in 3′ untranslated region (UTR) of DENV3 Slemann/78 strain [26], a rDEN4Δ30 vaccine strain obtained by Δ30 deletion in 3′ UTR of DENV4 Dominica/81 strain [27], and a Brazilian Zika virus strain (ZIKVBR) (Evandro Chagas Institute, Belem, PA, Brazil).

4.2. Dengue NS1 Monoclonal Antibody (mAb) Production

Four to six week-old female BALB/c mice were immunized via footpad route with 10 µg rNS1 adsorbed to 1 µg recombinant heat-labile toxin (rLT) [22] as adjuvant. The immunization protocols consisted of three booster injections of the rNS1 and rLT in 0.01 M phosphate buffered saline (PBS), pH 7.4 at 15 days intervals. The mouse with the highest antibody titer was boosted with 10 µg of rNS1 three days prior to cell fusion. The popliteal lymph node cells were fused to SP2/O-Ag14 mouse myeloma cells (2:1) using polyethylene glycol 1500 (Sigma Aldrich, St Louis, MO, USA) [28], with modifications [29]. The supernatant fluids were screened for specific antibodies by indirect ELISA in which 100 µL of hybridoma supernatant was added to a 96-well MaxiSorp microplates (Nunc®, Rochester, NY, USA) previously coated with 1 µg/mL of purified rNS1 to screen cultures for antibody production. Antibody-secreting cells were expanded and cloned at limiting dilution [29]. This study was carried out in accordance with the recommendations of Ethical Principles in Animal Research, adopted by the Brazilian College of Animal Experimentation. The protocol was approved by the Ethical Committee for Animal Research of Butantan Institute (995/12).

4.3. Dengue NS1 mAbs Characterization

Hybridoma supernatants were incubated with each of the anti-isotype (anti-IgG1, anti-IgG2a, anti-IgG2b, anti-IgG3, anti-IgA and anti-IgM antibodies) previously coated at MaxiSorp microplates followed by incubation with horseradish peroxidase-conjugated rabbit anti-mouse-IgG+A+M+ (1:1000) (Zymed, San Francisco, CA, USA) [27]. The supernatants from selected clones were filtered (0.45 µm) and purified by protein G affinity chromatography (GE-Healthcare, Freiburg, Germany). MAb purity was observed in a 12% polyacrylamide gel electrophoresis containing sodium dodecyl sulphate (SDS-PAGE) staining with Coomassie blue R-250.

The detection limit was established using rNS1 concentrations from 1 to 512 ng coated on microplates, followed incubation with 10 µg/mL of NS1 mAb and with goat anti-mouse peroxidase-conjugated antibody (Invitrogen, Carlsbad, CA, USA) diluted 1:5000. The three-step ELISA was employed to determine the dissociation constants (KD) of antigen-antibody interactions under equilibrium conditions [28].

ELISA assay was also applied in order to observe the reactivity of mAb NS1 against intact and denatured rNS1. For this, MaxiSorp microplates (Thermo Fischer Scientific, Waltham, MA, USA) were coated with 4 µg/mL of Dengue virus serotype 2 (DENV2) rNS1 heat-treated (100 °C for 10 min) or non-heated. The NS1 mAbs were serially diluted (log2) in an initial concentration of 2.5 µg/mL followed by incubation with goat anti-IgG mouse conjugated with horseradish peroxidase (1:10,000).

The reactivity of NS1 mAb against intact and denatured rNS1 was also analyzed by immunoblotting. Thus, 1 µg or 0.5 µg of rNS1 denatured (heat-treated for 10 min at 100 °C) or intact (non-heated) were separated by electrophoresis in denaturing condition polyacrylamide gel containing sodium dodecyl sulphate (SDS-PAGE) 15%. Nitrocellulose membranes (GE-Healthcare, Freiburg, Germany) containing the transferred proteins were tested with NS1 mAbs at a final concentration of 200 ng/mL. Thus, the membranes were incubated with goat anti-IgG mouse conjugated with peroxidase (1:10,000). The reactive protein bands were identified by exposing membranes to a solution of luminol-hydrogen peroxide according to the manufacturer's instructions (Sigma Aldrich, St Louis, MO, USA). Images were captured by Image Lab™ software (Bio-Rad, Hercules, CA, USA).

4.4. Epitope Characterization and Structure Analysis

Peptide mapping was performed using CelluSpot Peptide Array (Intavis, Heidelberg, Germany) following the manufacturer's recommendations. The slides were produced with dots containing 11 amino acids with overlapping of eight amino acids. Briefly, the slides were blocked, followed by incubation with 30 µg/mL mAb. Next, the slides were incubated with anti-mouse horseradish peroxidase conjugate (1:5000). After washing, diaminobenzidine and hydrogen peroxide were added and the reaction was stopped by the addition of distilled water.

We employed PyMol program (DeLano Scientific LLC, San Carlos, CA, USA, 2009) to predict the structure and the epitope of NS1. For the NS1 structure, we used the available PDB file from Protein Data Bank (code: 4O6B) [6]. For the structure of monoclonal antibodies, we first performed the prediction with Phyre [30].

4.5. NS1 Sequences Database Building

One database consisting of amino acid sequences of the NS1 protein in FASTA format was built. Sequences were retrieved from the National Center for Biotechnology Information (NCBI) protein database. Sequences from serotype 1 of DENV have the following accession numbers: ABG75766, ABG75761 and AFN54943. Sequences from serotype 2 of DENV have the following accession numbers: AIE17400, ABK51383 and AFZ40226. Sequences from serotype 3 of DENV have the following accession numbers: ADM63678, AAT79552 and ALI16137. Sequences from serotype 4 of DENV have the following accession numbers: AGI95993, ALB78116 and AFD53008. Sequences from ZIKV isolates have the following accession numbers: AMR39836, AMD61710, ASK51714, ARB07991, AMD16557 and ARB07967.

4.6. Conservancy Analysis

The IEDB conservancy analysis was used to determine the conservancy of epitopes for monoclonal antibodies 4F6, 4H2, 2H5 and 4H1BC. A sequence identity threshold of ≥20% was applied.

4.7. NS1 mAbs Reactivity to Dengue Virus Serotypes and Zika Virus

Vero cells grown on six-well plates were infected with the viral strains at a MOI of 0.5 for 48 h for DENV and at a MOI of 0.05 for 72 h for ZIKV. The cells were fixed with 1% formaldehyde for 10 min at 4 °C and then permeabilized with 0.5% saponin in PBS for DENV and with cold acetone at -20 °C for ZIKV. The cells were then blocked with PBS containing 10% bovine fetal serum for 30 min. Both cells were treated with mAbs diluted in permeabilization buffer at a concentration of 10 µg/mL for 1 h at room temperature for DENV and 37 °C for 30 min for ZIKV. After three washing steps with PBS, the cells were further treated with Alexa fluor488® (Thermo Fisher, Waltham, MA, USA) conjugated goat anti-mouse IgG at room temperature for 1 h. After another washing period (five times) with PBS, cells were examined using an EVOS digital inverted microscope. Mock infected and cell infected marked with just secondary antibody was the negative control. Also, Vero cells infected with ZIKVBR were tested in order to determine the cross reactivity of the mAbs.

5. Conclusions

In the present study we generated four monoclonal antibodies against the nonstructural protein 1 (NS1) of dengue virus serotype 2. One of them (4H2 mAb) recognizes by immunofluorescence the four-dengue virus serotype and did not cross react to zika virus. Thus, the selection and characterization of the epitope, specificity of anti-NS1 mAbs, may contribute to the development of diagnostic tools able to differentiate DENV and ZIKV infections.

Supplementary Materials: The following are available online at www.mdpi.com/2073-4468/6/3/14/s1, Figure S1: Characterization of NS1-specific mAbs reactivity by immunoblotting; Figure S2: Epitope mapping with NS1-derived synthetic peptides; Table S1: Epitope analysis of conservancy in DENV serotypes. Table S2: Epitope analysis of conservancy in different strains of DENV, ZIKV, YFV, JEV and WNV strains.

Acknowledgments: We would like to thank Pedro Vasconcelos, from Evandro Chagas Institute, Belém, PA, Brazil for providing a lyophilized ZIKVBR seed. This work was supported by grants from São Paulo Research Foundation (FAPESP): 2009/53894-3; 2011/51761-6; 2014/17595-0 to L.C.S.F. 2013/06589-6; 2013/50955-7 to R.M.F.P and from Conselho Nacional de Desenvolvimento Científico e Tecnológico (CNPq) awarded to R.M.F.P. LBR was a recipient of post-doctoral FAPESP fellowship (2012/09096-8).

Author Contributions: L.B.R., R.P.d.S.A., L.R.P., V.F.B., D.B.L.O., L.C.d.S.F. and R.M.F.P. conceived and designed the experiments; L.B.R., R.P.d.S.A., B.A.C., L.R.P., T.M. and J.M.P. performed the experiments; L.B.R., R.P.d.S.A., B.A.C., L.R.P., T.M., D.B.L.O., D.L., L.C.d.S.F. and R.M.F.P. analyzed the data; J.H.A., N.M.F.G., R.P., A.R.P., C.F.H.G., L.C.d.S.F. and R.M.F.P. contributed reagents/materials/analysis tools; L.B.R., R.P.d.S.A., B.A.C., D.L., L.C.d.S.F. and R.M.F.P. wrote the paper.

Conflicts of Interest: The authors declare no conflict of interest. The founding sponsors had no role in the design of the study; in the collection, analyses, or interpretation of data; in the writing of the manuscript, and in the decision to publish the results.

References

1. Guzman, M.G.; Halstead, S.B.; Artsob, H.; Buchy, P.; Farrar, J.; Gubler, D.J.; Hunsperger, E.; Kroeger, A.; Margolis, H.S.; Martínez, E.; et al. Dengue: A continuing global threat. *Nat. Rev. Microbiol.* **2010**, *8*, S7–S16. [CrossRef] [PubMed]
2. Bhatt, S.; Gething, P.W.; Brady, O.J.; Messina, J.P.; Farlow, A.W.; Moyes, C.L.; Drake, J.M.; Brownstein, J.S.; Hoen, A.G. The global distribution and burden of dengue. *Nature* **2013**, *496*, 504–507. [CrossRef] [PubMed]
3. Daep, C.A.; Muñoz-Jordán, J.L.; Eugenin, E.A. Flaviviruses, an expanding threat in public health: Focus on dengue, West Nile, and Japanese encephalitis virus. *J. Neurovirol.* **2014**, *6*, 539–560. [CrossRef] [PubMed]
4. Secretaria de Vigilância em Saúde, Ministério da Saúde. *Bol. Epidemiol.* **2017**, *48*, 1–9.
5. Chambers, T.J.; Hahn, C.S.; Galler, R.; Rice, C.M. Flavivirus genome organization, expression, and replication. *Annu. Rev. Microbiol.* **1990**, *44*, 649–688. [CrossRef] [PubMed]
6. Akey, D.L.; Brown, W.C.; Dutta, S.; Konwerski, J.; Jose, J.; Jurkiw, T.K.; DelProposto, J.; Ogata, C.W.; Skiniotis, G.; Kuhn, R.J.; et al. Flavivirus NS1 structures reveal surfaces for associations with membranes and the immune system. *Science* **2014**, *343*, 881–885. [CrossRef] [PubMed]

7. Mackenzie, J.M.; Jones, M.K.; Young, P.R. Immunolocalization of the dengue virus nonstructural glycoprotein NS1 suggests a role in viral RNA replication. *Virology* **1996**, *1*, 232–240. [CrossRef] [PubMed]
8. Falconar, A.K.; Young, P.R.; Miles, M.A. Precise location of sequential dengue virus subcomplex and complex B cell epitopes on the nonstructural-1 glycoprotein. *Arch. Virol.* **1994**, *137*, 315–326. [CrossRef] [PubMed]
9. Silva, E.M.; Conde, J.N.; Allonso, D.; Nogueira, M.L.; Mohana-Borges, R. Mapping the interactions of dengue virus NS1 protein with human liver proteins using a yeast two-hybrid system: Identification of C1q as an interacting partner. *PLoS ONE* **2013**, *8*, e57514. [CrossRef] [PubMed]
10. Muller, D.A.; Young, P.R. The flavivirus NS1 protein: Molecular and structural biology, immunology, role in pathogenesis and application as a diagnostic biomarker. *Antivir. Res.* **2013**, *98*, 192–208. [CrossRef] [PubMed]
11. Gyawali, N.; Bradbury, R.S.; Taylor-Robinson, A.W. The global spread of Zika virus: Is public and media concern justified in regions currently unaffected? *Infect. Dis. Poverty* **2016**, *19*, 5. [CrossRef] [PubMed]
12. Kochakarn, T.; Kotanan, N.; Kümpornsin, K.; Loesbanluechai, D.; Thammasatta, M.; Auewarakul, P.; Wilairat, P.; Chookajorn, T. Comparative genome analysis between Southeast Asian and South American Zika viruses. *Asian Pac. J. Trop. Med.* **2016**, *9*, 1048–1054. [CrossRef] [PubMed]
13. Zhang, X.; Sheng, J.; Plevka, P.; Kuhn, R.J.; Diamond, M.S.; Rossmann, M.G. Dengue structure differs at the temperatures of its human and mosquito hosts. *Proc. Natl. Acad. Sci. USA* **2013**, *110*, 6795–6799. [CrossRef] [PubMed]
14. Jiao, Y.; Legge, F.S.; Zeng, X.; Treutlein, H.R.; Zeng, J. Antibody recognition of Shiga toxins (Stxs): Computational identification of the epitopes of Stx2 subunit A to the antibodies 11E10 and S2C4. *PLoS ONE* **2014**, *2*, e88191. [CrossRef] [PubMed]
15. Jones, M.L.; Legge, F.S.; Lebani, K.; Mahler, S.M.; Young, P.R.; Watterson, D.; Treutlein, H.R.; Zeng, J. Computational Identification of Antibody Epitopes on the Dengue Virus NS1 Protein. *Molecules* **2017**, *22*, 607. [CrossRef] [PubMed]
16. Lebani, K.; Jones, M.L.; Watterson, D.; Ranzoni, A.; Traves, R.J.; Young, P.R.; Mahler, S.M. Isolation of serotype-specific antibodies against dengue virus non-structural protein 1 using phage display and application in a multiplexed serotyping assay. *PLoS ONE* **2017**, *12*, e0180669. [CrossRef] [PubMed]
17. Chaudhury, S.; Gromowski, G.D.; Ripoll, D.R.; Khavrutskii, I.V.; Desai, V.; Wallqvist, A. Dengue virus antibody database: Systematically linking serotype-specificity with epitope mapping in dengue virus. *PLoS Negl. Trop. Dis.* **2017**, *11*, e0005395. [CrossRef] [PubMed]
18. Friguet, B.; Chaffotte, A.F.; Djavadi-Ohaniance, L.; Goldberg, M.E. 1985. Measurements of the true affinity constant in solution of antigen-antibody complexes by enzyme-linked immunosorbent assay. *J. Immunol. Methods* **1985**, *77*, 305–319. [CrossRef]
19. Vaughn, D.W.; Green, S.; Kalayanarooj, S.; Innis, B.L.; Nimmannitya, S.; Suntayakorn, S.; Rothman, A.L.; Ennis, F.A.; Nisalak, A. Dengue in the early febrile phase: Viremia and antibody responses. *J. Infect. Dis.* **1997**, *76*, 322–330. [CrossRef]
20. Jiang, L.; Zhou, J.M.; Yin, Y.; Fang, D.Y.; Tang, Y.X.; Jiang, L.F. Selection and identification of B-cell epitope on NS1 protein of dengue virus type 2. *Virus Res.* **2010**, *150*, 49–55. [CrossRef] [PubMed]
21. Gelanew, T.; Poole-Smith, B.K.; Hunsperger, E. Development and characterization of mouse monoclonal antibodies against monomeric dengue virus non-structural glycoprotein 1 (NS1). *J. Virol. Methods* **2015**, *222*, 214–223. [CrossRef] [PubMed]
22. Amorim, J.H.; Porchia, B.F.M.M.; Balan, A.; Cavalcante, R.C.M.; da Costa, S.M.; de Barcelos Alves, A.M.; de Souza Ferreira, L.C. Refolded dengue virus type 2 NS1 protein expressed in Escherichia coli preserves structural and immunological properties of the native protein. *J. Virol. Methods* **2010**, *167*, 186–192. [CrossRef] [PubMed]
23. Salvador, F.S.; Amorim, J.H.; Alves, R.P.S.; Pereira, S.A.; Ferreira, L.C.S.; Romano, C.M. Complete Genome Sequence of an Atypical Dengue Virus Serotype 2 Lineage Isolated in Brazil. *Genome Announc.* **2015**, *3*, e00779-15. [CrossRef] [PubMed]
24. Amorim, J.H.; Pereira Bizerra, R.S.; dos Santos Alves, R.P.; Sbrogio-Almeida, M.E.; Levi, J.E.; Capurro, M.L.; de Souza Ferreira, L.C. A genetic and pathologic study of a DENV2 clinical isolate capable of inducing encephalitis and hematological disturbances in immunocompetent mice. *PLoS ONE* **2012**, *7*, e44984. [CrossRef] [PubMed]

25. Whitehead, S.S.; Falgout, B.; Hanley, K.A.; Blaney, J.E., Jr.; Markoff, L.; Murphy, B.R. A live, attenuated Dengue virus type 1 vaccine candidate with a 30-nucleotide deletion in the 3′ untranslated region is highly attenuated and immunogenic in monkeys. *J. Virol.* **2003**, *77*, 1653–1657. [CrossRef] [PubMed]

26. Blaney, J.E., Jr.; Sathe, N.S.; Goddard, L.; Hanson, C.T.; Romero, T.A.; Hanley, K.A.; Murphy, B.R.; Whitehead, S.S. Dengue virus type 3 vaccine candidates generated by introduction of deletions in the 3′ untranslated region (3′-UTR) or by exchange of the DENV-3 3′-UTR with that of DENV-4. *Vaccine* **2008**, *26*, 817–828. [CrossRef] [PubMed]

27. Blaney, J.E., Jr.; Johnson, D.H.; Manipon, G.G.; Firestone, C.Y.; Hanson, C.T.; Murphy, B.R.; Whitehead, S.S. Genetic basis of attenuation of Dengue virus type 4 small plaque mutants with restricted replication in suckling mice and in SCID mice transplanted with human liver cells. *Virology* **2002**, *300*, 125–139. [CrossRef] [PubMed]

28. Khöler, G.; Milstein, C. Continuous cultures of fused cells secreting antibody of predefined specificity. *Biotechnology* **1975**, *24*, 524–526.

29. Rocha, L.B.; Luz, D.E.; Moraes, C.T.P.; Caravelli, A.; Fernandes, I.; Guth, B.E.C.; Horton, D.S.P.Q.; Piazza, R.M.F. Interaction between Shiga toxin and monoclonal antibodies: Binding characteristics and in vitro neutralizing abilities. *Toxins (Basel)* **2012**, *4*, 729–747. [CrossRef] [PubMed]

30. Kelley, L.A.; Mezulis, S.; Yates, C.M.; Wass, M.N.; Sternberg, M.J.E. The Phyre2 web portal for protein modeling, prediction and analysis. *Nat. Protoc.* **2015**, *10*, 845–858. [CrossRef] [PubMed]

Review

Monoclonal Antibody: A New Treatment Strategy against Multiple Myeloma

Shih-Feng Cho [1,2,3], Liang Lin [3], Lijie Xing [3,4], Tengteng Yu [3], Kenneth Wen [3], Kenneth C. Anderson [3] and Yu-Tzu Tai [3,*]

[1] Division of Hematology & Oncology, Department of Internal Medicine, Kaohsiung Medical University Hospital, Kaohsiung Medical University, Kaohsiung 807, Taiwan; shih-feng_cho@dfci.harvard.edu
[2] Faculty of Medicine, College of Medicine, Kaohsiung Medical University, Kaohsiung 807, Taiwan
[3] LeBow Institute for Myeloma Therapeutics and Jerome Lipper Multiple Myeloma Center, Dana-Farber Cancer Institute, Harvard Medical School, Boston, MA 02215, USA; liang_lin@dfci.harvard.edu (L.L.); lijie_xing@dfci.harvard.edu (L.X.); tengteng_yu@dfci.harvard.edu (T.Y.); kenneth_wen@dfci.harvard.edu (K.W.); Kenneth_anderson@dfci.harvard.edu (K.C.A.)
[4] Department of Hematology, Shandong Provincial Hospital Affiliated to Shandong University, No. 324, Jingwu Road, Jinan 250021, China
*   Correspondence: yu-tzu_tai@dfci.harvard.edu; Tel.: +1-617-632-3875; Fax: +1-617-632-2140

Received: 20 October 2017; Accepted: 10 November 2017; Published: 14 November 2017

Abstract: 2015 was a groundbreaking year for the multiple myeloma community partly due to the breakthrough approval of the first two monoclonal antibodies in the treatment for patients with relapsed and refractory disease. Despite early disappointments, monoclonal antibodies targeting CD38 (daratumumab) and signaling lymphocytic activation molecule F7 (SLAMF7) (elotuzumab) have become available for patients with multiple myeloma in the same year. Specifically, phase 3 clinical trials of combination therapies incorporating daratumumab or elotuzumab indicate both efficacy and a very favorable toxicity profile. These therapeutic monoclonal antibodies for multiple myeloma can kill target cells via antibody-dependent cell-mediated cytotoxicity, complement-dependent cytotoxicity, and antibody-dependent phagocytosis, as well as by direct blockade of signaling cascades. In addition, their immunomodulatory effects may simultaneously inhibit the immunosuppressive bone marrow microenvironment and restore the key function of immune effector cells. In this review, we focus on monoclonal antibodies that have shown clinical efficacy or promising preclinical anti-multiple myeloma activities that warrant further clinical development. We summarize mechanisms that account for the in vitro and in vivo anti-myeloma effects of these monoclonal antibodies, as well as relevant preclinical and clinical results. Monoclonal antibody-based immunotherapies have already and will continue to transform the treatment landscape in multiple myeloma.

Keywords: multiple myeloma; monoclonal antibody; immunomodulatory activity; bone marrow microenvironment

1. Introduction

Multiple myeloma is the second most common hematologic malignancy, characterized by the proliferation of malignant plasma cells in the bone marrow and excessive production of immunoglobulins [1,2]. The clinical outcome of patients with multiple myeloma has been improved in recent decades due to the development of novel therapeutic agents such as the proteasome inhibitors bortezomib [3], carfilzomib [4,5], and ixazomib [6] or immunomodulatory drugs (IMiDs) including thalidomide [7], lenalidomide [8], and pomalidomide [9,10]. With the incorporation of these novel agents into myeloma treatment strategies, the response rate and extent, progression-free survival, and overall survival have also been significantly improved in newly diagnosed patients [11–17].

However, in most cases, it remains a chronic and incurable disease due to its typical pattern of remission and relapse [18,19]. In addition, patients with refractory disease or who relapse after treatment with proteasome inhibitors and IMiDs have a very poor prognosis [18,20]. Thus, exploring novel approaches targeting different mechanisms to overcome drug resistance and minimize disease relapse are urgently needed.

With increased understanding of the biology of the disease, the development and evolution of multiple myeloma has been closely linked to specific immune system impairments. Malignant plasma cells express lower levels of tumor antigens and human leukocyte antigen (HLA) molecules [21,22], as well as higher levels of programmed cell death ligand 1 (PD-L1), which have been linked to defects in the antigen-presenting capacity of dendritic cells and a state of immune tolerance, respectively [23,24]. In addition, the bone marrow microenvironment in multiple myeloma has been shown to be immunosuppressive, providing a protective niche for the proliferation, migration, survival, and acquisition of drug resistance by malignant plasma cells [25–29]. Previous studies revealed that secreted inflammatory cytokines support the growth of immunosuppressive cells such as myeloid derived suppressor cells (MDSCs), tumor-associated macrophages (TAMs), and regulatory T-cells (Treg). Bone marrow stromal cells (BMSCs), osteoclasts (OCs), and plasmacytoid dendritic cells (pDC), as well as cytokines, i.e., interleukin-6 (IL-6), Macrophage colony-stimulating factor (M-CSF), interleukin-10 (IL-10), tumor necrosis factor beta (TGFβ), C-C Motif Chemokine Ligand 2 (CCL2), and vascular endothelial growth factor (VEGF), also play important roles in maintaining an immunosuppressive environment in the bone marrow of multiple myeloma patients [25,27]. These findings suggest that an effective anti-myeloma treatment will require not only targeting the malignant plasma cell itself but also restoring the anti-tumor responses of immune effector cells via blockade of tumor evasion and disruption of inhibitory signals on effector cells.

Monoclonal antibody-based treatments which provide additional effector cell-mediated tumor killing mechanisms when compared with targeted small molecules are successful therapeutic strategies for cancer. Monoclonal antibodies targeting specific surface antigens on cancer cells can kill the targeted cell via various effector-dependent and -independent mechanisms. Thus far, therapeutic IgG1-based monoclonal antibodies are designed to induce effector-mediated tumor cell lysis, including antibody-dependent cellular cytotoxicity (ADCC), complement-dependent cytotoxicity (CDC), and/or antibody-dependent phagocytosis (ADPC). Dependent on target antigens, therapeutic antibodies also act via receptor blockade to inhibit cell growth, induce apoptosis, or specifically deliver drug, radiation, or cytotoxic agent. Furthermore, the Fc region of antibodies plays an important role in mediating the killing of cancer cells via activation of certain immune cells (NK cells or cytotoxic T cells) as well as the induction of phagocytosis, CDC or ADCC [30,31]. For the treatment of hematological malignancies, the development of the anti-CD20 monoclonal antibody rituximab represents an important landmark and opened new venues for targeted cancer immunotherapies. Due to the success of rituximab in the treatment of B-cell lymphomas, the search for novel monoclonal antibodies for myeloma treatment has been rigorously pursued. Only small percentages of MM patients express CD20, so rituximab is not generally useful in myeloma [32,33]. Following the demonstration of promising preclinical and clinical activities [34–36], two monoclonal antibodies targeting CD38 (daratumumab) [37] and SLAMF7 (elotuzumab) [38] were approved by the Food and Drug Administration (FDA) to treat patients with relapsed and refractory multiple myeloma in late 2015.

Here, we focus on monoclonal antibodies showing multiple anti-myeloma mechanisms, including those with immunomodulatory effects (Figure 1). The preclinical and clinical data of these monoclonal antibodies are summarized. We also discuss other molecular targets with therapeutic potential in multiple myeloma.

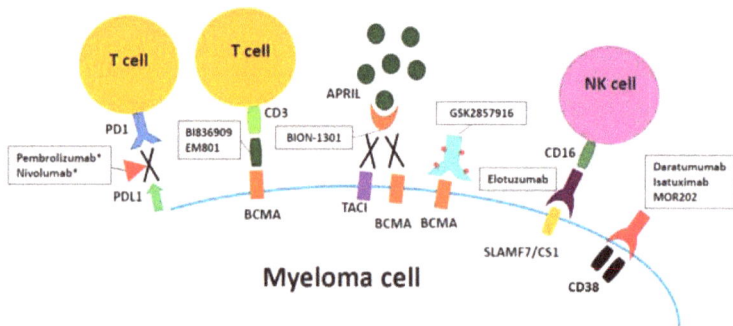

Figure 1. Therapeutic monoclonal antibodies for current multiple myeloma treatment. Daratumumab and elotuzumab, targeting CD38 and signaling lymphocytic activation molecule F7 (SLAMF7)/CS1, respectively, are naked IgG1 monoclonal antibodies which have been approved by FDA for treatment of relapsed/refractory multiple myeloma in late 2015. GSK2857916 is an antibody-drug conjugate composed of Fc-engineered IgG and a potent anti-tubulin drug MMAF. EM801 and BI836909 are bispecific T cell engagers targeting B-cell maturation antigen (BCMA) on myeloma cells and re-directing CD3+T cells to kill myeloma cells. BION-1301 exerts anti-myeloma activity by blocking the binding of APRIL to its cognate receptors BMCA and Transmembrane activator and CAML interactor (TACI), thereby abrogating growth, survival, and immuno-suppression signaling for myeloma cells. Clinical investigations of above agents are ongoing. * The clinical trials of PD-1 inhibitors (pembrolizumab and nivolumab) have been hold by Food and Drug Administration (FDA).

2. Targets and Monoclonal Antibodies

2.1. CD38

CD38 is a 46-kDa type II transmembrane glycoprotein with a short N-terminal cytoplasmic tail (20-aa) and a long extracellular domain (256-aa), which is identified on the surface of several cells of the immune system [39]. The intensity of CD38 expression is increased when lymphocytes are activated, and its expression is found in the majorities of hematopoietic linage cells. It is widely represented on lymphoid and myeloid cells, but absent from most mature resting lymphocytes. It catalyzes production of secondary messengers that affect Ca^{2+} mobilization. The biological function of CD38 has been linked to the regulation of calcium homeostasis in CD38-expressing lymphocytes [40]. Moreover, CD38 plays multiple but independent biologic roles since it acts both as a bifunctional enzyme responsible for the synthesis and hydrolysis of cyclic ADP-ribose, and as a signal-transducing surface receptor. CD38-/- mice are viable, without histological or pathological abnormalities. The natural ligands for CD38 are nicotinamide adenine dinucleotide (NAD)+, the substrate for its ecto-enzyme activity (ADP-ribosyl cyclase), and CD31/PECAM. Binding of CD31/PECAM and CD38 induces tyrosine phosphorylation and downstream signaling events regulating proliferation and cytokine release in lymphocytes. Regarding CD38 and multiple myeloma, previous studies have revealed that this glycoprotein is strongly and homogeneously expressed in terminally differentiated normal and malignant plasma cells [41,42]. A role of CD38 in the pathophysiology is postulated due to its high expression on a variety of hematological malignancies [43] including multiple myeloma [42,44], B- and T- acute lymphoblastic leukemia (ALL) [45,46], Non-Hodgkin lymphoma (NHL) [47], Acute myeloid leukemia (AML) [48] and Chronic Lymphocytic Leukemia (CLL) [49,50]. A recent study also suggested that CD38 enzymatic activity may be associated with immunosuppression in patients with multiple myeloma patients, due to its involvement in the production of immunosuppressive adenosine (ADO) [51].

2.1.1. Daratumumab

Daratumumab, a human IgG1-kappa monoclonal antibody, was the first naked CD38 monoclonal antibody to be further developed for clinical use following demonstration of promising anti-myeloma activity in preclinical studies with cell lines and animal models. In preclinical studies, it was found that daratumumab (formerly named HumaxCD38) kills CD38-expressing lymphoma and myeloma cells by various mechanisms including CDC, ADCC, ADCP, and induction of apoptosis after Fcγ receptor-mediated crosslinking with anti-human IgG1 secondary antibody [36,52,53]. In other CD38-expressing cells such as human NK cells, B and T cells, activated T cells or monocytes, the process of CDC cytotoxicity induced by daratumumab was not seen [36]. CD38 expression is relatively low on the cell membrane of these cells when compared with malignant plasma cells from multiple myeloma patients. Possible explanations for this therapeutic index include increased expression of complement regulatory proteins on the surface membrane of these cells or the need for a minimum threshold level of antigen expression needed to activate CDC [54,55]. Unlike CDC, daratumumab-mediated ADCC was seen in these cells and in primary tumor cells. In another study, ADCC was significantly enhanced if mononuclear effector cells derived from healthy peripheral blood donors were pretreated with lenalidomide, associated with activation of NK effector cells by lenalidomide [56]. Daratumumab also demonstrated potent antitumor activity in CD38-expressing xenografts in immune deficient mice [36], suggesting that daratumumab may mediate non-immune mediated anti-tumor activities in vivo.

A recent correlative study analyzed via flow cytometry on bone marrow and peripheral blood samples from participants in a clinical trial showed that daratumumab treatment rapidly depleted CD38 high-expressing immunosuppressive regulatory T cells (Treg) and B cells (Breg), as well as myeloid-derived suppressor cells (MDSC) [57]. In contrast, the numbers of immune effector cells such as helper and cytotoxic T cells increased [57,58]. CD38 levels are heterogenous among sub-populations of hematopoietic lineage cells. It is found to be expressed at significantly higher levels in Treg, Breg, and MDSCs, when compared with normal T, B, NK, and monocytes. This study therefore indicates that daratumumab quickly reduces these key immune inhibitory cellular components, thereby relieving their suppressive immune function and increasing effector cell-induced tumor cell lysis. Thus, in addition to its multiple FcR-dependent tumor cell killing mechanisms, daratumumab further blocks immunosuppressive cellular components, which may provide for more long-term responses [57].

Clinical Trials of Daratumumab

In phase 1–2 study, daratumumab monotherapy was administered to heavily pretreated patients with relapsed or refractory multiple myeloma (with a median of 5.5 lines of prior therapy, 75% refractory to lenalidomide and bortezomib) [37]. There were 32 patients enrolled in the dose-escalation phase. Daratumumab was administered weekly for 8 weeks, at doses ranging from 0.005 to 24 mg/kg. The maximum tolerated dose was not reached. In the expansion phase, 72 patients received 8 mg/kg or 16 mg/kg of daratumumab. The patients who received 16 mg/kg of daratumumab showed a better overall response rate (36% vs. 10%) and longer median progression-free survival (5.6 vs. 2.4 months) than patients treated with 8 mg/kg. Infusion-related reactions were the most frequently reported adverse effects, occurring in 71% of patients in the dose expansion group, mostly grade 1 or 2. In terms of hematologic adverse effects, neutropenia was most common, occurring in 12% of patients in the 16 mg/kg cohort [37]. Results of this small clinical study demonstrated the impressive activity of monotherapy with this agent in patients with no other available treatment options, leading the approval by the FDA in 2015 [37].

In the phase 2 SIRIUS study, 106 patients with multiple myeloma refractory to proteasome inhibitors and IMiDs (with a median of 5 lines of prior treatment) received daratumumab at 16 mg/kg. The overall response rate was 29.2% [59]. The time to response and median progression-free survival were 1.0 month and 3.7 months, respectively. The 12-month overall survival was 64.8%, and the median overall survival was 17.5 months. Infusion-related reactions were noted in 42%

of patients, mostly grade 1 or 2 The most common grade 3 or 4 adverse effects were anemia (24%), thrombocytopenia (19%) and neutropenia (12%) [59].

With respect to combination studies, the results of two phase 3 studies have been published [60,61]. In the CASTOR trial, 498 patients with relapsed or refractory multiple myeloma ($\geq$1 prior line of therapy) received bortezomib plus dexamethasone with or without daratumumab [60]. The incorporation of daratumumab in the treatment regimen significantly improved the overall response rate (82.9% vs. 63.2%, $p < 0.001$), the 12-month progression-free survival (60.7% vs. 26.9%), and the median progression-free survival (not reached vs. 7.2 months, $p < 0.001$). The most common grade 3 or 4 adverse events reported in the daratumumab group were thrombocytopenia (45.3%), anemia (14.4%), and neutropenia (12.8%). Infusion related reactions were noted in 45.3% of patients from the daratumumab group.

In another phase 3 trial, the POLLUX study, daratumumab proved to be a good therapeutic combination with lenalidomide and dexamethasone [61]. In this study, 569 patients who had received one or more lines of anti-myeloma treatment received lenalidomide with or without daratumumab. Adding daratumumab to lenalidomide and dexamethasone was associated with better response rates (93% vs. 76%, $p < 0.0001$), complete response rates (43.1% vs. 19.2%, $p < 0.0001$) and progression-free survival at 12 months (83.2% vs. 60.1%). The daratumumab group also showed a higher rate of minimal residual disease negativity (22.4% vs. 4.6%, $p < 0.001$). The most common grade 3 or 4 adverse effects in the daratumumab group were neutropenia (51.9%), thrombocytopenia (12.7%) and anemia (12.4%). Infusion-related reactions were noted in 47.7% of patients of the daratumumab group [61].

An important finding from both CASTOR and POLLUX was that the benefit of the addition of daratumumab to existing doublets persisted regardless of the number of prior lines of therapy. Greater benefit was seen when the triplet modality was used earlier in the disease course. Although close to half of the patients experienced daratumumab-related infusion reactions, >90% of these events occurred only upon the first infusion. This observation indicated that repeated dosing is safe. Both regimens were approved in November 2016 by the FDA for the treatment of multiple myeloma patients who have received at least one prior therapy. In addition, the unprecedented results stimulated studies for the detection of minimal residual disease (MRD) with next generation sequencing (NSG) and next generation flow-cytometry. The new MRD categories are currently being standardized to report across clinical trials in order to validate their importance as key prognostic markers and to guide treatment decisions.

2.1.2. Isatuximab (SAR650984)

Isatuximab, formerly called SAR650984 [62], is a novel humanized IgG1-kappa anti-CD38 monoclonal antibody currently under clinical development. Isatuximab was selected because of its direct induction of apoptosis in CD38-expressing lymphoma cell lines, in addition to its multiple effector cell-dependent cytotoxicity. In a preclinical study, isatuximab induced cell death in myeloma cell lines by ADCC, CDC, and ADCP, as well as the induction of tumor cell death in a CD38-dependent manner [62]. It is the latter activity which differentiates isatuximab from other therapeutic CD38 monoclonal antibodies because tumor cell death is directly induced by isatuximab in the absence of immune effector cells. It has similar half maximal effective concentrations (EC50 ~0.1 µg/mL) and maximal binding as daratumumab but MOR03087 (MOR202) (discussed later in this article) has a lower apparent affinity (EC50 ~0.3 µg/mL) [63]. These three CD38 monoclonal antibodies were equally potent at inducing ADCC against CD38-expressing tumor cells [63]. Daratumumab demonstrated superior induction of CDC in Daudi lymphoma cells as determined by flow cytometry, when compared with other CD38 antibodies in current clinical development. Specifically, isatuximab, more potently than daratumumab, inhibits ecto-enzyme function of CD38. It produced the largest inhibition of cyclic GDP-ribose (cGDPR) production, indicating a higher modulation of CD38 cyclase activity.

In in vivo studies using the same multiple myeloma cell lines xenografted in Severe combined immunodeficiency (SCID) mice, isatuximab showed more potent anti-myeloma activity than bortezomib [62]. Importantly, without the addition of Fc crosslinking agents or effector cells, isatuximab induced homotypic aggregation-associated multiple myeloma cell killing in a CD38-dependent manner [64].

In contrast, under similar conditions in ex vivo co-cultures, daratumumab shows no direct toxicity against multiple myeloma cells. Significantly, its F(ab)'2 fragments, just like the full-length version of isatuximab, could trigger lysosome-dependent cell death via upregulation of lysosome related protease cathepsin B and the translocation of lysosomal-associated membrane protein 1 (LAMP1) from lysosome to cell membrane, as well as increased reactive oxygen species. This effect was preferentially seen in myeloma cells expressing elevated levels of CD38 regardless of p53 mutation, which represents a key feature of most resistant patient group. Isatuximab specifically induced lysosome-dependent cell death by enlarging lysosomes and increasing lysosomal membrane permeability, despite the presence of protective IL-6 or bone marrow stromal cells. Furthermore, the addition of pomalidomide augments the direct and indirect killing effects of isatuximab in pomalidomide/lenalidomide-resistant myeloma cells [64]. Caspase 3/7-mediated apoptosis in drug-resistant myeloma cells was synergistically enhanced when both Isatuximab and pomalidomide were added together. Pomalidomide also increased ADCC mediated by isatuximab, further supporting ongoing phase III combination clinical trials.

Most recently, the effects of isatuximab on immune cell populations in the bone marrow microenvironment were investigated, since CD38 is widely expressed on hematopoietic cells [65]. First, significantly increased levels of CD38 were shown on the cell membrane of Tregs (CD4$^+$CD25^{high}Foxp3$^+$) and myeloma cells when compared with conventional T (Tcon, CD4+CD25$^-$) cells containing the majority of effector T cells. Higher CD38 expression and higher CD38$^+$ subsets were identified on the cell membrane of Tregs (CD4$^+$CD25^{high}Foxp3$^+$) versus Tcon, in accord with findings from a recent study of daratumumab trials [57]. Importantly, following isatuximab treatment, the percentage and function of CD38 high-expressing Tregs were decreased via the induction of apoptosis and decreased proliferation. In parallel, isatuximab blocked Treg-inhibited growth of conventional T cells in a dose-dependent manner. Longer ex vivo cocultures in the presence of low dose lenalidomide or pomalidomide further increased CD38 levels and percentages of the CD38$^+$ sub-population in the viable Tregs. This result indicates that IMiDs can enhance the sensitivity of this immune inhibitory subset to isatuximab, providing an additional mechanism to support combination trials of these reagents to sustain T effector cell function. Significantly, isatuximab augmented degranulation of NK and CD8+ T effector cells leading to increased multiple myeloma cell lysis, which was further enhanced by pomalidomide or lenalidomide. Isatuximab also reduced Foxp3 and IL10, associated with the inhibitory function of Tregs, and restored the proliferation of CD4$^+$CD25$^-$ naive T cells. Importantly, patients with multiple myeloma have elevated CD38high Tregs that block the proliferation of Tcons; and multiple myeloma cells convert Tcon into Tregs in ex vivo co-cultures. Myeloma cells induce the generation of Tregs (iTreg) which also highly express CD38 in addition to other known Treg markers. This study further showed that Tregs can be induced by cell-to-cell contact-dependent and -independent interactions between myeloma cells and Tcons, mimicking increased frequency of Tregs in MM patients versus normal donors. These iTregs generated in ex vivo co-cultures show significantly elevated CD38 and Foxp3 levels when compared with Tcons. They still demonstrate inhibitory function and significantly decrease proliferation of Tcons, which is overcome by isatuximab [65]. In a similar fashion as occurred in multiple myeloma cells overexpressing CD38, isatuximab preferentially targets CD38high-expressing Tregs of patient MM cells [65]. These results are in accord with a recent report showing that these immune inhibitory CD38high subsets were rapidly depleted by daratumumab in a recent correlative study [57]. These studies indicate that isatuximab can mitigate the immunosuppressive bone marrow microenvironment, thereby further restoring anti-multiple myeloma immunity.

Clinical Trials of Isatuximab

In a phase 1b dose-escalation study, isatuximab was administered in combination with lenalidomide and dexamethasone to treat patients with relapsed and refractory multiple myeloma (with a median of five prior regimens) [66]. Isatuximab was given at different doses and in two schedules (3, 5 or 10 mg/kg every 2 weeks (Q2W) or 10 or 20 mg/kg weekly for 4 weeks, then Q2W thereafter (QW/Q2W)). The maximum tolerated dose was not reached in this study. The analysis of efficacy revealed that the overall response rate was 56%. The median progression-free survival was 8.5 months. The most common treatment-related adverse events were fatigue and nausea, mostly grade 1 or 2. Infusion related reactions were noted in 56% of patients, mostly grade 1 or 2, and mainly occurred during the first infusion [66].

The combination of isatuximab and pomalidomide/dexamethasone is also generally well tolerated and clinically active in patients with heavily pre-treated relapsed and refractory multiple myeloma [67]. Of the eight patients who achieved at least PR up to 10 mg/kg, all continued to respond without confirmed disease progression at data cut-off. The pharmacokinetic (PK) parameters of isatuximab were not affected by co-administration with pomalidomide/dexamethasone. Data are continuing to be collected for longer-term follow up, including the 20 mg/kg cohort. A Phase III trial to evaluate isatuximab plus pomalidomide/dexamethasone is planned.

2.1.3. MOR03087 (MOR202)

MOR202 (HuCAL) is a novel fully human anti-CD38 IgG1 monoclonal antibody. In preclinical studies, MOR202 killed CD38-expressing cell lines and primary myeloma cells from patients by ADCC and ADCP. In SCID-mouse xenograft models, MOR202 inhibited tumor growth [68]. Furthermore, the addition of lenalidomide to MOR202 activated immune effector cells and augmented ADCP-mediated cytotoxicity [69]. Pomalidomide also enhances the cytotoxicity of MOR on MM cells [70].

Clinical trials of MOR03087 (MOR202)

In a phase I/II dose-escalation clinical trial, MOR202 was administered alone or combined with IMiDs (lenalidomide or pomalidomide) to treat 66 patients with relapsed or refractory multiple myeloma (more than 2 lines of prior therapy) [71]. MOR202 was infused for 2 h. The maximum tolerated dose was not reached in this study. Of 16 evaluable patients in the MOR202 monotherapy cohort, 3 patients (19%) showed partial responses and 2 patients (13%) showed very good partial responses. In the MOR202/lenalidomide cohort, 5 of 7 patients exhibited partial responses. In the MOR202/pomalidomide cohort, 3 of 5 patients showed a response to treatment, including 2 complete responses. Infusion-related reactions were seen in 3 patients (3/31, 10%), all occurring during the first infusion and less than grade 2.

2.1.4. Side Effects of CD38 Monoclonal Antibody

Some important clinical scenarios should be considered when using daratumumab to treat myeloma patients. First, daratumamab can be detected as an individual monoclonal band, which would interfere with serum immunofixation electrophoresis tests (IFE) [72,73]. To address this issue, daratumumab IFE reflex assay was developed to abrogate interference. Second, daratumumab has been shown to interfere with evaluation of bone marrow aspirates by multiparameter flow cytometry in a small study [74]. There were no CD38 or CD138 events detected in two patients treated with daratumumab. This was contrasting to the finding of the aspirate morphology and immunohistochemical study, which showed abnormal plasma cells positive for CD38 or CD138. Utilization of antibodies binding to different epitopes of CD38 may address this issue. Third, daratumumab also interferes with blood typing because CD38 is expressed on human red blood cells. Daratumumab binds to CD38 antigen on the reagent blood cells and leads to a positive indirect Coombs test. Treatment of dithiothreitol on reagent red cells with to denatures CD38 on the surface represents a potential method to circumvent this issue.

To prevent an infusion related reaction, medications such as glucocorticoid, antihistamine and acetaminophen can be administered before infusion during the initial cycles of treatment [75–77]. If an infusion related reaction occurs despite the administration of premedication, the infusion should be held until the appropriate symptom management is done and symptoms have resolved. After the resolution of symptoms, daratumumab can be restarted at a lower infusion rate.

2.2. SLAMF7/CS1

SLAMF7 (signaling lymphocytic activation molecule F7), previously known as CS1 (cell surface 1), is a cell surface glycoprotein that is a member of the signaling lymphocytic activation molecule family [78]. SLAMF7 is expressed on NK cells, activated monocytes, T cell subsets, and normal plasma cells. The biological function of SLAMF has been linked to the regulation and activation of NK cells [78,79]. Importantly, >90% of patients with myeloma cells expressed SLAMF7 messenger RNA (mRNA) and protein, regardless of disease status and treatments [34,35]. High SLAMF7 protein expression on the cell surface of multiple myeloma cell lines and patient myeloma cells spurred the development of monoclonal antibodies targeting SLAMF7 for the treatment of this cancer.

2.2.1. Elotuzumab

Elotuzumab is a humanized IgG1-kappa monoclonal antibody that targets SLAMF7. In a preclinical study, elotuzumab specifically bound to $CD138^+$ myeloma cells, natural killer (NK), NK-like T cells, and $CD8^+$ T cells, but not to hematopoietic $CD34^+$ stem cells. The major mechanism for the anti-myeloma activity of elotuzumab was the induction of dose-dependent NK cell-mediated ADCC [34,35], since elotuzumab did not induce CDC against myeloma cells. In a mouse xenograft model, elotuzumab showed in vivo efficacy mainly via NK effector cell-mediated toxicity [34]. In addition, it inhibited the adhesion of myeloma cells to bone marrow stromal cells, thereby blocking their proliferation and survival [34]. The anti-myeloma ADCC effect of elotuzumab was further enhanced by pretreatment with lenalidomide or bortezomib [34,80], providing the rationale for the combination trial of elotuzumab with lenalidomide/dexamethasone [38] or bortezomib [81].

Clinical Trials of Elotuzumab

In a phase I dose-escalation study, elotuzumab monotherapy was administered to thirty-five patients with relapsed or refractory myeloma ($\geq$2 prior therapies, median 4.5) [82]. Elotuzumab was administered in doses ranging from 0.5 to 20 mg/kg once every two weeks. The maximum tolerated dose was not reached. The treatment response, evaluated according to the European Group for Blood and Marrow Transplantation myeloma response criteria, showed that nine patients (26.5%) had stable disease. Common adverse events such as chills, fever and flushing were generally mild to moderate in severity (grade 1 or 2). Before the protocol amendment, 13 of 25 treated patients suffered from infusion-related reactions. This study found that CS1 on myeloma cells was highly saturated (>95%) with antibodies at dose levels of 10 and 20 mg/kg without dose-limiting toxicity. Two more two phases 1 studies were conducted to evaluate the efficacy of combining elotuzumab with other anti-myeloma agents [81,83]. In the first study, patients with relapsed and refractory multiple myeloma ($\geq$1 prior therapies (median 3), pretreated with lenalidomide were eligible) were treated with elotuzumab, lenalidomide and dexamethasone [83]. The overall response rate was 82%, with 29% showing at least a very good partial response (VGPR) [83]. In the second study, elotuzumab was combined with bortezomib to treat patients with relapsed and refractory multiple myeloma (1 to 3 prior treatments (median 2), pretreated with bortezomib were eligible) [81]. The overall response rate was 48%, and the median time to progression was 9.46 months.

In the pivotal phase III ELOQUENT-2 study, 646 patients with relapsed or refractory multiple myeloma (1 to 3 prior lines of treatment) received lenalidomide plus dexamethasone with or without elotuzumab [38]. Patients with prior lenalidomide treatment were enrolled if the best response seen was a partial response or better. The elotuzumab group showed a higher overall response rate than the control group (79% vs. 66%, $p < 0.001$). After a median follow-up of 24.5 months, the elotuzumab group showed better progression-free survival when compared with the control group (19.4 vs. 14.9 months, $p < 0.001$). Infection was noted in 81% and 74% of patients in the elotuzumab and control groups, respectively. The most common grade 3 or 4 adverse effects in the elotuzumab group were lymphocytopenia, neutropenia, fatigue, and pneumonia. With steroid premedication, infusion-related reactions occurred in 10% of patients of the elotuzumab group, mostly grade 1 or 2 [38].

2.3. PD-1/PD-L1

The immune checkpoint inhibitor programmed cell death protein 1 (PD-1)/programmed cell death ligand 1 (PD-L1) pathway plays a significant role in the evasion of host immunity by tumor cells [84]. PD-1 is a cell surface receptor of the immunoglobulin superfamily and is expressed on T cells, B cells, and NK cells [84,85]. Blockade of the PD-1 pathway may restore the cytotoxic function of T cells against myeloma cells in vitro [86]. Furthermore, T cells produce INFγ, which upregulates PD-L1 expression on tumor and infiltrating immune cells, forming a feedback loop that generates a PD-1 signal maintaining immunosuppression [87]. PD-L1 is highly expressed on malignant plasma cells from myeloma patients, especially in those with relapsed and refractory disease, but not on normal plasma cells [86,88]. In multiple myeloma, the expression of PD-L1 can be induced by IL-6, and higher levels of PD-L1 expression are associated with increased anti-apoptotic ability and more aggressive behavior of myeloma cells. The binding of PD-1 to PD-L1 has been associated with drug resistance by myeloma cells [89]. Significantly, PD-L1 is also expressed on cells with immunosuppressive effects that support the growth of myeloma cells in the bone marrow microenvironment, such as pDCs and MDSCs [86,90,91], leading to T cell anergy upon cellular contact. PD-L1 is further induced on multiple myeloma cells by a proliferation-inducing ligand (APRIL) [28,92] or the contact with bone marrow accessory cells [65,91,93]. It is also expressed on osteoclasts which support myeloma cell growth and survival in addition to the induction of bone lesions [28]. In preclinical studies, blockade of the PD-1/PD-L1 pathway was shown to inhibit myeloma cell growth mediated by bone marrow stromal cells [93]. These studies suggest that targeting PD-1/PD-L1 is promising immunotherapeutic strategy to overcome the ability of tumor cells to evade host immunity.

Since lenalidomide and pomalidomide (IMiDs) represent an efficient clinical approach in MM treatment to improve patients' survival, studies on the regulation of PD-1/PD-L1 pathway by IMiDs have been reported [93,94]. In addition to promoting tumor apoptosis, IMiDs activate T and NK cells, thereby increasing NK-mediated tumor recognition and killing. IMiDs stimulate T cell proliferation and cytokine secretion, decrease the expression of PD-1 on both T and NK cells in MM patients, as well as decrease both PD-1 and PD-L1 on MM cells. This leads to the inhibition of the negative signal induced by PD-1/PD-L1 axis on NK and T cells, restoring NK and T cell cytotoxic functions [91]. Thus, the combination of IMiDs with anti-PD-1/PD-L1 blocking strategies could represent a promising approach to re-establish the recognition of myeloma cells by exhausted NK and T cells to induce effective immune response. A recent study showed that PD-1/PD-L1 blockade induces anti-multiple myeloma immune response that can be enhanced by lenalidomide, which provides the framework for clinical evaluation of combination therapy [93]. Currently, there are at least two PD-1 inhibitors, nivolumab (IgG4-kappa) and pembrolizumab (IgG4-kappa) approved for solid cancer treatment.

Clinical trials of PD-1/PD-L1 Inhibitors

In a phase 1b study, a PD-1 inhibitor, nivolumab, was administered to patients with relapsed or refractory hematologic malignancies, including twenty-seven patients with multiple myeloma [95]. No significant disease regression was observed. Stable disease was noted in seventeen patients, with a median duration of 11.4 weeks.

With respect to the other PD-1 inhibitor, there are two trials showing promising results. In a phase I trial (Keynote-023), pembrolizumab was combined with lenalidomide and dexamethasone to treat relapsed or refractory myeloma patients (≥2 lines of prior therapy) [96]. The maximum tolerated dose for pembrolizumab was a fixed dose of 200 mg with 25 mg of lenalidomide and low-dose (40 mg) dexamethasone. Thrombocytopenia (47%), neutropenia (41%), and fatigue (29%) were the most common treatment-related adverse effects. With a median follow-up of 287 days, 13 out of 17 patients

(76%) responded to treatment, with 4 showing a very good partial response and 9 showing a partial response. The median duration of response was 9.7 months.

In a phase II study, pembrolizumab was administered with pomalidomide and dexamethasone to treat forty-eight patients with relapsed or refractory multiple myeloma (≥2 prior therapies) [97]. Autoimmune pneumonitis and hypothyroidism were seen in 13% and 10% of patients, respectively, mostly less than grade 2. The overall response rate was 60%. With a median follow-up of 15.6 months, the progression-free survival was 17.4 months. The median overall survival was not reached. A higher level of expression of PD-1 in bone marrow samples may correlate with improved progression-free survival [97]. However, two phase III trials comparing lenalidomide or pomalidomide with or without pembrolizumab have been put on hold due to excess deaths in the pembrolizumab cohort (http://www.ascopost.com/News/57813) [98].

Regarding antibodies targeting PD-L1, durvalumab is being studied alone and in combination with lenalidomide (NCT02685826) in patients with newly diagnosed multiple myeloma. Durvalumab, alone and in combination with pomalidomide (NCT02616640), is being evaluated in patients with relapsed/refractory disease. Durvalumab in combination with daratumumab or in combination with pomalidomide, dexamethasone, and daratumumab (NCT02807454). The other anti-PD-L1 antibody atezolizumab is currently being tested with daratumumab in patients with refractory multiple myeloma (NCT02431208) and in patients with asymptomatic multiple myeloma (NCT02784483). However, FDA recently placed a full clinical hold of checkpoint blockade in myeloma based on risks identified in other trials for an anti-PD-1 antibody, pembrolizumab.

2.4. B-Cell Maturation Antigen (BCMA)

BCMA (B-cell maturation antigen), a glycoprotein and non-tyrosine kinase receptor, is selectively expressed on the surface of mature B cells or plasma cells, but not naive B cells or most memory B cells [99–101]. BCMA can be induced by stimulation with cytokines during the differentiation of plasma cells [102] and plays an important role in the survival of long-lived plasma cells in the bone marrow [103,104]. It is not required for B cell homeostasis, and BCMA knockout mice is not lethal [104]. Importantly, BCMA is highly expressed in all multiple myeloma patient cells [99–101,105] and its expression levels correlates with disease status [105,106]. BCMA was being identified as a target of donor B-cell immunity in patients with myeloma who respond to donor lymphocyte infusion (DLI) [107]. Thus, in addition to donor T cells mediated graft-versus-myeloma response following allogenic hematopoietic stem-cell transplantation, the induction of specific antibodies against cell surface BCMA may directly contribute to tumor rejection in vivo [107]. Importantly, BCMA has a more specific expression pattern when compared with the other multiple myeloma antigens CD38 and SLAMF7. Malignant plasma cells have significantly increased BCMA when compared with normal plasma cells. Other than on plasma cells, BCMA is only detected on pDCs [105,108] which can promote myeloma cell growth, survival, and drug resistance. However, BCMA expression levels are significantly lower on pDCs when compared with plasma cells from the same individual, regardless of disease status. In contrast, all other cells including monocytic DCs, normal cells, and stem cells, do not express this antigen at mRNA and protein levels. The expression of BCMA on pDCs is significantly greater in myeloma patients than in normal individuals [105], further supporting BCMA as an ideal target for myeloma treatment. Moreover, BCMA upregulation in multiple myeloma cells has been associated with higher PD-L1 expression levels in addition to key survival proteins Mcl1, Bcl2, and Bcl-xL [92]. These data suggest that BCMA may modulate the immune response against multiple myeloma in the bone marrow microenvironment.

BCMA is shed by gamma-secretase, and soluble BCMA levels in serum are elevated and correlate with disease activity in systemic lupus erythematosus [102]. Gamma-secretase releases soluble BCMA that acts as a decoy, neutralizing its cognate ligand a proliferation-inducing ligand (APRIL). Like other multiple myeloma antigens, i.e., SLAMF7, CD138, CD38, soluble BCMA was detected in the elevated levels in the serum samples of multiple myeloma when compared with normal donors [106]. It was proposed that serum BCMA levels may be a new biomarker for monitoring disease status and overall survival of MM patients. Soluble BCMA levels may also play a pathophysiological role in multiple myeloma since they could inhibit the ligand (B cell activating factor, BAFF) binding to its membrane-bound BCMA to induce signaling and stimulate normal B-cell and plasma cell development, thereby resulting in reduced polyclonal antibody levels [109].

Immunotherapeutically Targeting BCMA

BCMA antibodies were developed with ligand blocking activity that could promote cytotoxicity of multiple myeloma cell lines as naked antibodies or as antibody-drug conjugates [110]. Recently, antagonistic humanized anti-BCMA antibody-drug conjugates via a noncleavable linker with auristatins (monomethyl auristatin E, MMAE or monomethyl auristatin F, MMAF) in preclinical studies demonstrated impressive in vitro and in vivo anti-multiple myeloma activity [105,111]. The BCMA antibody MMAF (GSK2857916) directly induces potent G2-M arrest, followed by apoptosis in multiple myeloma cell lines and patient cells [105]. Compared with its MMAE counterpart, GSK2857916 did not induce any bystander cytotoxicity when evaluated in co-cultures of MM cells with bone marrow stromal cells. Importantly, GSK2857916 also induces 1-log higher ADCC and ADPC against MM cells due to its afucosylation via FcR engineering, when compared with its homolog with normal FcR fragment. A phase 1 clinical study is ongoing, and preliminary data suggested a clinical activity at higher doses in patients with recurrent disease (NCT02064387) [112]. Thus far, maximal tolerated dose has not been reached. Adverse events were manageable, with ocular toxicity emerging as the most frequent reason for dose modifications.

One potential immunotherapeutic strategy is the development of T-cell bispecific antibodies, which bind simultaneously to a surface tumor cell antigen and a T-cell receptor to induce T cell-mediated killing of tumor cells harboring the target surface antigen. A BCMA/CD3 bispecific T-cell engager (BiTE®, Amgen, Thousand Oaks, CA, USA) antibody (BI836909), which was made based on blinatumomab (Anti-CD19/CD3 BiTE® antibody, Amgen, Thousand Oaks, CA, USA), was tested in a preclinical study using BCMA on myeloma cells as a target and CD3 on T cells as the other target [113]. Following treatment with BI836909, selective lysis of BCMA-positive myeloma cells, activation and proliferation of T cells, as well as release of multiple key cytokines related to effector T cell function (IFNγ, IL-2, IL-6, TNFα, IL-10) were noted. The anti-myeloma effect of BI836909 was not significantly affected by the presence of bone marrow stromal cells, soluble BCMA and APRIL (up to 150 and 100 ng/ml, respectively). BI836909 potently induces autologous patient myeloma cell lysis regardless of disease status. Clinical trial of BI836909 (NCT02514239) is ongoing in relapsed and refractory multiple myeloma.

An IgG-based BCMA-T cell bispecific antibody (EM801) also increased CD3$^+$ T cell/myeloma cell crosslinking, followed by CD4$^+$/CD8$^+$ T cell activation, and secretion of interferon-gamma, granzyme B, and perforin A [101]. It induced autologous T cell-mediated cell death in 34 of 43 bone marrow aspirates from patients with myeloma, including those with relapsed or refractory disease. Pharmacokinetics and pharmacodynamics indicate weekly intravenous/subcutaneous administration of EM801. Another bispecific antibody against BCMA (BiFab-BCMA) also potently and specifically redirects T cells to lyse malignant multiple myeloma cells [114]. BiFab-BCMA lysed BCMA-positive cell lines up to 20-fold more potently than a CS1-targeting bispecific antibody (BiFab-CS1) developed in an analogous fashion. In addition, the in vitro and in vivo activities of BiFab-BCMA are comparable to those of anti-BCMA chimeric antigen receptor T cell therapy (CAR-T-BCMA) [114], which has demonstrated impressive anti-myeloma activity in at least 4 recent clinical trials [115,116].

2.5. A Proliferation-Inducing Ligand (APRIL)

APRIL (a proliferation-inducing ligand), a member of the tumor necrosis factor family, is one of two ligands for BCMA [117–119]. Compared with the other ligand, BAFF, APRIL is more plasma cell-specific because it has stronger binding affinity towards receptors on plasma cells [120]. APRIL also binds to the receptor transmembrane activator, calcium modulator and cyclophilin ligand interactor (TACI), but the expression of TACI on myeloma cells is variable and lower than that of BCMA [99,121,122]. APRIL is produced by cells in the bone marrow, including myeloid-derived cells, osteoclasts, and DCs. APRIL can promote the survival of malignant plasma cells and rescue myeloma cell lines from apoptosis after IL-6 deprivation [92,121,123,124]. APRIL also promotes cell cycle progression in myeloma cells [92,125]. Based on these findings, targeting APRIL to prevent BCMA-mediated activation of myeloma cells constitutes a potential therapeutic strategy.

Blocking APRIL Biotherapeutics

An antagonistic anti-APRIL antibody hAPRIL01A (01A) was generated to block APRIL binding to BCMA and TACI [126]. It prevents in vitro proliferation and IgA production of APRIL-reactive B cells, and effectively impairs the chronic lymphocytic leukemia (CLL)-like phenotype of aging APRIL transgenic mice. Importantly, this antibody blocks APRIL binding to human B-cell lymphomas and prevents the survival effect induced by APRIL. Importantly, 01A inhibits APRIL- and osteoclast-induced myeloma cell proliferation and further induces apoptosis of myeloma cells in co-cultures [92]. 01A significantly blocks the growth of myeloma cells in a SCID-hu murine model, where multiple myeloma cells grow in the bone chips implanted in SCID mice. 01A augmented the cytotoxicity mediated by IMiDs and proteasome inhibitors in the co-cultures of myeloma cells with BCMA-negative bone marrow accessory cells and effector cells. Furthermore, following 01A treatment, APRIL-induced expression of genes involved in immunosuppression, i.e., PD-1, transforming growth factor beta (TGF-β), and interleukin 10 (IL-10), is decreased in multiple myeloma cells [28,92,127]. BION-1301, the clinical candidate for 01A, will be tested soon in multiple myeloma.

2.6. Potential Targets

Several additional cell surface antigens with therapeutic potential have been identified [128–131]. BAFF (B-cell activating factor) is a member of the TNFα superfamily that promotes the adhesion of myeloma cells to bone marrow stromal cells via activation of the AKT/NF-κB signaling pathway [122]. High levels of BAFF have been described in patients with multiple myeloma [121,132]. In an animal model of MM, mice treated with anti-BAFF antibody had significantly lower levels of soluble human IL-6 receptor and improved survival when compared with controls [132]. Tabalumab, a humanized monoclonal antibody targeting BAFF, was evaluated in a phase I clinical trial with bortezomib [133]. FcRH5, a B-cell lineage marker broadly expressed in myeloma, was targeted as one of the T-cell bispecific antibodies to induce T cell-mediated killing of FcRH5-expressing tumor cells [131]. PD-L1 blockade further enhanced the activity of FcRH5-CD3 T-cell bispecific antibody, suggesting the possibility for combination therapy in patients with multiple myeloma. Other therapeutic targets that are also being rigorously evaluated include IL-6 [134–137], CD40 [138], CD138 [139], MUC-1 [140] and Dickkopf-1(DKK-1) [141].

3. Conclusions

The development of monoclonal antibodies targeting selective multiple myeloma antigens represents an important advance in the improvement of effective immunotherapies for patients with multiple myeloma. In addition to various mechanisms mediated via FcR-expressing effector cells (ADCC, CDC or ADCP), monoclonal antibodies can produce immunomodulatory effects on immune cells in the bone marrow microenvironment by decreasing the function and number of immunosuppressive cells and restoring the tumor-killing activities of immune effector cells (Table 1). Such novel immunomodulatory effects may further lead to deepened clinical responses and improved efficacy, which is exemplified by daratumumab in recent large phase 3 clinical trials. Previous clinical trials have demonstrated that monoclonal antibodies constitute an efficacious therapeutic option even for heavily pretreated patients with relapsed and refractory multiple myeloma [142]. Either as a single agent or combined with other anti-myeloma drugs as well as immune checkpoint blockade and vaccination strategies, these antibodies will further improve the prognosis significantly. Furthermore, their acceptable safety profiles make monoclonal antibodies ideal partners to combine with other anti-myeloma agents in the search for better and more durable responses in patients with all stages, especially in early disease when the immune cells are still functional. In the case of monoclonal antibodies that have already been approved for the treatment of relapsed and refractory myeloma, their possible role as frontline treatments is being rigorously investigated. Ongoing and future studies are addressing the issue of which combinations are most effective at various stages of the disease. With the continued rapid development of novel monoclonal antibodies, we can expect transformation of the treatment landscape and associated improvement in patient outcome.

Table 1. Summary of monoclonal antibodies used in the treatment of myeloma.

Target	Name of the Antibody	Anti-Myeloma Mechanism	Immunomodulatory Effects
CD38	Daratumumab	CDC, ADCC, ADCP, induction of apoptosis when crosslinked, enzymatic modulation [36]	1. Deletion of CD38$^+$ Tregs and Bregs [57] 2. Expansion of CD8$^+$ cytotoxic T cells and CD4$^+$ helper T cells [57]
[34,35] CD38	Isatuximab	ADCC, CDC, ADCP, direct cell death via lysosome-mediated and apoptotic pathway [62]	1. Augmentation of NK and CD8$^+$ T effector cell-mediated anti-tumor immune responses [65] 2. Reduction of Foxp3 and IL10 in Tregs [65] 3. Restoration of proliferation and function of naive T cells [65]
CD38	MOR03087	ADCC, ADCP [69,70]	Activation of immune effector cells (Combined with IMID) [69,70]
SLAMF7/CS1	Elotuzumab	ADCC [34,35]	Activation of NK cells [79,143]
PD1	Pembrolizumab	Induction of apoptosis [86]	Activation and proliferation of T cells [86]
	Nivolumab		
BCMA	BI 836909	Potent induction of apoptosis [113]	BCMA- induced T-cell activation and cytokine release [113]
	GSK2857916	ADCC, ADCP, G2-M arrest followed by apoptosis [105]	1. Improved potency and efficacy of effector cell-mediated MM cell lysis [105] 2. G2-M growth arrest followed by apoptosis
	EM801	Induce myeloma cell death by autologous T cells [101]	Activation of CD4$^+$/CD8$^+$ T cells [101]
APRIL	BION-1301	Blockage of APRIL-induced growth and survival, induction of apoptosis [92]	Decreased expression of PD-1, TGF-β and IL-1 genes) [92]

CDC, complement-dependent cytotoxicity; ADCC, antibody-dependent cell-mediated cytotoxicity; ADCP, antibody-dependent cellular phagocytosis. IMID, immunomodulatory drugs; NK, natural killer cell; SLAMF, signaling lymphocytic activation molecule F7; BCMA, B-cell maturation antigen; APRIL, proliferation-inducing ligand.

Acknowledgments: Funding: The funding was provided by National Institutes of Health Grants RO1CA050947, RO1CA207237 and P50CA100707; KCA is an American Cancer Society Clinical Research.

Author Contributions: S.-F.C. and Y.-T.T. designed and wrote the manuscript. L.L., L.X., T.Y. and K.W. performed literature research. K.C.A. critically reviewed and edited the review.

Conflicts of Interest: K.C.A. serves on advisory boards to Millennium, Bristol Myers Squibb, Gilead, and is a scientific founder of Oncopep and C4 Therapeutics. The remaining authors declare no competing financial.

References

1. Kyle, R.A.; Gertz, M.A.; Witzig, T.E.; Lust, J.A.; Lacy, M.Q.; Dispenzieri, A.; Fonseca, R.; Rajkumar, S.V.; Offord, J.R.; Larson, D.R.; et al. Review of 1027 patients with newly diagnosed multiple myeloma. *Mayo Clin. Proc.* **2003**, *78*, 21–33. [CrossRef] [PubMed]
2. Palumbo, A.; Anderson, K. Multiple myeloma. *N. Engl. J. Med.* **2011**, *364*, 1046–1060. [CrossRef] [PubMed]
3. Richardson, P.G.; Sonneveld, P.; Schuster, M.W.; Irwin, D.; Stadtmauer, E.A.; Facon, T.; Harousseau, J.L.; Ben-Yehuda, D.; Lonial, S.; Goldschmidt, H.; et al. Bortezomib or high-dose dexamethasone for relapsed multiple myeloma. *N. Engl. J. Med.* **2005**, *352*, 2487–2498. [CrossRef] [PubMed]

4.   Stewart, A.K.; Rajkumar, S.V.; Dimopoulos, M.A.; Masszi, T.; Spicka, I.; Oriol, A.; Hajek, R.; Rosinol, L.; Siegel, D.S.; Mihaylov, G.G.; et al. Carfilzomib, lenalidomide, and dexamethasone for relapsed multiple myeloma. *N. Engl. J. Med.* **2015**, *372*, 142–152. [CrossRef] [PubMed]

5.   Dimopoulos, M.A.; Moreau, P.; Palumbo, A.; Joshua, D.; Pour, L.; Hajek, R.; Facon, T.; Ludwig, H.; Oriol, A.; Goldschmidt, H.; et al. Carfilzomib and dexamethasone versus bortezomib and dexamethasone for patients with relapsed or refractory multiple myeloma (endeavor): A randomised, phase 3, open-label, multicentre study. *Lancet Oncol.* **2016**, *17*, 27–38. [CrossRef]

6.   Moreau, P.; Masszi, T.; Grzasko, N.; Bahlis, N.J.; Hansson, M.; Pour, L.; Sandhu, I.; Ganly, P.; Baker, B.W.; Jackson, S.R.; et al. Oral ixazomib, lenalidomide, and dexamethasone for multiple myeloma. *N. Engl. J. Med.* **2016**, *374*, 1621–1634. [CrossRef] [PubMed]

7.   Singhal, S.; Mehta, J.; Desikan, R.; Ayers, D.; Roberson, P.; Eddlemon, P.; Munshi, N.; Anaissie, E.; Wilson, C.; Dhodapkar, M.; et al. Antitumor activity of thalidomide in refractory multiple myeloma. *N. Engl. J. Med.* **1999**, *341*, 1565–1571. [CrossRef] [PubMed]

8.   Dimopoulos, M.; Spencer, A.; Attal, M.; Prince, H.M.; Harousseau, J.L.; Dmoszynska, A.; San Miguel, J.; Hellmann, A.; Facon, T.; Foa, R.; et al. Lenalidomide plus dexamethasone for relapsed or refractory multiple myeloma. *N. Engl. J. Med.* **2007**, *357*, 2123–2132. [CrossRef] [PubMed]

9.   San Miguel, J.; Weisel, K.; Moreau, P.; Lacy, M.; Song, K.; Delforge, M.; Karlin, L.; Goldschmidt, H.; Banos, A.; Oriol, A.; et al. Pomalidomide plus low-dose dexamethasone versus high-dose dexamethasone alone for patients with relapsed and refractory multiple myeloma (mm-003): A randomised, open-label, phase 3 trial. *Lancet Oncol.* **2013**, *14*, 1055–1066. [CrossRef]

10.   Paludo, J.; Mikhael, J.R.; LaPlant, B.R.; Halvorson, A.E.; Kumar, S.; Gertz, M.A.; Hayman, S.R.; Buadi, F.K.; Dispenzieri, A.; Lust, J.A.; et al. Pomalidomide, bortezomib, and dexamethasone for patients with relapsed lenalidomide-refractory multiple myeloma. *Blood* **2017**, *130*, 1198–1204. [CrossRef] [PubMed]

11.   Rajkumar, S.V.; Blood, E.; Vesole, D.; Fonseca, R.; Greipp, P.R. Eastern Cooperative Oncology Group. Phase iii clinical trial of thalidomide plus dexamethasone compared with dexamethasone alone in newly diagnosed multiple myeloma: A clinical trial coordinated by the eastern cooperative oncology group. *J. Clin. Oncol.* **2006**, *24*, 431–436. [CrossRef] [PubMed]

12.   Palumbo, A.; Bringhen, S.; Caravita, T.; Merla, E.; Capparella, V.; Callea, V.; Cangialosi, C.; Grasso, M.; Rossini, F.; Galli, M.; et al. Oral melphalan and prednisone chemotherapy plus thalidomide compared with melphalan and prednisone alone in elderly patients with multiple myeloma: Randomised controlled trial. *Lancet* **2006**, *367*, 825–831. [CrossRef]

13.   San Miguel, J.F.; Schlag, R.; Khuageva, N.K.; Dimopoulos, M.A.; Shpilberg, O.; Kropff, M.; Spicka, I.; Petrucci, M.T.; Palumbo, A.; Samoilova, O.S.; et al. Bortezomib plus melphalan and prednisone for initial treatment of multiple myeloma. *N. Engl. J. Med.* **2008**, *359*, 906–917. [CrossRef] [PubMed]

14.   Richardson, P.G.; Weller, E.; Lonial, S.; Jakubowiak, A.J.; Jagannath, S.; Raje, N.S.; Avigan, D.E.; Xie, W.; Ghobrial, I.M.; Schlossman, R.L.; et al. Lenalidomide, bortezomib, and dexamethasone combination therapy in patients with newly diagnosed multiple myeloma. *Blood* **2010**, *116*, 679–686. [CrossRef] [PubMed]

15.   Cavo, M.; Tacchetti, P.; Patriarca, F.; Petrucci, M.T.; Pantani, L.; Galli, M.; Di Raimondo, F.; Crippa, C.; Zamagni, E.; Palumbo, A.; et al. Bortezomib with thalidomide plus dexamethasone compared with thalidomide plus dexamethasone as induction therapy before, and consolidation therapy after, double autologous stem-cell transplantation in newly diagnosed multiple myeloma: A randomised phase 3 study. *Lancet* **2010**, *376*, 2075–2085. [PubMed]

16.   Rosinol, L.; Oriol, A.; Teruel, A.I.; Hernandez, D.; Lopez-Jimenez, J.; de la Rubia, J.; Granell, M.; Besalduch, J.; Palomera, L.; Gonzalez, Y.; et al. Superiority of bortezomib, thalidomide, and dexamethasone (VTD) as induction pretransplantation therapy in multiple myeloma: A randomized phase 3 PETHEMA/GEM study. *Blood* **2012**, *120*, 1589–1596. [CrossRef] [PubMed]

17.   Benboubker, L.; Dimopoulos, M.A.; Dispenzieri, A.; Catalano, J.; Belch, A.R.; Cavo, M.; Pinto, A.; Weisel, K.; Ludwig, H.; Bahlis, N.; et al. Lenalidomide and dexamethasone in transplant-ineligible patients with myeloma. *N. Engl. J. Med.* **2014**, *371*, 906–917. [CrossRef] [PubMed]

18.   Kumar, S.K.; Lee, J.H.; Lahuerta, J.J.; Morgan, G.; Richardson, P.G.; Crowley, J.; Haessler, J.; Feather, J.; Hoering, A.; Moreau, P.; et al. Risk of progression and survival in multiple myeloma relapsing after therapy with imids and bortezomib: A multicenter international myeloma working group study. *Leukemia* **2012**, *26*, 149–157. [CrossRef] [PubMed]

19. Laubach, J.P.; Voorhees, P.M.; Hassoun, H.; Jakubowiak, A.; Lonial, S.; Richardson, P.G. Current strategies for treatment of relapsed/refractory multiple myeloma. *Expert Rev. Hematol.* **2014**, *7*, 97–111. [CrossRef] [PubMed]

20. Kumar, S.K.; Dimopoulos, M.A.; Kastritis, E.; Terpos, E.; Nahi, H.; Goldschmidt, H.; Hillengass, J.; Leleu, X.; Beksac, M.; Alsina, M.; et al. Natural history of relapsed myeloma, refractory to immunomodulatory drugs and proteasome inhibitors: A multicenter imwg study. *Leukemia* **2017**, *31*, 2443–2448. [CrossRef] [PubMed]

21. Walz, S.; Stickel, J.S.; Kowalewski, D.J.; Schuster, H.; Weisel, K.; Backert, L.; Kahn, S.; Nelde, A.; Stroh, T.; Handel, M.; et al. The antigenic landscape of multiple myeloma: Mass spectrometry (re)defines targets for T-cell-based immunotherapy. *Blood* **2015**, *126*, 1203–1213. [CrossRef] [PubMed]

22. Kumar, S.K.; Anderson, K.C. Immune therapies in multiple myeloma. *Clin. Cancer Res.* **2016**, *22*, 5453–5460. [CrossRef] [PubMed]

23. Liu, J.; Hamrouni, A.; Wolowiec, D.; Coiteux, V.; Kuliczkowski, K.; Hetuin, D.; Saudemont, A.; Quesnel, B. Plasma cells from multiple myeloma patients express B7-H1 (PD-L1) and increase expression after stimulation with IFN-{gamma} and TLR ligands via a MyD88-, TRAF6-, and MEK-dependent pathway. *Blood* **2007**, *110*, 296–304. [CrossRef] [PubMed]

24. Corthay, A.; Lundin, K.U.; Lorvik, K.B.; Hofgaard, P.O.; Bogen, B. Secretion of tumor-specific antigen by myeloma cells is required for cancer immunosurveillance by CD4+ T cells. *Cancer Res.* **2009**, *69*, 5901–5907. [CrossRef] [PubMed]

25. Hideshima, T.; Mitsiades, C.; Tonon, G.; Richardson, P.G.; Anderson, K.C. Understanding multiple myeloma pathogenesis in the bone marrow to identify new therapeutic targets. *Nat. Rev. Cancer* **2007**, *7*, 585–598. [CrossRef] [PubMed]

26. Zheng, Y.; Cai, Z.; Wang, S.; Zhang, X.; Qian, J.; Hong, S.; Li, H.; Wang, M.; Yang, J.; Yi, Q. Macrophages are an abundant component of myeloma microenvironment and protect myeloma cells from chemotherapy drug-induced apoptosis. *Blood* **2009**, *114*, 3625–3628. [CrossRef] [PubMed]

27. Kawano, Y.; Moschetta, M.; Manier, S.; Glavey, S.; Gorgun, G.T.; Roccaro, A.M.; Anderson, K.C.; Ghobrial, I.M. Targeting the bone marrow microenvironment in multiple myeloma. *Immunol. Rev.* **2015**, *263*, 160–172. [CrossRef] [PubMed]

28. An, G.; Acharya, C.; Feng, X.; Wen, K.; Zhong, M.; Zhang, L.; Munshi, N.C.; Qiu, L.; Tai, Y.T.; Anderson, K.C. Osteoclasts promote immune suppressive microenvironment in multiple myeloma: Therapeutic implication. *Blood* **2016**, *128*, 1590–1603. [CrossRef] [PubMed]

29. Leone, P.; Berardi, S.; Frassanito, M.A.; Ria, R.; De Re, V.; Cicco, S.; Battaglia, S.; Ditonno, P.; Dammacco, F.; Vacca, A.; et al. Dendritic cells accumulate in the bone marrow of myeloma patients where they protect tumor plasma cells from CD8+ T-cell killing. *Blood* **2015**, *126*, 1443–1451. [CrossRef] [PubMed]

30. Scott, A.M.; Wolchok, J.D.; Old, L.J. Antibody therapy of cancer. *Nat. Rev. Cancer* **2012**, *12*, 278–287. [CrossRef] [PubMed]

31. Scott, A.M.; Allison, J.P.; Wolchok, J.D. Monoclonal antibodies in cancer therapy. *Cancer Immunol.* **2012**, *12*, 14.

32. Mateo, G.; Castellanos, M.; Rasillo, A.; Gutierrez, N.C.; Montalban, M.A.; Martin, M.L.; Hernandez, J.M.; Lopez-Berges, M.C.; Montejano, L.; Blade, J.; et al. Genetic abnormalities and patterns of antigenic expression in multiple myeloma. *Clin. Cancer Res.* **2005**, *11*, 3661–3667. [CrossRef] [PubMed]

33. Kapoor, P.; Greipp, P.T.; Morice, W.G.; Rajkumar, S.V.; Witzig, T.E.; Greipp, P.R. Anti-CD20 monoclonal antibody therapy in multiple myeloma. *Br. J. Haematol.* **2008**, *141*, 135–148. [CrossRef] [PubMed]

34. Tai, Y.T.; Dillon, M.; Song, W.; Leiba, M.; Li, X.F.; Burger, P.; Lee, A.I.; Podar, K.; Hideshima, T.; Rice, A.G.; et al. Anti-CS1 humanized monoclonal antibody HuLuc63 inhibits myeloma cell adhesion and induces antibody-dependent cellular cytotoxicity in the bone marrow milieu. *Blood* **2008**, *112*, 1329–1337. [CrossRef] [PubMed]

35. Hsi, E.D.; Steinle, R.; Balasa, B.; Szmania, S.; Draksharapu, A.; Shum, B.P.; Huseni, M.; Powers, D.; Nanisetti, A.; Zhang, Y.; et al. CS1, a potential new therapeutic antibody target for the treatment of multiple myeloma. *Clin. Cancer Res.* **2008**, *14*, 2775–2784. [CrossRef] [PubMed]

36. De Weers, M.; Tai, Y.T.; van der Veer, M.S.; Bakker, J.M.; Vink, T.; Jacobs, D.C.; Oomen, L.A.; Peipp, M.; Valerius, T.; Slootstra, J.W.; et al. Daratumumab, a novel therapeutic human CD38 monoclonal antibody, induces killing of multiple myeloma and other hematological tumors. *J. Immunol.* **2011**, *186*, 1840–1848. [CrossRef] [PubMed]

37. Lokhorst, H.M.; Plesner, T.; Laubach, J.P.; Nahi, H.; Gimsing, P.; Hansson, M.; Minnema, M.C.; Lassen, U.; Krejcik, J.; Palumbo, A.; et al. Targeting CD38 with daratumumab monotherapy in multiple myeloma. *N. Engl. J. Med.* **2015**, *373*, 1207–1219. [CrossRef] [PubMed]

38. Lonial, S.; Dimopoulos, M.; Palumbo, A.; White, D.; Grosicki, S.; Spicka, I.; Walter-Croneck, A.; Moreau, P.; Mateos, M.V.; Magen, H.; et al. Elotuzumab therapy for relapsed or refractory multiple myeloma. *N. Engl. J. Med.* **2015**, *373*, 621–631. [CrossRef] [PubMed]

39. Deaglio, S.; Mehta, K.; Malavasi, F. Human CD38: A (r)evolutionary story of enzymes and receptors. *Leuk. Res.* **2001**, *25*, 1–12. [CrossRef]

40. Lee, H.C. Structure and enzymatic functions of human CD38. *Mol. Med.* **2006**, *12*, 317–323. [PubMed]

41. Fernandez, J.E.; Deaglio, S.; Donati, D.; Beusan, I.S.; Corno, F.; Aranega, A.; Forni, M.; Falini, B.; Malavasi, F. Analysis of the distribution of human CD38 and of its ligand CD31 in normal tissues. *J. Biol. Regul. Homeost. Agents* **1998**, *12*, 81–91. [PubMed]

42. Lin, P.; Owens, R.; Tricot, G.; Wilson, C.S. Flow cytometric immunophenotypic analysis of 306 cases of multiple myeloma. *Am. J. Clin. Pathol.* **2004**, *121*, 482–488. [CrossRef] [PubMed]

43. Van de Donk, N.W.; Janmaat, M.L.; Mutis, T.; Lammerts van Bueren, J.J.; Ahmadi, T.; Sasser, A.K.; Lokhorst, H.M.; Parren, P.W. Monoclonal antibodies targeting CD38 in hematological malignancies and beyond. *Immunol. Rev.* **2016**, *270*, 95–112. [CrossRef] [PubMed]

44. Stevenson, F.K.; Bell, A.J.; Cusack, R.; Hamblin, T.J.; Slade, C.J.; Spellerberg, M.B.; Stevenson, G.T. Preliminary studies for an immunotherapeutic approach to the treatment of human myeloma using chimeric anti-CD38 antibody. *Blood* **1991**, *77*, 1071–1079. [PubMed]

45. Seegmiller, A.C.; Kroft, S.H.; Karandikar, N.J.; McKenna, R.W. Characterization of immunophenotypic aberrancies in 200 cases of B acute lymphoblastic leukemia. *Am. J. Clin. Pathol.* **2009**, *132*, 940–949. [CrossRef] [PubMed]

46. Atanackovic, D.; Steinbach, M.; Radhakrishnan, S.V.; Luetkens, T. Immunotherapies targeting CD38 in multiple myeloma. *Oncoimmunology* **2016**, *5*, e1217374. [CrossRef] [PubMed]

47. Mihara, K.; Yanagihara, K.; Takigahira, M.; Kitanaka, A.; Imai, C.; Bhattacharyya, J.; Kubo, T.; Takei, Y.; Yasunaga, S.; Takihara, Y.; et al. Synergistic and persistent effect of T-cell immunotherapy with anti-CD19 or anti-CD38 chimeric receptor in conjunction with rituximab on B-cell non-hodgkin lymphoma. *Br. J. Haematol.* **2010**, *151*, 37–46. [CrossRef] [PubMed]

48. Keyhani, A.; Huh, Y.O.; Jendiroba, D.; Pagliaro, L.; Cortez, J.; Pierce, S.; Pearlman, M.; Estey, E.; Kantarjian, H.; Freireich, E.J. Increased CD38 expression is associated with favorable prognosis in adult acute leukemia. *Leuk. Res.* **2000**, *24*, 153–159. [CrossRef]

49. Pittner, B.T.; Shanafelt, T.D.; Kay, N.E.; Jelinek, D.F. CD38 expression levels in chronic lymphocytic leukemia B cells are associated with activation marker expression and differential responses to interferon stimulation. *Leukemia* **2005**, *19*, 2264–2272. [CrossRef] [PubMed]

50. Malavasi, F.; Deaglio, S.; Damle, R.; Cutrona, G.; Ferrarini, M.; Chiorazzi, N. CD38 and chronic lymphocytic leukemia: A decade later. *Blood* **2011**, *118*, 3470–3478. [CrossRef] [PubMed]

51. Horenstein, A.L.; Chillemi, A.; Quarona, V.; Zito, A.; Roato, I.; Morandi, F.; Marimpietri, D.; Bolzoni, M.; Toscani, D.; Oldham, R.J.; et al. NAD(+)-metabolizing ectoenzymes in remodeling tumor-host interactions: The human myeloma model. *Cells* **2015**, *4*, 520–537. [CrossRef] [PubMed]

52. Overdijk, M.B.; Verploegen, S.; Bogels, M.; van Egmond, M.; Lammerts van Bueren, J.J.; Mutis, T.; Groen, R.W.; Breij, E.; Martens, A.C.; Bleeker, W.K.; et al. Antibody-mediated phagocytosis contributes to the anti-tumor activity of the therapeutic antibody daratumumab in lymphoma and multiple myeloma. *mAbs* **2015**, *7*, 311–321. [CrossRef] [PubMed]

53. Overdijk, M.B.; Jansen, J.H.; Nederend, M.; Lammerts van Bueren, J.J.; Groen, R.W.; Parren, P.W.; Leusen, J.H.; Boross, P. The therapeutic CD38 monoclonal antibody daratumumab induces programmed cell death via fcgamma receptor-mediated cross-linking. *J. Immunol.* **2016**, *197*, 807–813. [CrossRef] [PubMed]

54. Terui, Y.; Sakurai, T.; Mishima, Y.; Mishima, Y.; Sugimura, N.; Sasaoka, C.; Kojima, K.; Yokoyama, M.; Mizunuma, N.; Takahashi, S.; et al. Blockade of bulky lymphoma-associated CD55 expression by RNA interference overcomes resistance to complement-dependent cytotoxicity with rituximab. *Cancer Sci.* **2006**, *97*, 72–79. [CrossRef] [PubMed]

55. Dzietczenia, J.; Wrobel, T.; Mazur, G.; Poreba, R.; Jazwiec, B.; Kuliczkowski, K. Expression of complement regulatory proteins: CD46, CD55, and CD59 and response to rituximab in patients with CD20+ non-hodgkin's lymphoma. *Med. Oncol.* **2010**, *27*, 743–746. [CrossRef] [PubMed]

56. Van der Veer, M.S.; de Weers, M.; van Kessel, B.; Bakker, J.M.; Wittebol, S.; Parren, P.W.; Lokhorst, H.M.; Mutis, T. Towards effective immunotherapy of myeloma: Enhanced elimination of myeloma cells by combination of lenalidomide with the human CD38 monoclonal antibody daratumumab. *Haematologica* **2011**, *96*, 284–290. [CrossRef] [PubMed]

57. Krejcik, J.; Casneuf, T.; Nijhof, I.S.; Verbist, B.; Bald, J.; Plesner, T.; Syed, K.; Liu, K.; van de Donk, N.W.; Weiss, B.M.; et al. Daratumumab depletes CD38+ immune regulatory cells, promotes t-cell expansion, and skews t-cell repertoire in multiple myeloma. *Blood* **2016**, *128*, 384–394. [CrossRef] [PubMed]

58. Tai, Y.T.; Anderson, K.C. A new era of immune therapy in multiple myeloma. *Blood* **2016**, *128*, 318–319. [CrossRef] [PubMed]

59. Lonial, S.; Weiss, B.M.; Usmani, S.Z.; Singhal, S.; Chari, A.; Bahlis, N.J.; Belch, A.; Krishnan, A.; Vescio, R.A.; Mateos, M.V.; et al. Daratumumab monotherapy in patients with treatment-refractory multiple myeloma (sirius): An open-label, randomised, phase 2 trial. *Lancet* **2016**, *387*, 1551–1560. [CrossRef]

60. Palumbo, A.; Chanan-Khan, A.; Weisel, K.; Nooka, A.K.; Masszi, T.; Beksac, M.; Spicka, I.; Hungria, V.; Munder, M.; Mateos, M.V.; et al. Daratumumab, bortezomib, and dexamethasone for multiple myeloma. *N. Engl. J. Med.* **2016**, *375*, 754–766. [CrossRef] [PubMed]

61. Dimopoulos, M.A.; Oriol, A.; Nahi, H.; San-Miguel, J.; Bahlis, N.J.; Usmani, S.Z.; Rabin, N.; Orlowski, R.Z.; Komarnicki, M.; Suzuki, K.; et al. Daratumumab, lenalidomide, and dexamethasone for multiple myeloma. *N. Engl. J. Med.* **2016**, *375*, 1319–1331. [CrossRef] [PubMed]

62. Deckert, J.; Wetzel, M.C.; Bartle, L.M.; Skaletskaya, A.; Goldmacher, V.S.; Vallee, F.; Zhou-Liu, Q.; Ferrari, P.; Pouzieux, S.; Lahoute, C.; et al. SAR650984, a novel humanized CD38-targeting antibody, demonstrates potent antitumor activity in models of multiple myeloma and other CD38+ hematologic malignancies. *Clin. Cancer Res.* **2014**, *20*, 4574–4583. [CrossRef] [PubMed]

63. Van Bueren, J.L.; Jakobs, D.; Kaldenhoven, N.; Roza, M.; Hiddingh, S.; Meesters, J.; Voorhorst, M.; Gresnigt, E.; Wiegman, L.; Buijsse, A.O.; et al. Direct in vitro comparison of daratumumab with surrogate analogs of CD38 antibodies mor03087, SAR650984 and Ab79. *Blood* **2014**, *124*, 3474.

64. Jiang, H.; Acharya, C.; An, G.; Zhong, M.; Feng, X.; Wang, L.; Dasilva, N.; Song, Z.; Yang, G.; Adrian, F.; et al. SAR650984 directly induces multiple myeloma cell death via lysosomal-associated and apoptotic pathways, which is further enhanced by pomalidomide. *Leukemia* **2016**, *30*, 399–408. [CrossRef] [PubMed]

65. Feng, X.; Zhang, L.; Acharya, C.; An, G.; Wen, K.; Qiu, L.; Munshi, N.C.; Tai, Y.T.; Anderson, K.C. Targeting CD38 suppresses induction and function of t regulatory cells to mitigate immunosuppression in multiple myeloma. *Clin. Cancer Res.* **2017**, *23*, 4290–4300. [CrossRef] [PubMed]

66. Martin, T.; Baz, R.; Benson, D.M.; Lendvai, N.; Wolf, J.; Munster, P.; Lesokhin, A.M.; Wack, C.; Charpentier, E.; Campana, F.; et al. A phase 1b study of isatuximab plus lenalidomide and dexamethasone for relapsed/refractory multiple myeloma. *Blood* **2017**, *129*, 3294–3303. [CrossRef] [PubMed]

67. Richardson, P.G.; Mikhael, J.R.; Usmani, S.Z.; Raje, N.; Bensinger, W.; Campana, F.; Gao, L.; Dubin, F.; Wack, C.; Anderson, K. Preliminary results from a phase Ib study of isatuximab in combination with pomalidomide and dexamethasone in relapsed and refractory multiple myeloma. *Blood* **2016**, *128*, 2123. [CrossRef]

68. Endell, J.; Samuelsson, C.; Boxhammer, R.; Strauss, S.; Steidl, S. Effect of MOR202, a human CD38 antibody, in combination with lenalidomide and bortezomib on bone lysis and tumor load in a physiological model of myeloma. *J. Clin. Oncol.* **2011**, *33*, 8588. [CrossRef]

69. Endell, J.; Boxhammer, R.; Wurzenberger, C.; Ness, D.; Steidl, S. The activity of MOR202, a fully human anti-CD38 antibody, is complemented by ADCP and is synergistically enhanced by lenalidomide in vitro and in vivo. *Blood* **2012**, *120*, 4018.

70. Boxhammer, R.; Steidl, S.; Endell, J. Effect of imid compounds on CD38 expression on multiple myeloma cells: MOR202, a human CD38 antibody in combination with pomalidomide. *J. Clin. Oncol.* **2015**, *33*, 8588.

71. Raab, M.S.; Chatterjee, M.; Goldshmidt, H.; Agis, H.; Blau, I.; Einsele, H.; Engelhardt, M.; Ferstl, B.; Gramatzki, M.; Röllig, C.; et al. A phase i/iia study of the CD38 antibody MOR202 alone and in combination with pomalidomide or lenalidomide in patients with relapsed or refractory multiple myeloma. *Blood* **2016**, *128*, 1152.

72. Van de Donk, N.W.; Otten, H.G.; El Haddad, O.; Axel, A.; Sasser, A.K.; Croockewit, S.; Jacobs, J.F. Interference of daratumumab in monitoring multiple myeloma patients using serum immunofixation electrophoresis can be abrogated using the daratumumab IFE reflex assay (DIRA). *Clin. Chem. Lab. Med.* **2016**, *54*, 1105–1109. [CrossRef] [PubMed]

73. McCudden, C.; Axel, A.E.; Slaets, D.; Dejoie, T.; Clemens, P.L.; Frans, S.; Bald, J.; Plesner, T.; Jacobs, J.F.; van de Donk, N.W.; et al. Monitoring multiple myeloma patients treated with daratumumab: Teasing out monoclonal antibody interference. *Clin. Chem. Lab. Med.* **2016**, *54*, 1095–1104. [CrossRef] [PubMed]

74. Perincheri, S.; Torres, R.; Tormey, C.A.; Smith, B.R.; Rinder, H.M.; Siddon, A.J. Daratumumab interferes with flow cytometric evaluation of multiple myeloma. *Blood* **2016**, *128*, 5630.

75. Chapuy, C.I.; Nicholson, R.T.; Aguad, M.D.; Chapuy, B.; Laubach, J.P.; Richardson, P.G.; Doshi, P.; Kaufman, R.M. Resolving the daratumumab interference with blood compatibility testing. *Transfusion* **2015**, *55*, 1545–1554. [CrossRef] [PubMed]

76. Chapuy, C.I.; Aguad, M.D.; Nicholson, R.T.; AuBuchon, J.P.; Cohn, C.S.; Delaney, M.; Fung, M.K.; Unger, M.; Doshi, P.; Murphy, M.F.; et al. International validation of a dithiothreitol (DTT)-based method to resolve the daratumumab interference with blood compatibility testing. *Transfusion* **2016**, *56*, 2964–2972. [CrossRef] [PubMed]

77. Costello, C. An update on the role of daratumumab in the treatment of multiple myeloma. *Ther. Adv. Hematol.* **2017**, *8*, 28–37. [CrossRef] [PubMed]

78. Kumaresan, P.R.; Lai, W.C.; Chuang, S.S.; Bennett, M.; Mathew, P.A. Cs1, a novel member of the CD2 family, is homophilic and regulates NK cell function. *Mol. Immunol.* **2002**, *39*, 1–8. [CrossRef]

79. Collins, S.M.; Bakan, C.E.; Swartzel, G.D.; Hofmeister, C.C.; Efebera, Y.A.; Kwon, H.; Starling, G.C.; Ciarlariello, D.; Bhaskar, S.; Briercheck, E.L.; et al. Elotuzumab directly enhances NK cell cytotoxicity against myeloma via CS1 ligation: Evidence for augmented NK cell function complementing ADCC. *Cancer Immunol. Immunother.* **2013**, *62*, 1841–1849. [CrossRef] [PubMed]

80. Van Rhee, F.; Szmania, S.M.; Dillon, M.; van Abbema, A.M.; Li, X.; Stone, M.K.; Garg, T.K.; Shi, J.; Moreno-Bost, A.M.; Yun, R.; et al. Combinatorial efficacy of anti-CS1 monoclonal antibody elotuzumab (HuLuc63) and bortezomib against multiple myeloma. *Mol. Cancer Ther.* **2009**, *8*, 2616–2624. [CrossRef] [PubMed]

81. Jakubowiak, A.J.; Benson, D.M.; Bensinger, W.; Siegel, D.S.; Zimmerman, T.M.; Mohrbacher, A.; Richardson, P.G.; Afar, D.E.; Singhal, A.K.; Anderson, K.C. Phase I trial of anti-CS1 monoclonal antibody elotuzumab in combination with bortezomib in the treatment of relapsed/refractory multiple myeloma. *J. Clin. Oncol.* **2012**, *30*, 1960–1965. [CrossRef] [PubMed]

82. Zonder, J.A.; Mohrbacher, A.F.; Singhal, S.; van Rhee, F.; Bensinger, W.I.; Ding, H.; Fry, J.; Afar, D.E.; Singhal, A.K. A phase 1, multicenter, open-label, dose escalation study of elotuzumab in patients with advanced multiple myeloma. *Blood* **2012**, *120*, 552–559. [CrossRef] [PubMed]

83. Lonial, S.; Vij, R.; Harousseau, J.L.; Facon, T.; Moreau, P.; Mazumder, A.; Kaufman, J.L.; Leleu, X.; Tsao, L.C.; Westland, C.; et al. Elotuzumab in combination with lenalidomide and low-dose dexamethasone in relapsed or refractory multiple myeloma. *J. Clin. Oncol.* **2012**, *30*, 1953–1959. [CrossRef] [PubMed]

84. Freeman, G.J.; Long, A.J.; Iwai, Y.; Bourque, K.; Chernova, T.; Nishimura, H.; Fitz, L.J.; Malenkovich, N.; Okazaki, T.; Byrne, M.C.; et al. Engagement of the PD-1 immunoinhibitory receptor by a novel B7 family member leads to negative regulation of lymphocyte activation. *J. Exp. Med.* **2000**, *192*, 1027–1034. [CrossRef] [PubMed]

85. Parry, R.V.; Chemnitz, J.M.; Frauwirth, K.A.; Lanfranco, A.R.; Braunstein, I.; Kobayashi, S.V.; Linsley, P.S.; Thompson, C.B.; Riley, J.L. CTLA-4 and PD-1 receptors inhibit T-cell activation by distinct mechanisms. *Mol. Cell. Biol.* **2005**, *25*, 9543–9553. [CrossRef] [PubMed]

86. Tamura, H.; Ishibashi, M.; Yamashita, T.; Tanosaki, S.; Okuyama, N.; Kondo, A.; Hyodo, H.; Shinya, E.; Takahashi, H.; Dong, H.; et al. Marrow stromal cells induce B7-H1 expression on myeloma cells, generating aggressive characteristics in multiple myeloma. *Leukemia* **2013**, *27*, 464–472. [CrossRef] [PubMed]

87. Spranger, S.; Gajewski, T. Rational combinations of immunotherapeutics that target discrete pathways. *J. Immunother. Cancer* **2013**, *1*, 16. [CrossRef] [PubMed]

88. Yousef, S.; Marvin, J.; Steinbach, M.; Langemo, A.; Kovacsovics, T.; Binder, M.; Kroger, N.; Luetkens, T.; Atanackovic, D. Immunomodulatory molecule PD-l1 is expressed on malignant plasma cells and myeloma-propagating pre-plasma cells in the bone marrow of multiple myeloma patients. *Blood Cancer J.* **2015**, *5*, e285. [CrossRef] [PubMed]

89. Ishibashi, M.; Tamura, H.; Sunakawa, M.; Kondo-Onodera, A.; Okuyama, N.; Hamada, Y.; Moriya, K.; Choi, I.; Tamada, K.; Inokuchi, K. Myeloma drug resistance induced by binding of myeloma B7-H1 (PD-l1) to PD-1. *Cancer Immunol. Res.* **2016**, *4*, 779–788. [CrossRef] [PubMed]

90. Favaloro, J.; Liyadipitiya, T.; Brown, R.; Yang, S.; Suen, H.; Woodland, N.; Nassif, N.; Hart, D.; Fromm, P.; Weatherburn, C.; et al. Myeloid derived suppressor cells are numerically, functionally and phenotypically different in patients with multiple myeloma. *Leuk. Lymphoma* **2014**, *55*, 2893–2900. [CrossRef] [PubMed]

91. Ray, A.; Das, D.S.; Song, Y.; Richardson, P.; Munshi, N.C.; Chauhan, D.; Anderson, K.C. Targeting PD1-PDL1 immune checkpoint in plasmacytoid dendritic cell interactions with T cells, natural killer cells and multiple myeloma cells. *Leukemia* **2015**, *29*, 1441–1444. [CrossRef] [PubMed]

92. Tai, Y.T.; Acharya, C.; An, G.; Moschetta, M.; Zhong, M.Y.; Feng, X.; Cea, M.; Cagnetta, A.; Wen, K.; van Eenennaam, H.; et al. APRIL and BCMA promote human multiple myeloma growth and immunosuppression in the bone marrow microenvironment. *Blood* **2016**, *127*, 3225–3236. [CrossRef] [PubMed]

93. Gorgun, G.; Samur, M.K.; Cowens, K.B.; Paula, S.; Bianchi, G.; Anderson, J.E.; White, R.E.; Singh, A.; Ohguchi, H.; Suzuki, R.; et al. Lenalidomide enhances immune checkpoint blockade-induced immune response in multiple myeloma. *Clin. Cancer Res.* **2015**, *21*, 4607–4618. [CrossRef] [PubMed]

94. Giuliani, M.; Janji, B.; Berchem, G. Activation of NK cells and disruption of PD-l1/PD-1 axis: Two different ways for lenalidomide to block myeloma progression. *Oncotarget* **2017**, *8*, 24031–24044. [CrossRef] [PubMed]

95. Lesokhin, A.M.; Ansell, S.M.; Armand, P.; Scott, E.C.; Halwani, A.; Gutierrez, M.; Millenson, M.M.; Cohen, A.D.; Schuster, S.J.; Lebovic, D.; et al. Nivolumab in patients with relapsed or refractory hematologic malignancy: Preliminary results of a phase Ib study. *J. Clin. Oncol.* **2016**, *34*, 2698–2704. [CrossRef] [PubMed]

96. Mateos, M.-V.; Orlowski, R.Z.; Siegel, D.S.D.C.; Reece, D.E.; Moreau, P.; Ocio, E.M. Pembrolizumab in combination with lenalidomide and low-dose dexamethasone for relapsed/refractory multiple myeloma (RRMM): Final efficacy and safety analysis. *J. Clin. Oncol.* **2016**, *34* (Suppl. 15), 8010. [CrossRef]

97. Badros, A.; Hyjek, E.; Ma, N.; Lesokhin, A.; Dogan, A.; Rapoport, A.P.; Kocoglu, M.; Lederer, E.; Philip, S.; Milliron, T.; et al. Pembrolizumab, pomalidomide and low dose dexamethasone for relapsed/refractory multiple myeloma. *Blood* **2017**, *130*, 1189–1197. [CrossRef] [PubMed]

98. The ASCO Post. FDA Places Clinical Hold on Three Studies Evaluating Pembrolizumab in Multiple Myeloma. Available online: http://www.ascopost.com/News/57813 (accessed on 20 October 2017).

99. Claudio, J.O.; Masih-Khan, E.; Tang, H.; Goncalves, J.; Voralia, M.; Li, Z.H.; Nadeem, V.; Cukerman, E.; Francisco-Pabalan, O.; Liew, C.C.; et al. A molecular compendium of genes expressed in multiple myeloma. *Blood* **2002**, *100*, 2175–2186. [CrossRef] [PubMed]

100. Carpenter, R.O.; Evbuomwan, M.O.; Pittaluga, S.; Rose, J.J.; Raffeld, M.; Yang, S.; Gress, R.E.; Hakim, F.T.; Kochenderfer, J.N. B-cell maturation antigen is a promising target for adoptive T-cell therapy of multiple myeloma. *Clin. Cancer Res.* **2013**, *19*, 2048–2060. [CrossRef] [PubMed]

101. Seckinger, A.; Delgado, J.A.; Moser, S.; Moreno, L.; Neuber, B.; Grab, A.; Lipp, S.; Merino, J.; Prosper, F.; Emde, M.; et al. Target expression, generation, preclinical activity, and pharmacokinetics of the BCMA-T cell bispecific antibody EM801 for multiple myeloma treatment. *Cancer Cell* **2017**, *31*, 396–410. [CrossRef] [PubMed]

102. Laurent, S.A.; Hoffmann, F.S.; Kuhn, P.H.; Cheng, Q.; Chu, Y.; Schmidt-Supprian, M.; Hauck, S.M.; Schuh, E.; Krumbholz, M.; Rubsamen, H.; et al. Gamma-secretase directly sheds the survival receptor BCMA from plasma cells. *Nat. Commun.* **2015**, *6*, 7333. [CrossRef] [PubMed]

103. Avery, D.T.; Kalled, S.L.; Ellyard, J.I.; Ambrose, C.; Bixler, S.A.; Thien, M.; Brink, R.; Mackay, F.; Hodgkin, P.D.; Tangye, S.G. BAFF selectively enhances the survival of plasmablasts generated from human memory B cells. *J. Clin. Investig.* **2003**, *112*, 286–297. [CrossRef] [PubMed]

104. O'Connor, B.P.; Raman, V.S.; Erickson, L.D.; Cook, W.J.; Weaver, L.K.; Ahonen, C.; Lin, L.L.; Mantchev, G.T.; Bram, R.J.; Noelle, R.J. BCMA is essential for the survival of long-lived bone marrow plasma cells. *J. Exp. Med.* **2004**, *199*, 91–98. [CrossRef] [PubMed]

105. Tai, Y.T.; Mayes, P.A.; Acharya, C.; Zhong, M.Y.; Cea, M.; Cagnetta, A.; Craigen, J.; Yates, J.; Gliddon, L.; Fieles, W.; et al. Novel anti-B-cell maturation antigen antibody-drug conjugate (GSK2857916) selectively induces killing of multiple myeloma. *Blood* **2014**, *123*, 3128–3138. [CrossRef] [PubMed]

106. Sanchez, E.; Li, M.; Kitto, A.; Li, J.; Wang, C.S.; Kirk, D.T.; Yellin, O.; Nichols, C.M.; Dreyer, M.P.; Ahles, C.P.; et al. Serum B-cell maturation antigen is elevated in multiple myeloma and correlates with disease status and survival. *Br. J. Haematol.* **2012**, *158*, 727–738. [CrossRef] [PubMed]

107. Bellucci, R.; Alyea, E.P.; Chiaretti, S.; Wu, C.J.; Zorn, E.; Weller, E.; Wu, B.; Canning, C.; Schlossman, R.; Munshi, N.C.; et al. Graft-versus-tumor response in patients with multiple myeloma is associated with antibody response to BCMA, a plasma-cell membrane receptor. *Blood* **2005**, *105*, 3945–3950. [CrossRef] [PubMed]

108. Schuh, E.; Musumeci, A.; Thaler, F.S.; Laurent, S.; Ellwart, J.W.; Hohlfeld, R.; Krug, A.; Meinl, E. Human plasmacytoid dendritic cells display and shed B cell maturation antigen upon TLR engagement. *J. Immunol.* **2017**, *198*, 3081–3088. [CrossRef] [PubMed]

109. Sanchez, E.; Gillespie, A.; Tang, G.; Ferros, M.; Harutyunyan, N.M.; Vardanyan, S.; Gottlieb, J.; Li, M.; Wang, C.S.; Chen, H.; et al. Soluble B-cell maturation antigen mediates tumor-induced immune deficiency in multiple myeloma. *Clin. Cancer Res.* **2016**, *22*, 3383–3397. [CrossRef] [PubMed]

110. Ryan, M.C.; Hering, M.; Peckham, D.; McDonagh, C.F.; Brown, L.; Kim, K.M.; Meyer, D.L.; Zabinski, R.F.; Grewal, I.S.; Carter, P.J. Antibody targeting of B-cell maturation antigen on malignant plasma cells. *Mol. Cancer Ther.* **2007**, *6*, 3009–3018. [CrossRef] [PubMed]

111. Lee, L.; Bounds, D.; Paterson, J.; Herledan, G.; Sully, K.; Seestaller-Wehr, L.M.; Fieles, W.E.; Tunstead, J.; McCahon, L.; Germaschewski, F.M.; et al. Evaluation of B cell maturation antigen as a target for antibody drug conjugate mediated cytotoxicity in multiple myeloma. *Br. J. Haematol.* **2016**, *174*, 911–922. [CrossRef] [PubMed]

112. Cohen, A.D.; Popat, R.; Trudel, S.; Richardson, P.G.; Libby, E.N., III; Lendvai, N.; Anderson, L.D., Jr.; Sutherland, H.J.; DeWall, S.; Ellis, C.E.; et al. First in human study with GSK2857916, an antibody drug conjugated to microtubule-disrupting agent directed against B-cell maturation antigen (BCMA) in patients with relapsed/refractory multiple myeloma (MM): Results from study BMA117159 part 1 dose escalation. *Blood* **2016**, *128*, 1148.

113. Hipp, S.; Tai, Y.T.; Blanset, D.; Deegen, P.; Wahl, J.; Thomas, O.; Rattel, B.; Adam, P.J.; Anderson, K.C.; Friedrich, M. A novel BCMA/CD3 bispecific T-cell engager for the treatment of multiple myeloma induces selective lysis in vitro and in vivo. *Leukemia* **2017**, *31*, 1743–1751. [CrossRef] [PubMed]

114. Ramadoss, N.S.; Schulman, A.D.; Choi, S.H.; Rodgers, D.T.; Kazane, S.A.; Kim, C.H.; Lawson, B.R.; Young, T.S. An anti-B cell maturation antigen bispecific antibody for multiple myeloma. *J. Am. Chem. Soc.* **2015**, *137*, 5288–5291. [CrossRef] [PubMed]

115. Ali, S.A.; Shi, V.; Maric, I.; Wang, M.; Stroncek, D.F.; Rose, J.J.; Brudno, J.N.; Stetler-Stevenson, M.; Feldman, S.A.; Hansen, B.G.; et al. T cells expressing an anti-B-cell maturation antigen chimeric antigen receptor cause remissions of multiple myeloma. *Blood* **2016**, *128*, 1688–1700. [CrossRef] [PubMed]

116. Mikkilineni, L.; Kochenderfer, J.N. Chimeric antigen receptor T-cell therapies for multiple myeloma. *Blood* **2017**. [CrossRef] [PubMed]

117. Yu, G.; Boone, T.; Delaney, J.; Hawkins, N.; Kelley, M.; Ramakrishnan, M.; McCabe, S.; Qiu, W.R.; Kornuc, M.; Xia, X.Z.; et al. APRIL and TALL-1 and receptors BCMA and TACI: System for regulating humoral immunity. *Nat. Immunol.* **2000**, *1*, 252–256. [CrossRef] [PubMed]

118. Marsters, S.A.; Yan, M.; Pitti, R.M.; Haas, P.E.; Dixit, V.M.; Ashkenazi, A. Interaction of the TNF homologues BLyS and APRIL with the TNF receptor homologues BCMA and TACI. *Curr. Biol.* **2000**, *10*, 785–788. [CrossRef]

119. Medema, J.P.; Planelles-Carazo, L.; Hardenberg, G.; Hahne, M. The uncertain glory of APRIL. *Cell Death Differ.* **2003**, *10*, 1121–1125. [CrossRef] [PubMed]

120. Schneider, P. The role of APRIL and BAFF in lymphocyte activation. *Curr. Opin. Immunol.* **2005**, *17*, 282–289. [CrossRef] [PubMed]

121. Moreaux, J.; Legouffe, E.; Jourdan, E.; Quittet, P.; Reme, T.; Lugagne, C.; Moine, P.; Rossi, J.F.; Klein, B.; Tarte, K. BAFF and APRIL protect myeloma cells from apoptosis induced by interleukin 6 deprivation and dexamethasone. *Blood* **2004**, *103*, 3148–3157. [CrossRef] [PubMed]

122. Tai, Y.T.; Li, X.F.; Breitkreutz, I.; Song, W.; Neri, P.; Catley, L.; Podar, K.; Hideshima, T.; Chauhan, D.; Raje, N.; et al. Role of B-cell-activating factor in adhesion and growth of human multiple myeloma cells in the bone marrow microenvironment. *Cancer Res.* **2006**, *66*, 6675–6682. [CrossRef] [PubMed]

123. Moreaux, J.; Sprynski, A.C.; Dillon, S.R.; Mahtouk, K.; Jourdan, M.; Ythier, A.; Moine, P.; Robert, N.; Jourdan, E.; Rossi, J.F.; et al. April and TACI interact with syndecan-1 on the surface of multiple myeloma cells to form an essential survival loop. *Eur. J. Haematol.* **2009**, *83*, 119–129. [CrossRef] [PubMed]

124. Matthes, T.; Dunand-Sauthier, I.; Santiago-Raber, M.L.; Krause, K.H.; Donze, O.; Passweg, J.; McKee, T.; Huard, B. Production of the plasma-cell survival factor a proliferation-inducing ligand (APRIL) peaks in myeloid precursor cells from human bone marrow. *Blood* **2011**, *118*, 1838–1844. [CrossRef] [PubMed]

125. Quinn, J.; Glassford, J.; Percy, L.; Munson, P.; Marafioti, T.; Rodriguez-Justo, M.; Yong, K. APRIL promotes cell-cycle progression in primary multiple myeloma cells: Influence of D-type cyclin group and translocation status. *Blood* **2011**, *117*, 890–901. [CrossRef] [PubMed]

126. Guadagnoli, M.; Kimberley, F.C.; Phan, U.; Cameron, K.; Vink, P.M.; Rodermond, H.; Eldering, E.; Kater, A.P.; van Eenennaam, H.; Medema, J.P. Development and characterization of APRIL antagonistic monoclonal antibodies for treatment of B-cell lymphomas. *Blood* **2011**, *117*, 6856–6865. [CrossRef] [PubMed]

127. Dulos, J.; Driessen, L.; Snippert, M.; Guadagnoli, M.; Bertens, A.; Hulsik, D.L.; Tai, Y.T.; Anderson, K.; Medema, J.P.; Cameron, K.; et al. Development of a first in class APRIL fully blocking antibody BION-1301 for the treatment of multiple myeloma. *Cancer Res.* **2017**, *77* (Suppl. 13), 2645. [CrossRef]

128. Zagouri, F.; Terpos, E.; Kastritis, E.; Dimopoulos, M.A. Emerging antibodies for the treatment of multiple myeloma. *Expert Opin. Emerg. Drugs* **2016**, *21*, 225–237. [CrossRef] [PubMed]

129. Touzeau, C.; Moreau, P.; Dumontet, C. Monoclonal antibody therapy in multiple myeloma. *Leukemia* **2017**, *31*, 1039–1047. [CrossRef] [PubMed]

130. Laubach, J.P.; Paba Prada, C.E.; Richardson, P.G.; Longo, D.L. Daratumumab, elotuzumab, and the development of therapeutic monoclonal antibodies in multiple myeloma. *Clin. Pharmacol. Ther.* **2017**, *101*, 81–88. [CrossRef] [PubMed]

131. Li, J.; Stagg, N.J.; Johnston, J.; Harris, M.J.; Menzies, S.A.; DiCara, D.; Clark, V.; Hristopoulos, M.; Cook, R.; Slaga, D.; et al. Membrane-proximal epitope facilitates efficient T cell synapse formation by anti-FcRH5/CD3 and is a requirement for myeloma cell killing. *Cancer Cell* **2017**, *31*, 383–395. [CrossRef] [PubMed]

132. Neri, P.; Kumar, S.; Fulciniti, M.T.; Vallet, S.; Chhetri, S.; Mukherjee, S.; Tai, Y.; Chauhan, D.; Tassone, P.; Venuta, S.; et al. Neutralizing B-cell activating factor antibody improves survival and inhibits osteoclastogenesis in a severe combined immunodeficient human multiple myeloma model. *Clin. Cancer Res.* **2007**, *13*, 5903–5909. [CrossRef] [PubMed]

133. Raje, N.S.; Faber, E.A., Jr.; Richardson, P.G.; Schiller, G.; Hohl, R.J.; Cohen, A.D.; Forero, A.; Carpenter, S.; Nguyen, T.S.; Conti, I.; et al. Phase 1 study of tabalumab, a human anti-B-cell activating factor antibody, and bortezomib in patients with relapsed/refractory multiple myeloma. *Clin. Cancer Res.* **2016**, *22*, 5688–5695. [CrossRef] [PubMed]

134. Suzuki, K.; Ogura, M.; Abe, Y.; Suzuki, T.; Tobinai, K.; Ando, K.; Taniwaki, M.; Maruyama, D.; Kojima, M.; Kuroda, J.; et al. Phase 1 study in Japan of siltuximab, an anti-IL-6 monoclonal antibody, in relapsed/refractory multiple myeloma. *Int. J. Hematol.* **2015**, *101*, 286–294. [CrossRef] [PubMed]

135. Voorhees, P.M.; Manges, R.F.; Sonneveld, P.; Jagannath, S.; Somlo, G.; Krishnan, A.; Lentzsch, S.; Frank, R.C.; Zweegman, S.; Wijermans, P.W.; et al. A phase 2 multicentre study of siltuximab, an anti-interleukin-6 monoclonal antibody, in patients with relapsed or refractory multiple myeloma. *Br. J. Haematol.* **2013**, *161*, 357–366. [CrossRef] [PubMed]

136. San-Miguel, J.; Blade, J.; Shpilberg, O.; Grosicki, S.; Maloisel, F.; Min, C.K.; Polo Zarzuela, M.; Robak, T.; Prasad, S.V.; Tee Goh, Y.; et al. Phase 2 randomized study of bortezomib-melphalan-prednisone with or without siltuximab (anti-IL-6) in multiple myeloma. *Blood* **2014**, *123*, 4136–4142. [CrossRef] [PubMed]

137. Matthes, T.; Manfroi, B.; Huard, B. Revisiting IL-6 antagonism in multiple myeloma. *Crit. Rev. Oncol. Hematol.* **2016**, *105*, 1–4. [CrossRef] [PubMed]

138. Bensinger, W.; Maziarz, R.T.; Jagannath, S.; Spencer, A.; Durrant, S.; Becker, P.S.; Ewald, B.; Bilic, S.; Rediske, J.; Baeck, J.; et al. A phase 1 study of lucatumumab, a fully human anti-CD40 antagonist monoclonal antibody administered intravenously to patients with relapsed or refractory multiple myeloma. *Br. J. Haematol.* **2012**, *159*, 58–66. [CrossRef] [PubMed]

139. Ikeda, H.; Hideshima, T.; Fulciniti, M.; Lutz, R.J.; Yasui, H.; Okawa, Y.; Kiziltepe, T.; Vallet, S.; Pozzi, S.; Santo, L.; et al. The monoclonal antibody nBT062 conjugated to cytotoxic maytansinoids has selective cytotoxicity against CD138-positive multiple myeloma cells in vitro and in vivo. *Clin. Cancer Res.* **2009**, *15*, 4028–4037. [CrossRef] [PubMed]

140. Burton, J.; Mishina, D.; Cardillo, T.; Lew, K.; Rubin, A.; Goldenberg, D.M.; Gold, D.V. Epithelial mucin-1 (MUC1) expression and MA5 anti-MUC1 monoclonal antibody targeting in multiple myeloma. *Clin. Cancer Res.* **1999**, *5*, 3065S–3072S. [PubMed]

141. Iyer, S.P.; Beck, J.T.; Stewart, A.K.; Shah, J.; Kelly, K.R.; Isaacs, R.; Bilic, S.; Sen, S.; Munshi, N.C. A phase IB multicentre dose-determination study of BHQ880 in combination with anti-myeloma therapy and zoledronic acid in patients with relapsed or refractory multiple myeloma and prior skeletal-related events. *Br. J. Haematol.* **2014**, *167*, 366–375. [CrossRef] [PubMed]

142. Zhang, T.; Wang, S.; Lin, T.; Xie, J.; Zhao, L.; Liang, Z.; Li, Y.; Jiang, J. Systematic review and meta-analysis of the efficacy and safety of novel monoclonal antibodies for treatment of relapsed/refractory multiple myeloma. *Oncotarget* **2017**, *8*, 34001–34017. [CrossRef] [PubMed]

143. Balasa, B.; Yun, R.; Belmar, N.A.; Fox, M.; Chao, D.T.; Robbins, M.D.; Starling, G.C.; Rice, A.G. Elotuzumab enhances natural killer cell activation and myeloma cell killing through interleukin-2 and TNF-alpha pathways. *Cancer Immunol. Immunother.* **2015**, *64*, 61–73. [CrossRef] [PubMed]

antibodies

MDPI

Article

Generation and Performance of R132H Mutant IDH1 Rabbit Monoclonal Antibody

Juliet Rashidian, Raul Copaciu [†], Qin Su [†], Brett Merritt, Claire Johnson, Aril Yahyabeik, Ella French and Kelsea Cummings *

MilliporeSigma, 6600 Sierra College Blvd, Rocklin, CA 95677, USA; jrashidian@sial.com (J.R.); raul.copaciu@sial.com (R.C.); qin.su@sial.com (Q.S.); bmerritt@sial.com (B.M.); claire.johnson@sial.com (C.J.); ayahyabeik@sial.com (A.Y.); ella.french@sial.com (E.F.)
* Correspondence: kcummings@sial.com; Tel.: +1-916-472-4664
† These authors contributed equally to this paper.

Received: 31 October 2017; Accepted: 22 November 2017; Published: 1 December 2017

Abstract: Isocitrate dehydrogenase 1 (IDH1) gene mutations have been observed in a majority of diffuse astrocytomas, oligodendrogliomas, and secondary glioblastomas, and the mutant IDH1 R132H is detectable in most of these lesions. By specifically targeting the R132H mutation through B-cell cloning, a novel rabbit monoclonal antibody, MRQ-67, was produced that can recognize mutant IDH1 R132H and does not react with the wild type protein as demonstrated by Enzyme-linked immunosorbent assay (ELISA) and Western blotting. Through immunohistochemistry, the antibody is able to highlight neoplastic cells in glioma tissue specimens, and can be used as a tool in glioma subtyping. Immunohistochemistry (IHC) detection of IDH1 mutant protein may also be used to visualize single infiltrating tumor cells in surrounding brain tissue with an otherwise normal appearance.

Keywords: IDH1; R132H; novel rabbit monoclonal antibody; B-cell cloning; immunohistochemistry

1. Introduction

Isocitrate dehydrogenase 1 (IDH1) functions as an enzyme in the Krebs (citric acid) cycle and is biologically active in the cytoplasmic and peroxisomal compartments under normal conditions [1]. Somatic mutations in the gene that encodes IDH1 have been reported to be present in some glioma subtypes in high frequencies. The majority of these particular tumors have been found to harbor heterozygous point mutations in codon 132, with a missense amino acid substitution of arginine to histidine (R132H) being observed to have highest rate of occurrence [2]. The high incidence of glioma-specific IDH1 mutations has implicated them as an early event that occurs during gliomagenesis and provides utility in distinguishing low grade astrocytomas and oligodendrogliomas, as well as secondary glioblastomas from reactive gliosis and primary glioblastomas [3]. The value of this mutant marker is further illustrated by the 2016 World Health Organization (WHO) Classification of Tumors of the Central Nervous System that newly incorporates IDH1 mutation status as a parameter for sub-classifying diffuse astrocytic and oligodendroglial tumors [4]. While genetic testing can be burdensome, a clinically established routine procedure like immunohistochemistry (IHC), using a specific monoclonal antibody directed against IDH1 R132H mutant protein, represents a useful tool for overcoming this diagnostic challenge.

This study describes the generation and performance of a novel rabbit monoclonal IDH1 R132H antibody (MRQ-67) by single B-cell cloning technology, a recently emerging strategy for monoclonal antibody development [5]. The capacity of MRQ-67 to identify mutant IDH1 R132H without reacting with wild type protein is demonstrated through binding specificity assays. The functional utility of the antibody to specifically detect mutant IDH1 protein in astrocytomas, oligodendrogliomas, and glioblastomas is also examined through immunohistochemical analysis.

2. Materials and Methods

2.1. Tissue Specimens

Immunohistochemical evaluation of MRQ-67 performance was assessed using formalin-fixed, paraffin-embedded (FFPE) tissue specimens, which included 18 cases of astrocytoma, 7 cases of oligodendroglioma, 7 cases of glioblastoma, 12 cases of meningioma, and 15 cases of non-neoplastic brain tissue. The FFPE tissue specimens used in this study were procured, qualified, and tested in accordance with the U.S. Food and Drug Administration (FDA) "Guidance on Informed Consent for In Vitro Diagnostic Device Studies Using Leftover Human Specimens that are Not Individually Identifiable". This study exclusively used leftover tissue specimens that are not individually identifiable for conducting IHC testing. More specifically, these are remnants of human specimens collected for routine clinical care or analysis that would have otherwise been discarded and where the identity of the subject is not known to, or may not be readily ascertained by, any individual associated with this study.

2.2. Immunization

New Zealand White Rabbits were immunized with synthetic peptide CKPIIIGHHAYGD coupled to Keyhole limpet hemocyanin (KLH) corresponding the amino acids 126 to 137 of the human IDH1 containing R132H mutation. All of the housing and immunization procedures were performed by Antibodies Incorporated (Antibodies Inc., Davis, CA, USA), according to the approved protocols and guidelines of the Institutional Animal Care and Use Committee (IACUC). The project numbers were 5834 and 5835 under IACUC protocol 0298-9 "Custom Polyclonal Antibody Production in Rabbits", approved on 1 July 2016 (PHS Assurance number A4064-01).

2.3. Isolation and Sorting of Rabbit B-Cells

Peripheral blood mononuclear cells (PBMCs) were isolated from the ethylenediaminetetraacetic acid (EDTA) containing peripheral blood by density-gradient centrifugation with Lympholyte-Mammal (Cedarlane, Burlington, NC, USA), as described in the manual. The isolated PBMCs (10^7 cells) were next washed with RPMI (Life Technologies, Carlsbad, CA, USA) containing DNase I (Roche, Basel, Switzerland) and re-suspended in phosphate-buffered saline (PBS) containing 0.5% bovine serum albumin (BSA). Then, the IgG expressing B-cells were isolated using Anti-Rabbit IgG Microbeads (Miltenyi Biotech, Auburn, CA, USA).

The isolated B-cells were stained for viability, incubated with a cocktail of anti-rabbit IgG and fluorochrome-conjugated specific peptide in the dark for 30 min at 4 °C, and washed with ice-cold PBS. Finally, cells were re-suspended in PBS and subjected to Fluorescence Activated Cell Sorting (FACS) analyses. Sorting was carried out using BD Influx Cell Sorter and BD FACS DIVA software (UC Davis Medical Center, Sacramento, CA, USA). Single B-cells expressing IgG were sorted into a 96-well plate (omitting row H).

2.4. Cloning Antibody Variable Regions

Single-cell reverse transcription polymerase chain reaction (RT-PCR) was performed with reverse transcriptase Superscript III First-Strand Synthesis system (Life Technologies, Carlsbad, CA, USA). Next, the RT-PCR reaction mixtures were used for subsequent polymerase chain reaction (PCR) reactions to amplify variable regions of IgG heavy chain (V_H) and light chain (V_L). The V_H and V_L were separately cloned in pTrans-CMV-MCS expression vectors (CEVEC, Köln, Germany) containing constant region coding sequences for rabbit IgG γ and IgG κ.

2.5. Transfection

Primary human amniocytes CAP-T cells (CEVEC, Köln, Germany) were transiently co-transfected with vectors containing the codon sequences of heavy chain and light chain originating from the same sorted cell using NovaCHOice® transfection kit (MilliporeSigma, Billerica, MA, USA), and the supernatants were harvested after seven days for evaluation by Enzyme-linked immunosorbent assay (ELISA), Western blotting, and IHC.

2.6. ELISA

The concentration of the IgG released by transfected cells was measured using Rabbit IgG ELISA kit (ZeptoMetrix Corp., Franklin, MA, USA). Serial dilutions of a rabbit IgG antibody (60, 30, 15, 7.5, 3.75, and 0 ng/mL), provided by the kit, were used to set up a standard curve. The specificity of the antibody was determined by immobilizing biotinylated synthetic peptides (provided by Antibodies Inc., Davis, CA, USA) KPIIIGHHAYGD (mutant) or KPIIIGRHAYGD (wild type) on a 96-well plate. The plate was coated with 2 µg/mL streptavidin (MilliporeSigma, Billerica, MA, USA) overnight at 4 °C and then the peptides were immobilized at 1 µg/mL for one hour at room temperature. After blocking with 5% skim milk for an hour, the plate was probed with the supernatant of transfected cells or a commercially available mouse monoclonal IDH1 R132H H09 antibody (Dianova, Hamburg, Germany) at different concentrations, starting from 1 µg/mL and lower and incubated for an hour at 37 °C. Next, a peroxidase-conjugated anti-rabbit IgG anibody (Jackson ImmunoResearch Lab, West Grove, PA, USA) (1:1000) was added to the plate and incubated for an hour at room temperature. The enzymatic reaction was conducted with Tetramethylbenzidine (TMB) (MilliporeSigma, Billerica, MA, USA) at room temperature and stopped by 0.25 M Sulfuric acid (MilliporeSigma, Billerica, MA, USA). The optical density was measured at 450 nm. The plate was washed four times after every step with PBS containing 0.05% Tween 20 (MilliporeSigma, Billerica, MA, USA).

2.7. Western Blotting

100 ng of recombinant human IDH1 R132H protein and wild type IDH1 protein (Abcam, Cambridge, MA, USA) were loaded onto 4–12% Bis-Tris mini gels (Life Technologies, Carlsbad, CA, USA) and blotted to PVDF membranes (iBlot™ Transfer Stack, PVDF, Invitrogen, Carlsbad, CA, USA). The Western blot was carried out using WesternBreeze Chromogenic kit (Invitrogen, Carlsbad, CA, USA) and the supernatant of the transfected cells (0.5 µg/mL IgG) or H09 (0.5 µg/mL antibody) were used to probe the membranes.

2.8. Immunohistochemistry

FFPE tissue samples were sectioned at a thickness of 4 µm and were prepared on Superfrost™ Plus microscope slides (Fisherbrand™, Pittsburgh, PA, USA). Prepared slides were stained by routine IHC on a BenchMark ULTRA automated staining instrument (Ventana Medical Systems Inc., Tucson, AZ, USA). Tissue slides were incubated for 64 min at 95 °C using an EDTA-based epitope retrieval solution, followed by a 32 min primary antibody incubation at 36 °C. Staining signal was visualized through a horseradish peroxidase (HRP)-based multimer detection system and 3-3'-Diaminobenzidine (DAB) chromogen. Counterstaining was performed by incubating tissue slides for 4 min with Hematoxylin II, followed by a 4 min incubation with bluing reagent. Stained slides were evaluated using light microscopy for target signal intensity and background signal by a qualified pathologist. Tumor cells exhibiting a strong, diffuse cytoplasmic staining pattern, as well as weaker nuclear labeling, were scored as positive for IDH1 mutant protein. The H09 antibody was used as a reference comparison during performance testing of the antisera from immunized rabbits and optimization of the selected MRQ-67 clone. The H09 clone further served in establishing the IDH1 R132H mutation status of glioma samples that were used in this study. The same automated staining conditions were used for both the MRQ-67 and H09 clones, with optimal antibody titers having been experimentally determined for each clone individually. The MRQ-67 and H09 clones were determined to perform optimally in IHC at concentrations of 2.54 µg/mL and 3.25 µg/mL, respectively.

3. Results

Four rabbits were immunized with the synthetic IDH1 R132H mutant peptide. Sera from immunized rabbits were tested by IHC. The PBMCs from the rabbit with the best immune response were isolated for cloning of V_H and V_L, followed by co-transfection in CAP-T cells.

ELISA analyses of the transfection reactions confirmed the production of rabbit IgG by 73% of the clones (Figure 1a). The positive clones were further screened for specificity of the antibodies and among them, antibody MRQ-67 generated by clone C5 was selected for further evaluations. As shown in Figure 1b, the MRQ-67 antibody specifically reacted with mutant peptide IDH1 R132H in a dose-dependent manner, but not with the wild type peptide IDH1 in ELISA assay, indicating that MRQ-67 specifically recognized IDH1 R132H. This result was consistent with the result obtained using the control antibody, H09 (Figure 1b).

(a)

	1	2	3	4	5	6	7	8	9	10	11	12
A	22.2	16.4	0.0	11.0	22.9	0.0	17.7	18.3	19.2	11.9	1.7	6.5
B	6.5	15.1	11.8	12.2	18.5	0.0	21.1	31.3	20.1	22.7	0.0	16.8
C	29.7	0.0	24.6	15.7	23.3	21.2	36.9	14.8	19.4	0.0	0.0	11.9
D	23.4	18.2	0.0	0.0	17.9	22.9	7.8	30.5	13.8	13.6	22.4	0.0
E	4.0	37.2	5.5	0.0	9.6	0.0	34.6	9.7	21.4	19.2	0.0	18.9
F	18.2	30.6	22.8	34.7	2.9	40.2	10.6	0.0	0.0	0.0	20.1	11.9
G	10.2	0.0	12	0.0	29.6	23.8	0.0	20.5	0.0	0.0	0.0	13
H	EW	EW	EW	EW	EW	EW	EW	EW	EW	EW	EW	EW

Concentration of rabbit IgG (µg/mL)
EW: Empty Well

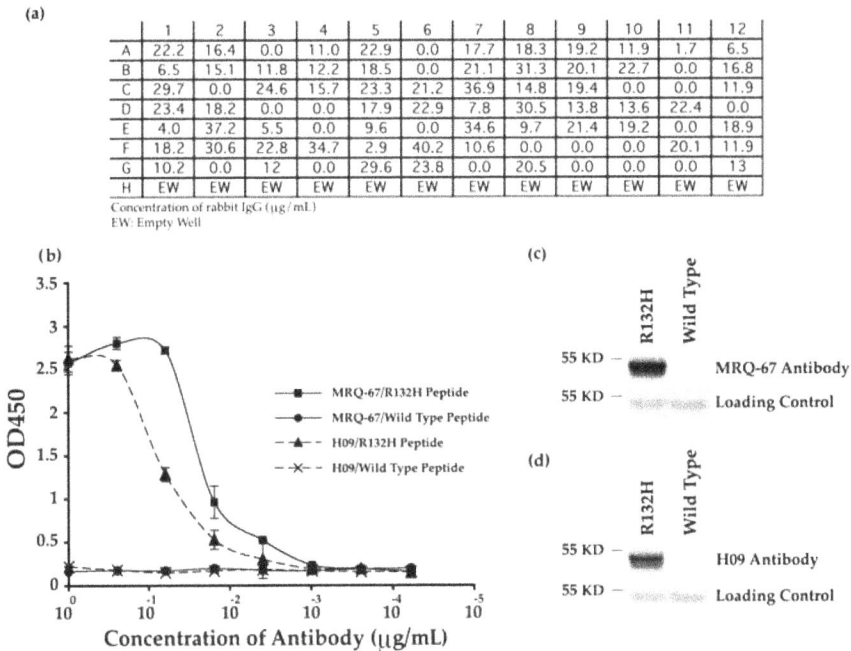

Figure 1. Generation and performance of rabbit monoclonal MRQ-67 antibody against IDH1 R132H and comparing its function with the H09 antibody by Enzyme-linked immunosorbent assay (ELISA) and Western blotting assays: (**a**) The concentration of antibody generated by clones was measured in µg/mL by rabbit IgG ELISA kit; (**b**) Peptide binding ELISA. MRQ-67 antibody specifically recognized only mutant IDH1 R132H peptide and not wild type IDH1 peptide. The ELISA binding assay result for the H09 antibody has been included as a control; (**c,d**) Analysis of MRQ-67 and H09 for detecting recombinant human IDH1 R132H and recombinant human IDH1 wild type proteins in Western blotting. MRQ-67 (**c**) and H09 (**d**) antibodies (0.5 µg/mL) were used to probe the membranes. Coomassie blue stainings of the proteins are shown as loading control.

The MRQ-67 antibody's specificity was further analyzed by Western blotting and compared with the specificity of H09 antibody. As shown in Figure 1c,d, MRQ-67 and H09 both detected the recombinant IDH1 R132H protein at a predicted molecular weight of 48 kDa for human IDH1 protein. Notably, while MRQ-67 antibody did not detect any band on the blot with wild type recombinant IDH1 protein (Figure 1c), the H09 antibody showed a weak reaction with this protein (Figure 1d). Overall, this data indicates that MRQ-67 is also useful in detecting not only IDH1 R132H peptide, but also the mutant IDH1 R132H protein.

To further characterize the rabbit IDH1 R132H monoclonal antibody, the capacity of the MRQ-67 antibody to immunohistochemically identify IDH1 mutant protein in FFPE tissues was investigated. The IHC staining results summarized in Table 1 demonstrate equivalent sensitivity and specificity performance of the MRQ-67 clone in comparison to H09 for distinguishing low grade gliomas from glioblastomas, meningiomas, and benign brain samples. The MRQ-67 clone generated strong, diffuse cytoplasmic staining with weaker nuclear reactivity in 50% of the diffuse and anaplastic astrocytomas (Figure 2a) that was equivalent to performance observed with H09 (Figure 2b). Positive tumor staining in 71% of oligodendroglioma samples by MRQ-67 was primarily demonstrated by cytoplasmic reactivity (Figure 3a) and was comparable to observed results from H09 testing (Figure 3b). From the seven high grade glioblastoma cases that were tested, only one reacted positively with MRQ-67 as indicated by weakly diffuse cytoplasmic and nuclear staining (Figure 4a), while no tumor cells were labeled in the rest of the tested cases (Figure 4b).

Table 1. IDH1 R132H immunohistochemistry (IHC) staining data. Summary of IHC staining results comparing the number of cases stained positive out of the total number of cases tested with the MRQ-67 and H09 clones.

Tissue	Cases Stained (MRQ-67)	Cases Stained (H09)
Astrocytoma	9/18 (50%)	9/18 (50%)
Oligodendroglioma	5/7 (71%)	5/7 (71%)
Gliobastoma	1/7 (14%)	1/7 (14%)
Meningioma	0/12	0/12
Non-neoplastic Brain	0/15	0/15

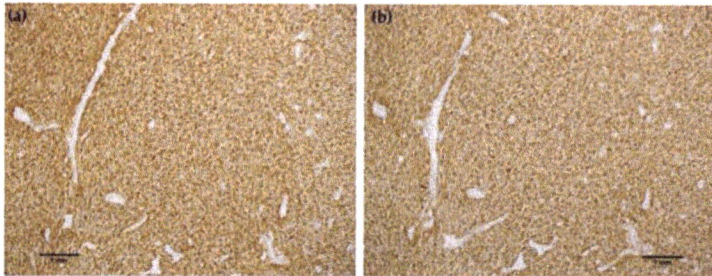

Figure 2. Comparison immunohistochemistry (IHC) staining results with MRQ-67 and H09 in astrocytoma: (**a**) Strong, diffuse cytoplasmic and weak nuclear labeling of tumor cells with MRQ-67 in a case of anaplastic astrocytoma (100×); (**b**) Equivalent cytoplasmic and nuclear staining of tumor cells with H09 in the same case of astrocytoma (100×).

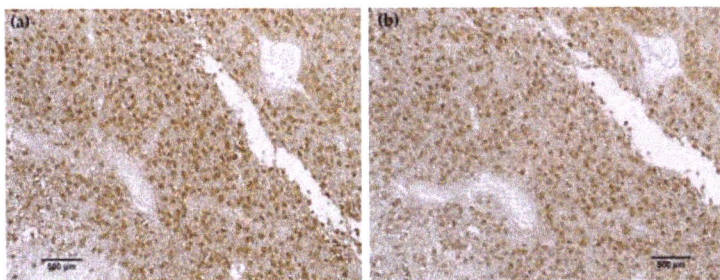

Figure 3. Comparison immunohistochemistry (IHC) staining results with MRQ-67 and H09 in oligodendroglioma: (**a**) Strong cytoplasmic and weak nuclear labeling of tumor cells with MRQ-67 in a case of WHO grade II oligodendroglioma (200×); (**b**) Comparable cytoplasmic and nuclear staining of tumor cells with H09 in the same case of oligodendroglioma (200×).

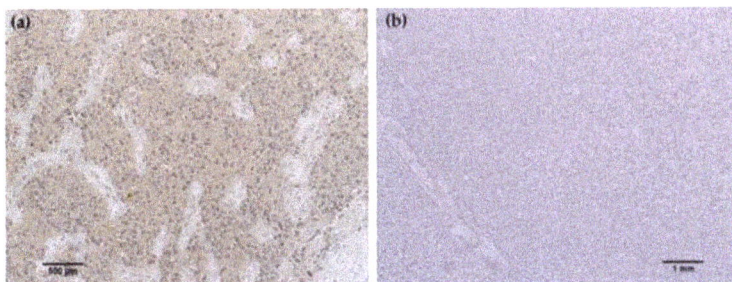

Figure 4. MRQ-67 immunohistochemistry (IHC) staining results in high grade glioblastoma: (**a**) Weak, diffuse cytoplasmic and nuclear labeling of tumor cells in a case of WHO grade IV glioblastoma with anaplastic astrocytoma involvement (200×); (**b**) No observed reactivity in tumor cells in another case of WHO grade IV glioblastoma (100×).

In all cancerous and benign brain samples tested, the MRQ-67 antibody did not react with endothelial cells, lymphocytic cells, or normal glial cells. The 15 cases of benign brain (Figure 5a,b) and 12 cases of meningioma (Figure 6c,d) that were assessed displayed no observable cross-reactivity, with only two cases (16%) of meningioma demonstrating equivocal background staining signal when tested using MRQ-67. However, meningioma samples stained with the H09 clone generated nonspecific background signal in fibrillar and spindle cell components in 11 out of 12 cases (92%), with 4 cases (33%) in particular developing notably strong background staining (Figure 6a,b). One particular case of benign brain was observed to have a small focus of tumor cells that was identified by anti-IDH1 R132H (Figure 7a). Importantly, the surrounding distal nervous tissue was identified to consist of single infiltrative tumor cells that exhibited strong positive reactivity (Figure 7b).

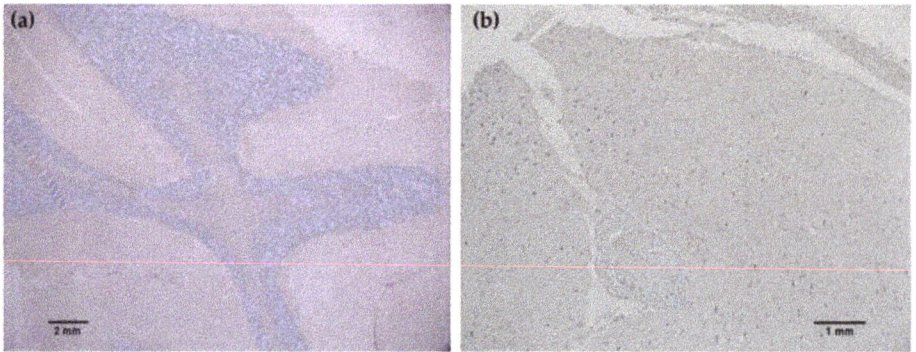

Figure 5. MRQ-67 immunohistochemistry (IHC) staining results in normal brain: (**a**) No observed reactivity in the normal cell types that constitute the gray and white matter in the cerebellar region of the brain (40×); (**b**) No observed reactivity in normal cerebral nervous tissue (100×).

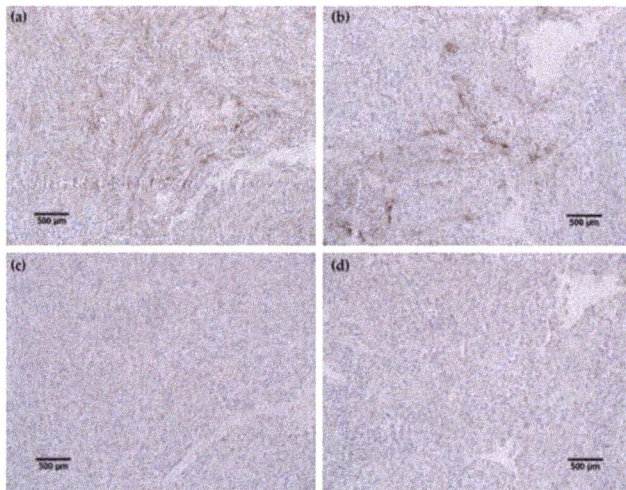

Figure 6. Comparison immunohistochemistry (IHC) staining results of MRQ-67 and H09 in meningioma specimens: (**a**) The H09 clone displayed positive reactivity in spindle cells of a meningioma sample (200×); (**b**) Positive labeling of fine fibrous elements by the H09 clone in another case of meningioma (200×); (**c**) The MRQ-67 clone demonstrated no cross-reactivity with spindle cells in the same meningioma case stained with the H09 clone (200×); (**d**) MRQ-67 also did not generate cross-reaction with fibrous elements as was observed with H09 staining in the same case of meningioma (200×).

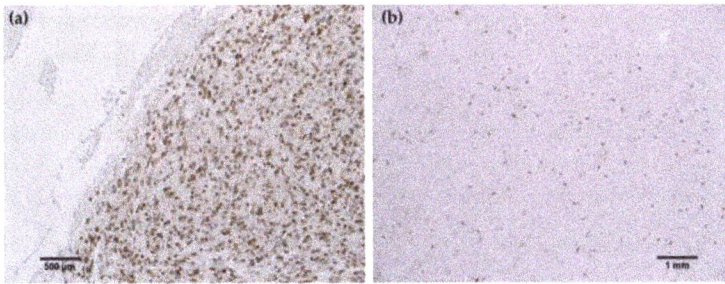

Figure 7. MRQ-67 immunohistochemistry (IHC) staining results in a primarily benign brain sample with a small tumor focus and some scattered tumor cells: (**a**) Cytoplasmic and weak nuclear labeling of cells within a small tumor focus of a benign brain sample (200×); (**b**) Scattered reactivity in single infiltrative tumor cells of the same benign brain case that have dispersed from the focal tumor site (100×).

4. Discussion

In the specific targeting of the IDH1 R132H mutation through B-cell cloning, a novel rabbit monoclonal antibody was produced that promises to be a useful tool for overcoming the diagnostic challenge of differentiating between different subtypes of glioma. The single B-cell cloning technology used to generate MRQ-67 is an attractive method for developing high quality monoclonal antibodies that has several advantages over previously established techniques. This technology is more efficient than cell fusion in hybridoma technique [6], and unlike display methods (e.g., phage display [7] and yeast surface display [8]), both heavy and light chains originate from a single sorted cell. This allows for the natural cognate pairing of the heavy and light chains to be preserved during synthesis and maturation of the antibody [5]. Moreover, culturing of isolated B-cells is not required, which removes a potential source of technical complications from the process.

In ELISA assays and Western blotting analyses, the MRQ-67 rabbit monoclonal antibody reacted with mutant IDH1 R132H, but not with wild type IDH1. This data indicates that MRQ-67 is capable of specifically detecting not only IDH1 R132H peptide, but also the mutant IDH1 R132H protein without cross-reacting with wild type IDH1. The specificity of MRQ-67 to the R132H mutation had not been tested against other less frequent mutations of IDH1 at the time of this report.

Identification of IDH1 mutation status by IHC presents a useful tool for many diagnostic institutions where genetic testing can be burdensome and may be inaccurate, especially regarding tissue samples with low tumor cell content. Detection by IHC provides the opportunity to visualize even single infiltrating tumor cells in otherwise normal appearing brain tissue. The IDH1 status of FFPE tissue samples used in this study had not been determined by sequencing at the time of this report. IDH1 mutation status of these tissues was therefore established through the use of the commercially available mouse monoclonal IDH1 R132H (H09) antibody.

Rabbit monoclonal anti-IDH1 R132H (MRQ-67) labeled each of the same cases as H09, indicating comparable ability to detect IDH1 mutant protein by IHC. However, the H09 clone generated nonspecific background signals in the fibrillar and spindle cell components in nearly all meningioma cases tested, with a few cases in particular developing notably strong background staining. The observed cross-reaction with fibrous elements in meningioma cases has been previously identified as nonspecific binding by anti-IDH1 R132H to extracellular matrix protein or a subtype of collagen fiber [9]. Staining with the MRQ-67 clone demonstrated only two instances of weak, nonspecific background signal in meningioma samples, indicating a particular advantage in specificity compared to the H09 clone. Since no immunoreaction was observed in meningioma tumor cells or normal brain samples, all tumor cells that stained are considered to be IDH1 mutant-positive, including the single infiltrating cells that dispersed from the primary tumor focus as seen in Figure 7.

Antibodies **2017**, *6*, 22

The presence of IDH1 mutations has been indicated to be much more frequent in secondary glioblastomas compared to primary glioblastomas [10]. Clinical information regarding the progression of the single case of glioblastoma in this study with identified mutant IDH1 staining signal was not available, but histopathological evidence of lower grade glioma involvement was observed. Further, population-based data indicates a considerably greater incidence rate of primary glioblastoma compared to that of secondary glioblastoma [11]. These observations, together with H09 comparison staining data in glioblastoma samples, suggest that MRQ-67 functions as intended for IHC applications.

Overall, the results from ELISA assays, Western blotting, and IHC analyses support the proposed utility of the novel rabbit monoclonal MRQ-67 antibody in the identification of mutated human IDH1 protein. IDH1 (MRQ-67) rabbit monoclonal antibody is a specific marker for immunohistochemical detection of IDH1 mutant protein in glioma subtypes and may have value as a tool in distinguishing between diffuse astrocytomas and oligodendrogliomas from secondary glioblastoma.

Acknowledgments: We would like to acknowledge Maricela Linhares for her support in preparing cell cultures.

Author Contributions: J.R. conceived and designed the experiments; J.R., R.C., B.M., A.Y., and E.F. performed the experiments; J.R., Q.S., and R.C. analyzed the data; J.R., R.C., K.C., and C.J. wrote the paper.

Conflicts of Interest: Each of the authors is employed by MilliporeSigma.

References

1. Guo, C.; Pirozzi, C.J.; Lopez, G.Y.; Yan, H. Isocitrate dehydrogenase mutations in gliomas: Mechanisms, biomarkers and therapeutic target. *Curr. Opin. Neurol.* **2011**, *24*, 648–652. [CrossRef] [PubMed]
2. Balss, J.; Meyer, J.; Mueller, W.; Korshunov, A.; Hartmann, C.; von Deimling, A. Analysis of the IDH1 codon 132 mutation in brain tumors. *Acta Neuropathol.* **2008**, *116*, 597–602. [CrossRef] [PubMed]
3. Watanabe, T.; Nobusawa, S.; Kleihues, P.; Ohgaki, H. IDH1 mutations are early events in the development of astrocytomas and oligodendrogliomas. *Am. J. Pathol.* **2009**, *174*, 1149–1153. [CrossRef] [PubMed]
4. Louis, D.N.; Perry, A.; Reifenberger, G.; von Deimling, A.; Figarella-Branger, D.; Cavenee, W.K.; Ohgaki, H.; Wiestler, O.D.; Kleihues, P.; Ellison, D.W. The 2016 World Health Organization Classification of Tumors of the Central Nervous System: A summary. *Acta Neuropathol.* **2016**, *131*, 803–820. [CrossRef] [PubMed]
5. Zhang, Z.; Liu, H.; Guan, Q.; Wang, L.; Yuan, H. Advances in the Isolation of Specific Monoclonal Rabbit Antibodies. *Front. Immunol.* **2017**, *8*, 494. [CrossRef] [PubMed]
6. Kohler, G.; Milstein, C. Continuous cultures of fused cells secreting antibody of predefined specificity. *Nature* **1975**, *256*, 495–497. [CrossRef] [PubMed]
7. Smith, G.P. Filamentous fusion phage: Novel expression vectors that display cloned antigens on the virion surface. *Science* **1985**, *228*, 1315–1317. [CrossRef] [PubMed]
8. Murai, T.; Ueda, M.; Yamamura, M.; Atomi, H.; Shibasaki, Y.; Kamasawa, N.; Osumi, M.; Amachi, T.; Tanaka, A. Construction of a starch-utilizing yeast by cell surface engineering. *Appl. Environ. Microbiol.* **1997**, *63*, 1362–1366. [PubMed]
9. Capper, D.; Weissert, S.; Balss, J.; Habel, A.; Meyer, J.; Jager, D.; Ackermann, U.; Tessmer, C.; Korshunov, A.; Zentgraf, H.; et al. Characterization of R132H mutation-specific IDH1 antibody binding in brain tumors. *Brain Pathol.* **2010**, *20*, 245–254. [CrossRef] [PubMed]
10. Nobusawa, S.; Watanabe, T.; Kleihues, P.; Ohgaki, H. IDH1 mutations as molecular signature and predictive factor of secondary glioblastomas. *Clin. Cancer Res.* **2009**, *15*, 6002–6007. [CrossRef] [PubMed]
11. Ohgaki, H.; Dessen, P.; Jourde, B.; Horstmann, S.; Nishikawa, T.; Di Patre, P.L.; Burkhard, C.; Schuler, D.; Probst-Hensch, N.M.; Maiorka, P.C.; et al. Genetic pathways to glioblastoma: A population-based study. *Cancer Res.* **2004**, *64*, 6892–6899. [CrossRef] [PubMed]

![antibodies logo] *antibodies*

MDPI

Article

Integration of Aqueous Two-Phase Extraction as Cell Harvest and Capture Operation in the Manufacturing Process of Monoclonal Antibodies

Axel Schmidt [1], Michael Richter [2], Frederik Rudolph [2] and Jochen Strube [1,*]

[1] Institute for Separation and Process Technology, Clausthal University of Technology, Leibnizstraße 15, 38678 Clausthal-Zellerfeld, Germany; schmidt@itv.tu-clausthal.de

[2] Boehringer Ingelheim Pharma GmbH & Co. KG, Bioprocess + Pharma. Dev. Biologicals, Birkendorfer Strasse 65, 88397 Biberach an der Riss, Germany; michael.richter@boehringer-ingelheim.com (M.R.); frederik.rudolph@boehringer-ingelheim.com (F.R.)

* Correspondence: strube@itv.tu-clausthal.de; Tel.: +49-5323-72-2200

Received: 30 October 2017; Accepted: 20 November 2017; Published: 1 December 2017

Abstract: Substantial improvements have been made to cell culturing processes (e.g., higher product titer) in recent years by raising cell densities and optimizing cultivation time. However, this has been accompanied by an increase in product-related impurities and therefore greater challenges in subsequent clarification and capture operations. Considering the paradigm shift towards the design of continuously operating dedicated plants at smaller scales—with or without disposable technology—for treating smaller patient populations due to new indications or personalized medicine approaches, the rising need for new, innovative strategies for both clarification and capture technology becomes evident. Aqueous two-phase extraction (ATPE) is now considered to be a feasible unit operation, e.g., for the capture of monoclonal antibodies or recombinant proteins. However, most of the published work so far investigates the applicability of ATPE in antibody-manufacturing processes at the lab-scale and for the most part, only during the capture step. This work shows the integration of ATPE as a combined harvest and capture step into a downstream process. Additionally, a model is applied that allows early prediction of settler dimensions with high prediction accuracy. Finally, a reliable process development concept, which guides through the necessary steps, starting from the definition of the separation task to the final stages of integration and scale-up, is presented.

Keywords: interfacial partitioning; cell harvest; capture; aqueous two-phase extraction; horizontal settler; static mixer; process integration; continuous phase separation

1. Introduction

Clarification processes and their corresponding devices, which are already well established in other industry sectors, e.g., flocculation, precipitation, or flotation, are increasingly being taken into consideration as alternatives in monoclonal antibodies (mAb) production in spite of individual limitations like process robustness, process cost, toxicity of flocculation or precipitation agents, and easy scale-up [1]. A quite promising concept is the application of aqueous two-phase extraction (ATPE) as a combined harvest and capture step, especially since this approach deals with all the aforementioned issues.

Interfacial partitioning of cells, cell debris, and other bioparticles occurs when a mixed system composed of phase-forming components begins to separate into its specific light and heavy phases [2–4]. During the process of settling and coalescence, particles and small solid objects accumulate at the surface of the dispersed phase (Figure 1). This phenomenon can be explained by the fact that the partitioning behavior of small particles is strongly dependent on surface forces [5].

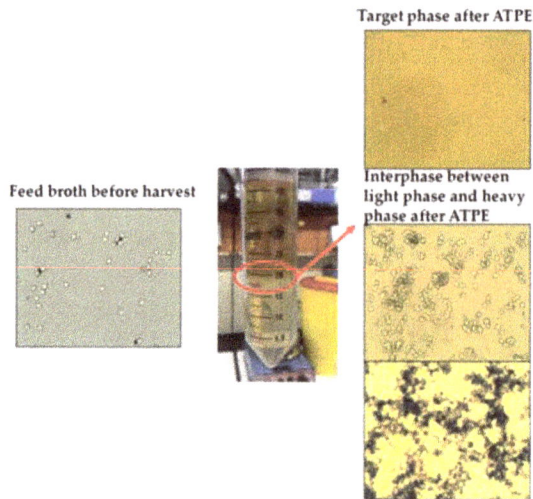

Figure 1. Cell clearance through interfacial partitioning. ATPE: Aqueous two-phase extraction.

Approaches for the combination of liquid–liquid extraction (LLE) and the separation of small particles by adsorption towards a dispersed phase are well investigated. These include interfacial partitioning for the recovery of bioparticles in general, three-phase partitioning (TPP) where interfacial partitioning is combined with the precipitation of proteins, or more recently, aqueous two-phase flotation, where gas bubbles are introduced as a dispersed third phase into an already separated aqueous two-phase system (ATPS) [6–8]. However, most approaches so far design the process around the particle-loaded phase because the solid phase contains the product. Some research has also been conducted towards interfacial partitioning for the removal of unwanted particles from a feed stream, which is especially necessary during the clarification of cultivation broths in the manufacturing process of monoclonal antibodies. However, to establish this technology in industry, there is also the need for a process development strategy that guides through the necessary steps, starting from the definition of the exact separation task, to the experiments necessary for the determination of crucial process and model parameters, up to the design considerations for optimal equipment dimensions. In this work, an ATPE process is outlined for the clarification of up to 12,000 L of cultivation broth in a time window of less than 3 h.

2. Theory

2.1. General Considerations

Unlike in the conventional production process of mAb, centrifugation and microfiltration as harvest and clearance operations are replaced with the outlined ATPE process. After the filtration trains, the product stream can, depending on the future process strategy, be further purified by precipitation or by integrated counter-current chromatography (iCCC) for continuous capture, replacing traditional protein A chromatography, as wells as ion-exchange and hydrophobic interaction chromatography for further purification and polishing [9–14].

The flowsheet of the discussed process, including the two purification and polishing operations, is shown in Figure 2. It provides a precise and straightforward view of how ATPE is integrated into the overall purification strategy. To quickly get an overview of all process-relevant effects, it is helpful to construct a cause-and-effect diagram, also known as an Ishikawa or fish bone diagram. It is meant to illustrate the different possible sources of reduced or insufficient process performance. The diagram is constructed of primary branches, which organize specific groups of effects. They can be fanned out further into major branches, representing major causes. Minor branches can be integrated as well to show the relationship between cause and effect in even more detail. Most diagrams have environment, people, materials, equipment, measurement systems, and methods as the primary branches. Though the diagram can be constructed like this, the causes and effects specific to the outlined ATPE targets is shown in Figure 3.

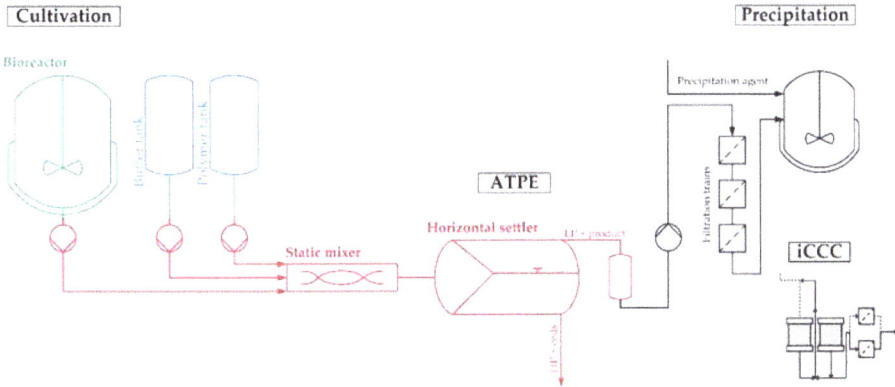

Figure 2. Process segment as discussed in the risk assessment. LP: Light phase. HP: Heavy phase. iCCC: integrated counter-current chromatography.

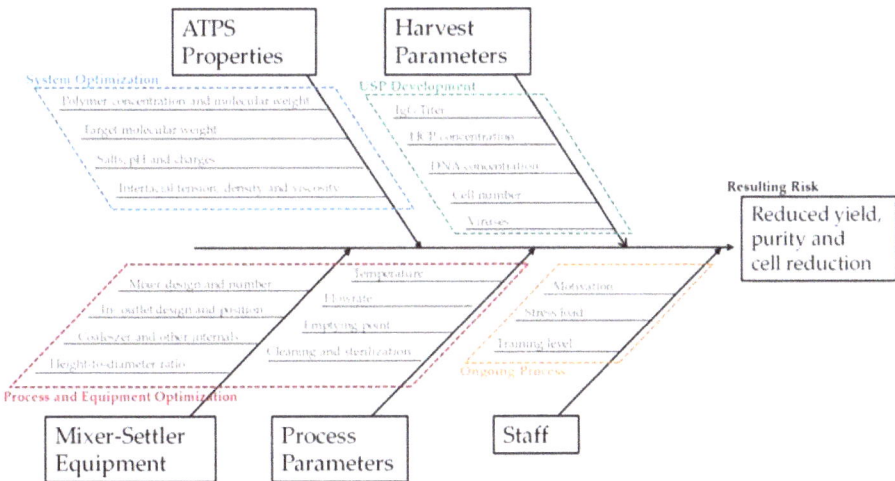

Figure 3. Ishikawa diagram illustrating the main groups of causes affecting yield, purity, and cell reduction in the outlined ATPE process. ATPS: separated aqueous two-phase system; HCP: host cell proteins.

The groups of causes in the diagram are set up by considering so-called prior knowledge. The green branch summarizes the factors that arise from upstream processing (USP) during cultivation and harvesting of the target parameters in this process step. There are several factors that can be optimized during upstream process development, like cultivation time, feed media, and supplement compositions, etc. An extensive overview and methodology for USP optimization has been published by Gronemeyer et al. [15]. The overall cell number or cell concentration is very important to consider, since ATPE as a harvest operation aims to remove most of the particles to lower the burden for the subsequent filtration trains. In addition to that, higher cell numbers decrease the settling velocity inside the apparatus as during interfacial partitioning the cells adsorb on the dispersed phase and hinder the coalescence of the droplets [3,16,17]. The effect of overall protein content, which is, for the most part, the sum of the mAb and the host cell proteins (HCP), needs to be considered as well. In most cases, protein partitioning between the specific light phase and heavy phase is a function of protein concentration, where total partitioning is only observable for low protein concentrations [18,19]. However, by increasing the polarity of the heavy phase, protein solubility can be significantly lowered due to salting-out effects, resulting in very high partitioning coefficients for the target component towards the light phase [18,20,21]. There are two main side-effects of this: firstly, salting-out very often affects HCP in the same way as mAb, which means that achieving high yield comes at the cost of low protein-based purity. Secondly, total partitioning of all proteins, in combination with purposely lowering the volume of the light phase in order to concentrate the target component, can lead to protein saturation and precipitation [22,23].

The considered effects are also included in the second primary branch, which organizes the effects of ATPS composition and phase properties (marked blue in Figure 3). The effects of ATPS type, composition, and phase properties are discussed widely in literature [21,24–28]. The identification of a potential operation space should, at this point, already have occurred in a system optimization study [29].

The third primary branch organizes the influences of the applied equipment (marked red in Figure 3). There are several potential devices available for LLE in general, and ATPE specifically. An extensive overview was published by Espitia-Saloma et al. [30]. In the outlined ATPE process, the combination of a static mixing pipe and a horizontal settler is applied. Although gravity settling is a relatively robust process, even slight changes to inlet geometry or the installation of a baffle have been shown to have a noticeable impact on the settling behavior [31].

As a further step, a risk assessment considering the relative impact and the relative occurrence of the different branches is applied. This assessment considers factors like temperature, which influences the settling behavior, and changing flow rates, which influences the hydrodynamic residence time and the specific power input.

Both parameters—temperature and changing flow rates—can potentially negatively impact the yield, purity, and cell reduction performance of the ATPE as illustrated in Figure 4. Increasing process temperatures can alter the intrinsic material properties of the phase-forming components which, like reducing the viscosities, could result in the creation of smaller droplets and, in the presence of cells, cell debris, and other solids, very large separation times [31]. However, coalescence kinetics in general are accelerated at higher temperatures, such that particle-free ATPS separation times can be reduced. Furthermore, the phase equilibrium itself is dependent on the temperature, since it affects the composition of the specific light and heavy phases, and therefore also changes the partitioning behavior of the target and side components [32–34]. The total flow rate determines the hydrodynamic residence time within the apparatus, but can result in an unnecessary firm dispersion. It is even possible to change the point of phase inversion of the system, which drastically affects the settling rates [35,36].

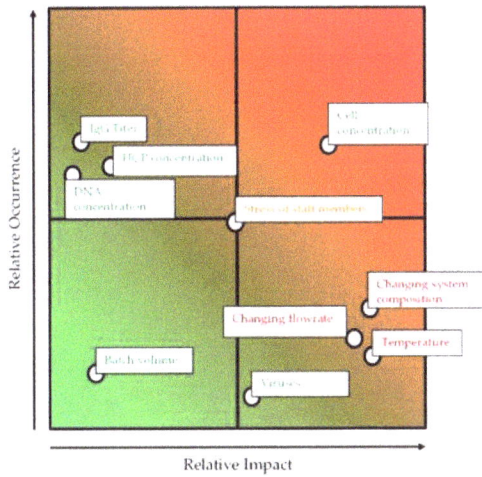

Figure 4. Occurrence–impact diagram for the outlined ATPE process.

2.2. Process Development

Process development starts by defining the design space that is feasible from an economic point of view, but also safe in terms of reliable process performance. Since ATPE in this work is applied mainly as a clarification operation, the parameter which characterizes the process performance of this step, is the reduction in total cell number. One important quality measure for the monitoring of subsequent filtration, precipitation, or chromatography steps is the particle load of intermediate pools. This is because, other than the cells themselves, cell debris (precipitated material, etc.) can also lead, above certain values, to either a blockage or the underperformance of these further purification and polishing steps. Particle load can be characterized by turbidity measurements or state-of-the-art DLS (dynamic light scattering) methods. Based on so-called prior knowledge, the equipment chosen for mixing the phase-forming components is a static mixing pipe, which offers gentle and homogeneous mixing at the same time. ATPS loaded with bio-particles tends to form strong emulsions when too much power is introduced into the system [31]. The specific light and heavy phases are separated in a horizontal settler unit. This apparatus is ideal for single stage LLE since it is easy to scale-up and is proven to show reliable separation performance even when confronted with high particle burdens.

2.3. Batch-Settling Behavior

To enable the early prediction of settler dimensions, other than physical properties, settling and coalescence rates must also be determined.

The settling behavior of an ATPS can be mathematically described by three model parameters. The aim is to describe the influence of different material parameters on the separation behavior. The system is separated when the majority of the phases have settled and an interphase has formed.

$$\frac{dh}{dt} = \frac{\sigma}{\mu_c} \, const. \left(\frac{\mu_d}{\mu_c}\right)^a \left(\frac{\sigma_w}{\sigma}\right)^b \left(\frac{\Delta\rho}{\rho_c}\right)^c \tag{1}$$

The settling time is dependent on the ratio of the interfacial tension between the phases (σ) and the viscosity of the continuous phase (μ_c). The values are composed of the dispersed (μ_d) and continuous viscosities, the interfacial tension of water (σ_w) and the two phases, as well as the density difference of the dispersed and continuous phases ($\Delta\rho$) and the density of the continuous phase (ρ_c). The exponents a, b, and c are calculated numerically based on experimentally-determined settling times [37].

2.4. Early Prediction of Settler Dimensions

To get a rough estimation of the dimensions of the apparatus, volume flow ($\dot{V}$) is calculated, which is dependent on the entire volume (V_{total}) of the ATPS in relation to the predetermined process duration (t_{total}):

$$\dot{V} = \frac{V_{total}}{t_{total}} \tag{2}$$

The length can then be calculated using volume flow and the predetermined residence time (t_{RDT}), as well as the cross-sectional area of the device (A_S):

$$L_S = \frac{\dot{V} \times t_{RDT}}{A_S} \tag{3}$$

Further influencing variables, such as the drop size, sedimentation rates, or other parameters influencing coalescence, are not taken into account. This quick estimation is often referred to as minimum-apparatus-volume (MAV).

2.5. Henschke Method for Settler Dimensioning

A procedure to calculate the dimensions of horizontal settler units was introduced in 1994 [38]. It was derived from aqueous organic systems and is investigated here in terms of its suitability for bio-particle loaded aqueous two-phase systems. In this model, the power input, sedimentation rate, and coalescence behavior of the single droplets are taken into account. To examine these influencing factors more precisely, the length of the separator (L_S) is divided into two regions: the inlet area of the dispersion (L_{in}) and the coalescence region (L_c) (Figure 5).

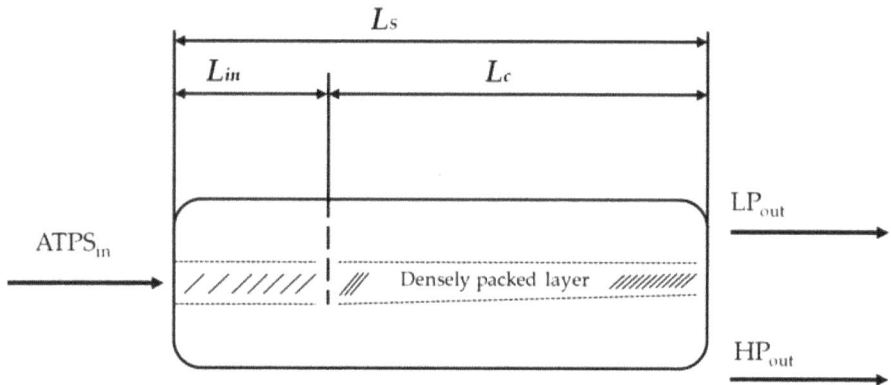

Figure 5. Structure of the settler unit [38]. ATPS: Aqueous two-phase system. L_S: Length of the separator. L_{in}: Inlet length. L_c: Coalescence length.

In the inlet area, occurring turbulences and the formation of a droplet layer are considered. Within the apparatus, drop coalescence and the formation of the densely packed layer (h_p) are of greater interest. The change in the densely packed height can be calculated as follows:

$$\frac{dh_p}{dl} = -\frac{\dot{V}_{Dis} \times (11.3 \times s \times \eta_{Dis} + 126 \times (\eta_c + \eta_d))}{h_p \times D_S^3 \times \bar{\varepsilon}_p \times (1 - \bar{\varepsilon}_p) \times \Delta\rho \times g} \tag{3}$$

It is dependent on the dispersion volume flow ($\dot{V}_{Dis}$), the dispersion viscosity (η_{Dis}), a slippage parameter (s), the diameter of the settler (D_S) and the dispersed volume fraction in the densely packed layer ($\bar{\varepsilon}_P$). The settling time (t_s) is normalized to level out the influence of long coalescence times at the boundary surface. The compensation parameter is (C^*). With the aid of these two parameters, the coalescence length of the separator can be determined as follows:

$$L_c = \frac{t_S \times \dot{V}_{d,0}}{D_S \times \sqrt[1,3]{\frac{H_p \times H_{d,0}^{0,3}}{\left(\frac{\Phi_{32,0,settle}}{\Phi_{32,0,settler}}\right)^{0,5} \times C^*}}} \tag{4}$$

It is dependent on the final height of dispersed phase in a batch-settling experiment ($H_{d,0}$), the volume flow of dispersed phase ($\dot{V}_{d,0}$) and the starting droplet diameter in a batch-settling experiment ($\Phi_{32,0,settle}$) and inside the settler ($\Phi_{32,0,settler}$). The height of the densely packed layer inside the settler (H_p) should be predefined. The inlet length must also be determined. It is dependent on the velocity inside the settler (v_S) the volume of the inlet (V_{in}), the diameter of the inlet (D_{in}), earth gravity (g) and the dispersion band height inside the settler (H_{Dis}). In this case, the mean density ($\bar{\rho}$) and the hold-up (ε_0), which corresponds to the volume fraction of the dispersed phase in the total volume, are determined:

$$L_{in} = \frac{43,7 \bar{\rho}^{0,3} v_S^{0,5} D_S^{0,4} V_{in}^{0,5} D_{in}^{0,5}}{\Delta \rho^{0,3} (\Phi_{32,0} + H_{Dis})^{0,4} (1 - \varepsilon_0)^{0,2} g^{0,5}} \tag{5}$$

$$\bar{\rho} = \varepsilon_0 \cdot \rho_d + (1 - \varepsilon_0) \cdot \rho_c \tag{6}$$

$$\varepsilon_0 = \frac{H_{d,0}}{H_{c,0} + H_{d,0}} \tag{7}$$

The total length of the separator (L_S) is composed of the inlet length (L_{in}) and the coalescence length (L_c):

$$L_S = L_{in} + L_c \tag{8}$$

The geometry of the dispersion layer is not further considered in the simplified model. For further explanations, reference is made to Henschke et al. [38].

3. Results and Discussion

3.1. System Selection and Batch-Settling Experiments

Based on the shaking flask results, focus is placed on system point 1 (SP1) in the following experiments, since the achievable yield is the highest of the investigated compositions at around 95%, while also showing high cell reduction capabilities. Concentration of IgG is nearly doubled in SP4, due to its phase-split, however, this comes at the cost of insufficient cell reduction (Table 1).

Table 1. Yield, purity, and cell reduction of the investigated ATPS.

System Point	Yield (%)	SEC-Purity (%)	Log Cell Reduction (−)
SP1	95 ± 2.33	80.93 ± 0.02	2.08 ± 0.1
SP2	80 ± 0.39	81.97 ± 1.7	2.47 ± 0.02
SP3	63 ± 3.00	67.98 ± 1.8	3.45 ± 0.2
SP4	80 ± 1.62	80.98 ± 0.03	0.61 ± 0.02

ATPS: aqueous two-phase system; SEC: size exclusion chromatography; SP: system point.

In Figure 6, the height profile for the specific light and heavy phases observed in the batch-settling experiments is plotted. The specific light phase begins at a reactor height of approximately 47 mm with settling, and is complete after approximately 9 min at a reactor height of approximately 21 mm. The heavy phase begins to visibly coalesce with a delay of 5 min and has settled after about 20 min at a height of about 17 mm. This is related to the delayed coalescence of the dispersed droplets. In Table 2 the resulting sedimentation and coalescence rates, including the standard errors, are listed.

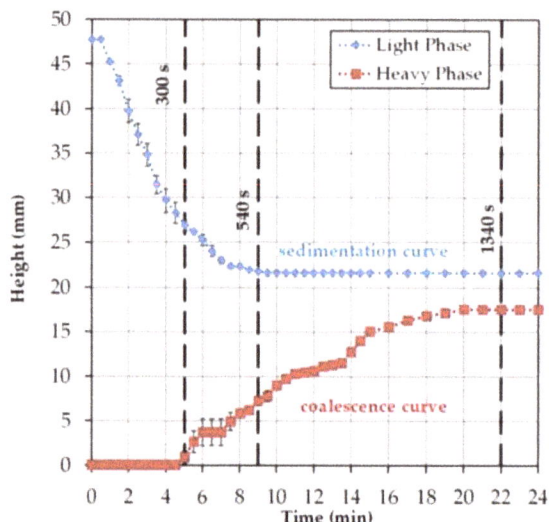

Figure 6. Settling curves obtained by triplicate batch-settling experiments.

The results from the settling experiments are used below to calculate the exponents of the Mistry model. This model only considers the settling behavior of the interphase. This is divided into two areas. In the first section, up to 9 min sedimentation of the specific light phase and coalescence of the heavy phase are observable. After that, only the heavy phase coalesces. The height change of the interphase can be determined from the batch-settling experiments. The results are shown in Table 2.

Table 2. Sedimentation and coalescence rates in mm/min obtained by triplicate batch-settling experiments.

Time Interval	dh_p/dt (mm/min)		
	1.1	1.2	1.3
540 s	3.8	3.87	3.62
1340 s	2.09	2.07	2.17

The interphase, up to the time of 9 min, settles down at approximately 3.7 mm per minute. With the range up to 22 min, a sedimentation rate of 2 mm per minute is obtained. The results of the settling behavior of the interphase can be used in Equation (1) to determine the exponents a, b, and c (Table 3). The other parameters for the calculation are listed in Table 4.

Table 3. Parameter set for the calculation of phase separation rates.

Time Interval	a	b	c
540 s	0.6230	0.1604	3.1096
1340 s	0.6276	0.0824	3.1644

Table 4. Material data for the calculation of phase separation rate.

σ (mN/m)	σ_w (mN/m)	ρ (kg/m^3)	$\Delta\rho$ (kg/m^3)	η_c (Pa·s)	η_d (Pa·s)
0.0702	72.75	1102.5	0.1337	0.0074	0.0048

For 9 min, a correlation between the calculated settling time and the experimental settling time can be seen. The linear course of the interphase, which results from the deposition of both phases, is well represented by Equation (1). The exponential course of the settling curve results when the sedimentation behavior is considered up to complete separation. This results from the much longer settling time of the dispersed, heavy phase (Figure 7). The applied approach can not reflect this trend. In order to further verify the results, the settling times of other ATPS would have to be compared experimentally and theoretically. Furthermore, cell-loading and power input must be taken into account, because they influence the settling behavior of the system.

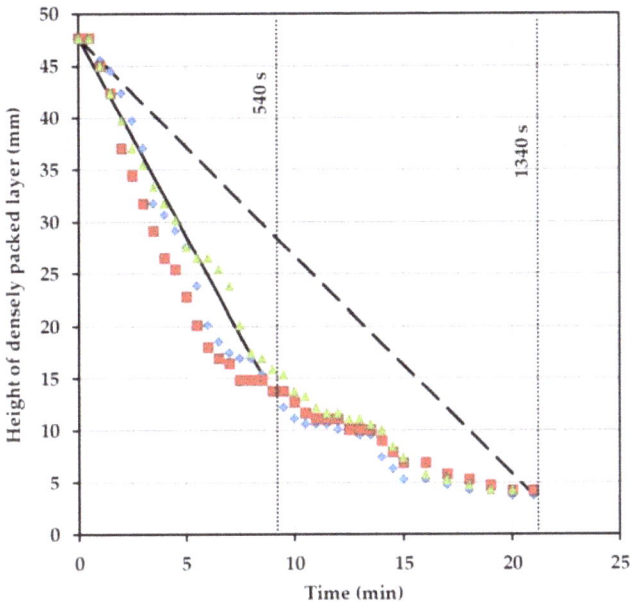

Figure 7. Height of the densely packed layer from triplicate batch-settling experiments (red, green, blue) as well as from the calculated separation rates (straight lines).

3.2. Model-Based Design

For the calculation of the separator length some parameters were fixed in advance. The hydrodynamic residence time for the MAV calculations equals the settling times (t_s) from the batch-settling experiments. In contrast, these settling times do not equal the residence times in the more detailed calculation because this model considers not only the inlet length, but also the actual height of the densely packed layer at the apparatus end, as well as the mean droplet size. The separation times used in both calculations are 300, 540, and 1340 s.

In the calculations, the ratio $D_S/L_S = 1/5$ is chosen to achieve comparability for the MAV and the detailed calculations of the laboratory settler devices used in the experiments. The specifications for the inlet diameter (D_{in}) and the inlet length of the separator (L_{in}) are based on the dimensions of the laboratory separator for small volumes as well. If the system volume exceeds 160 L, the specifications are scaled-up to the dimensions of the pilot-scale device. The results of the laboratory experiments also reveal a slightly higher dispersion level (H_{Dis}) than the densely packed dispersion layer (H_p) at the end of the settler. The height of the densely packed dispersion layer is set in relation to the diameter of the settler. In addition, the height of the densely packed dispersion layer varies depending on the predetermined separation time. For this reason, a factor is used in the calculation linking the height of the densely packed dispersion layer to the diameter of the separator. This results in the following ratios:

- $H_p/D_S = 0.54$ at a cut-off time of 300 s,
- $H_p/D_S = 0.303$ at a separation time of 540 s,
- $H_p/D_S = 0.1$ with a separation time of 1340 s.

The ratio is obtained from the filling height of the batch-settling reactor and the height of the experimentally-determined densely packed layer (Figure 8).

Firstly, a cut-off time of 300 s (hereinafter called H300) is considered. At this point, the light phase has not yet completely settled. However, there is sufficient separation to retrieve clear phase at the top. The interphase is strongly pronounced at this point and it is possible that a large part of the cells are withdrawn together with the heavy phase. However, there is also the risk of losing a small amount of the specific light phase.

A separation time of 540 s is considered next (hereinafter, H540). At this point, the specific light phase in the settling test has completely settled down. The specific heavy phase can be withdrawn together with cells between the dispersed droplets without obtaining significant losses in the yield.

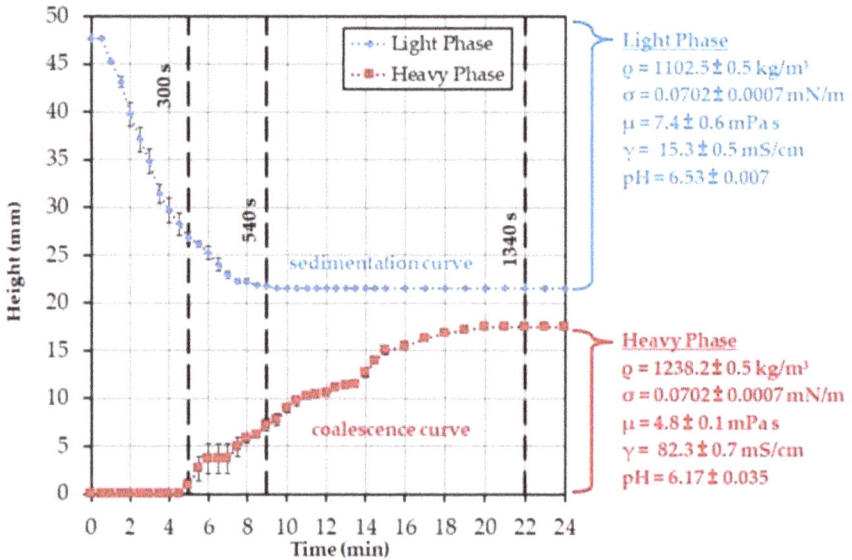

Figure 8. Marked separation times for the calculation from the settling experiments.

Complete separation of both phases is now observed at 1340 s (hereinafter called H1340). Both phases are completely separated now and it is possible to remove them separately from the separator. Removal of the densely packed layer in combination with the cells in between is only possible in this case if the phase is withdrawn very close to the phase boundary.

The length of the separator is calculated in each case for different specifications at a fixed system volume, and only for variation in the diameter of the separators to a ratio of 0.2. The investigated system volumes vary between 2 L and the industrially realizable volume of 24,000 L (roughly 12,000 L cultivation volume). As before, the systems should be separated within 3 h.

At low system volumes between 1 L and 10 L, the sizes of the separators according to H1340 are significantly larger than the designs according to the MAV-method (Figure 9) and H300 and H540, respectively (Figure 10). This is due to the sm all, predetermined height of the densely packed layer at the end of the separator. This results in a very large separator length, necessary to realize such a complete separation. In comparison, the settler volumes at H300 are very small due to the early withdrawal of the light phase and the accompanying larger densely packed layer. In the designs for H540, the settler dimensions are approximately 20 times smaller than the system volume. Between the increasing system volumes, a linear profile of the separator volume is indicated.

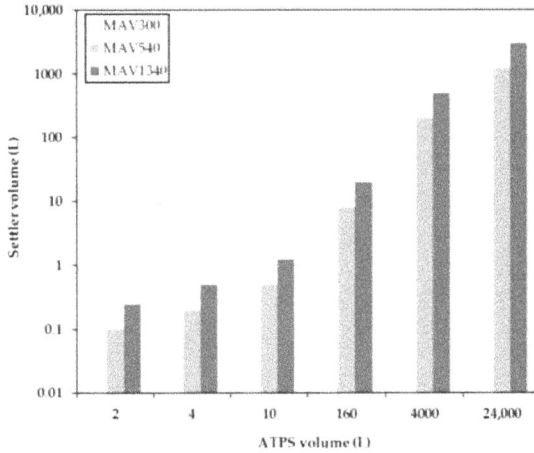

Figure 9. Settler volume as a function of residence time for different system scales obtained for the MAV method. MAV: minimum-apparatus-volume. MAV300, MAV540 and MAV1340 represent separation times of 300, 540 and 1340 s.

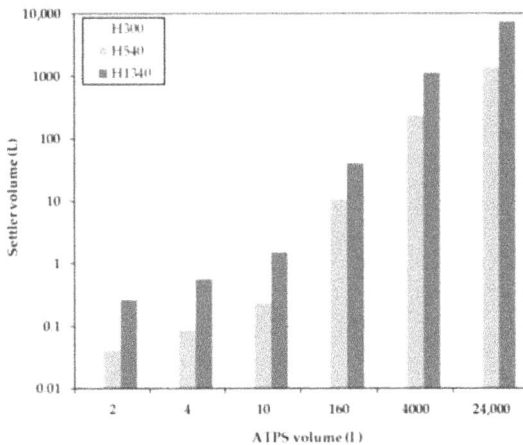

Figure 10. Settler volume as a function of residence time for different system scales obtained for the simplified Henschke method. H300, H540 and H1340 represent separation times of 300, 540 and 1340 s.

3.3. Process Performance

To evaluate the validity of the early prediction of settler dimensions, experimental studies were performed at lab- and pilot-scale.

Figure 11 shows the results of eight continuous settling experiments that were conducted utilizing a settler with a nominal diameter (DN) of 50 mm (750 mL hold-up). Product yield was, across the experiments, higher than 80%. Cell reduction was at least 1 log or higher, except for experiments S6 and S7. The reason for this was a change in flowrate. S6 and S7 were part of a phase inversion study. By increasing the volume flow, effects on yield, phase separation, and cell clearance should be determined. The relatively high yields of 90–100% indicate that mass transfer, as well as phase separation, is not negatively affected by the increase in flowrate.

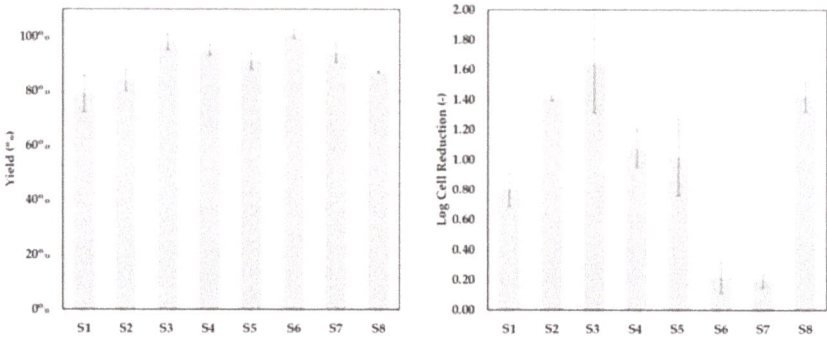

Figure 11. Yield and cell clearance for eight continuous settling experiments (S1–S8).

However, as can be seen in Figure 12, since the length of the static mixing pipe is fixed, the residence times in S6 and S7 are decreased not only in the settler, but more importantly in the mixing pipe as well. As a result, the phase-forming components are insufficiently mixed and no interfacial partitioning occurs. This system behavior can be reproduced in the batch-settling experiments as well. Leaving the standard set-up conditions constant, but reducing the mixing time from 5 to less than 1 min, results in an absence of interfacial partitioning (Figure 13).

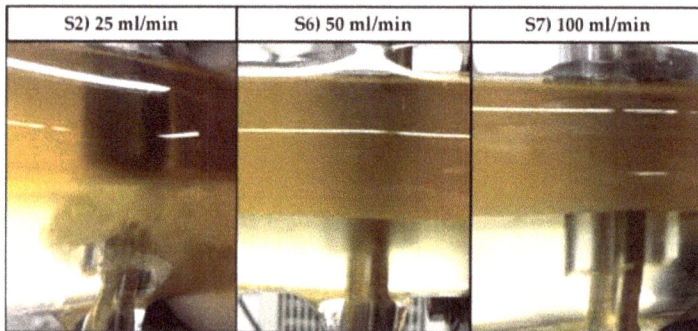

Figure 12. Exit area of the DN50 settler at three different flowrates but an otherwise identical set-up (DN: Diameter nominal). In experiment S2, cells can be retrieved together with heavy phase through bottom phase exiting (left).

Figure 13. SP1 after batch-settling. Clear light phase due to interfacial cell partitioning (**left**). High turbidity in light phase, because of insufficient mixing (**right**).

3.4. Comparison of Experiment and Model

In Figure 14, the phase separation profile inside a laboratory settler during continuous settling operation is shown. A volume flow of 35 mL/min is constantly separated during the procedure. For 3 h process time, this results in a processed ATPS volume of around 6.3 L. Complete phase separation of light and heavy phases is achieved.

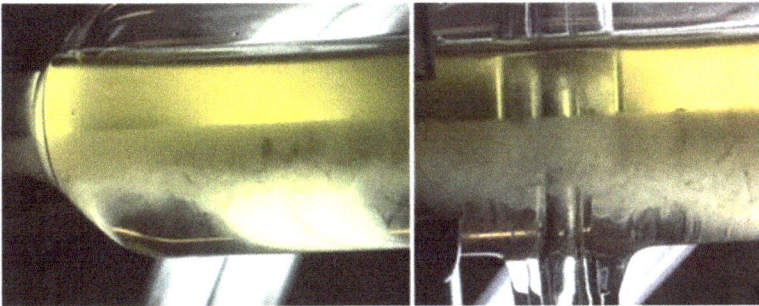

Figure 14. Side view of the laboratory settler (DN35, 175 mL hold-up. DN: diameter nominal) during continuous settling operation (300 s residence time). The image shows the inlet area (**left**) and the heavy phase outlet (**right**).

Also, the pronounced interphase, composed of cells and densely packed dispersed phase, meets the coalescence profile from the batch-settling experiment (Figure 8). The continuous settling experiments executed in the DN50 settler, except for S6 and S7 for the above-discussed reasons, show very narrow interphase bands (Figure 12, left), which is also in accordance with the batch-settling profile, considering that the residence times are beyond the 1340 s marking line.

Finally, a pilot-scale operation of the continuous settling process is discussed. Figure 15 shows the experimental set-up. Like in the lab-scale studies, phase-forming components, after passing through a static mixer pipe, continuously enter the apparatus. To ensure total sedimentation of the light phase, a residence time of 900 s was adjusted. Due to a small measurement inaccuracy of the flow meters, the PEG content in the system was lower (13 instead of 15.5 w%) and the percentage of phosphate rose from 15.5 to 18.7 w%. Thus, the operating point is closer to SP4 than to SP1. This results in phase inversion, which is also visible in Figure 15. In order to obtain a predominantly light phase for the filtration train, large parts of the light phase with the cells were withdrawn through the immersion tube during operation, which led to a product yield of 80.4%. For better comparability, the same process was carried out in batch-settling mode simultaneously. The comparisons in yield, cell reduction, and subsequent filterability are listed in Table 5.

Both operation modes clarified approximately 160 L total volume within the aimed-at time-frame of less than 3 h. No filter blockage occurred, however, due to the discussed shift in system composition, the product phase from the continuous settling process showed a higher yield loss. Also, the necessary filtration pressure was significantly higher.

Figure 15. Side view of the pilot-scale settler (DN150, 12.6 L hold-up) during continuous settling operation (900 s residence time).

Table 5. Comparison of batch and continuous ATPE of 160 L system volume. The pressure increase refers to 117 L/m^2 area-specific filtration volumes.

Mode	Yield IgG (%)	Log Cell Reduction (−)	Pressure Increase (bar)
Batch	95	1.29	0.1
Continuous	80.4	0.44	0.9

Compared to the settler dimensions, modeled only on the basis of batch-settling experiments (Figure 16, dashed lines indicate tested scenarios), the used DN150 settler (12.6 L hold-up), is very close to the calculated 10.9 L settler volume (H540), and in accordance to the thick dispersion layer significantly smaller than 40 L (H1340), which would be necessary for a narrow dispersion band.

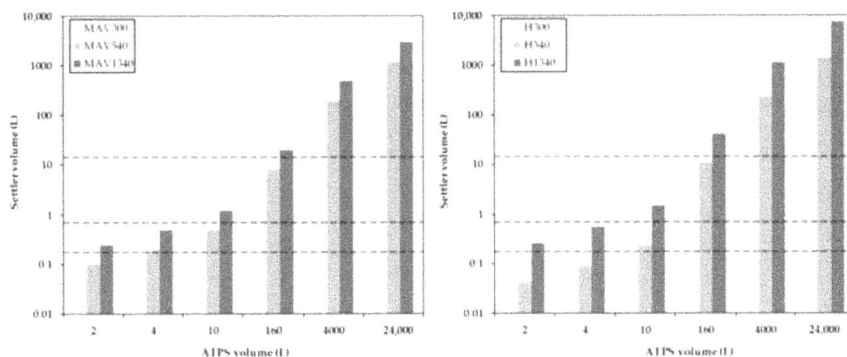

Figure 16. Settler volume as a function of residence time for different system scales obtained for the MAV (**left**) and simplified Henschke method (**right**). The dashed lines indicate the experimentally-tested scenarios (from top to bottom): 12.6 L, 0.75 L, and 0.175 L settler volumes for 160 L, 10 L, and 4 L ATPS volumes.

4. Materials and Methods

4.1. Aqueous Two-Phase Systems

In this work, four ATPS are considered. The components of the individual systems with their approximate phase splits are listed in Table 6.

Table 6. Composition of the investigated ATPS.

	SP1	SP2	SP3	SP4
PEG 400 (w%)	15.5	17	20	9
Buffer (w%)	40	43.75	45	53
Broth (w%)	44.5	39.75	35	38
Volume split LP/HP	1:1	1:1	1:1	1:4

4.2. Cultivation

The fed-batch cultivation of Chinese hamster ovary cells (cell line CHO DG44), used for mAb production, is carried out in a commercial serum-free medium. The cells were cultivated one week at 37 °C, 5% CO_2, and 130 rpm in shaking flasks.

4.3. Analytical Procedure

Protein A chromatography (PA ID Sensor Cartridge, Applied Biosystems, Bedford, MA, USA) was carried out for the determination of IgG concentrations. The buffers were PBS A (pH 7.4) for binding and PBS B (pH 2.6) for elution. For the analysis, a flow of 1.6 mL/min was applied. The injection mass for the calibration varied between 10 µg and 95 µg. Sample quantities of 50 µg are applied. The measured signals (280 nm) were evaluated via peak areas.

For size exclusion chromatography (SEC), a 1 M disodium hydrogen/sodium dihydrogen phosphate buffer containing 1 M sodium sulfate was used (Merck KGaA, Darmstadt, Germany). The column was a Yarra™ 3 µm SEC-3000 (Phenomenex Ltd., Aschaffenburg, Germany). The analysis was carried out at a flow rate of 0.35 mL/min and a duration of 23 min.

Cells are counted using a Motic BA 310 microscope (Motic Deutschland GmbH, Wetzlar, Germany). A sample is taken and dyed with trypan blue ((Sigma-Aldrich, St. Louis, MO, USA) in order to screen dead cells. The cells are applied to a Neubauer counting chamber (Brandt, 0.1 mm depth and 0.0025 mm^2).

Density (DMA 500, Anton Paar GmbH, Graz, Austria), pH (InoLab pH 720, WTW, Weilheim, Germany), and conductivity (InoLab pH 720, WTW, Weilheim, Germany) were measured for each ATPS. Dynamic viscosity (Rotary viscometer of the type HAAKE™ ViscoTester™ 550, ThermoFisher Scientific™, Waltham, MA, USA) and interfacial tension (Spinning Drop Tensiometer (SDT), Krüss GmbH, Hamburg, Germany) were determined for model calculations as well.

4.4. Procedures

4.4.1. Shaking Flask Experiments

The system was weighed in tubes and was sealed. PEG 400 is added first and afterwards the broth and buffer must be added quickly. The samples rest for at least 15 min. A sample, approx. 8–10 mL, was taken from the light phase and heavy phase and centrifuged in 15 mL tubes at 3000 rpm for 10 min to separate phase residues and solids from the product.

4.4.2. Batch-Settling Experiments

At the beginning of the experiment, the cell count in the cultivation flask was determined. The cell count was adjusted via dilution with a cell free broth or a concentration of the cells using a centrifuge. The predetermined stirrer speed of 300 rpm was set up with a tachometer in order to obtain a defined power input. Furthermore, 100 g of ATPS was weighed into the reactor at a cell number of around 200E5 cells/mL and then mixed for 5 min with a blade stirrer (4 blades, 45° blade angle, 29 mm diameter, 5 mm height). These parameters were sufficient for mixing the system. The settling height of the light phase and the heavy phase was recorded in 30 s intervals. From a settling time of 15 min, the values were recorded every minute. After separation of the light phase, heavy phase, and interphase, a sample of each phase was taken. The samples were centrifuged at 3000 rpm for 10 min to separate cell and phase residues.

4.4.3. Continuous-Settling Experiments

In order to prevent sedimentation of the cells in the feed vessel, the feed vessel was stirred on a magnetic stirring plate at about 290 rpm. The components were combined in two Y-pieces by three peristaltic pumps (Ismatec IP65, Cole-Parmer GmbH, Wertheim, Germany) according to mass-flow measurements (Mini Cori-Flow, Bronkhorst Deutschland Nord GmbH, Kamen, Germany) and mixed by static helix mixers (ESSKA.de GmbH, Hamburg, Germany). The mass-flow was recorded during the test by a Labbox (Labbox 3 M, HiTec Zang GmbH, Herzogenrath, Germany). At the outlet of the separator, the phases were collected separately from each other. During the experiment, samples were taken in order to detect changes in product yield or cell reduction over time.

5. Conclusions

The presented work displays development and implementation strategies for aqueous two-phase extraction as a cell harvest operation (Figure 17). After defining the operation space for the specific separation task in terms of cell reduction, yield, and process time by a limited number of shaking flask experiments to obtain equilibrium composition and material data by lab-scale batch and continuous settling tests, early predictions of necessary settler dimensions are possible.

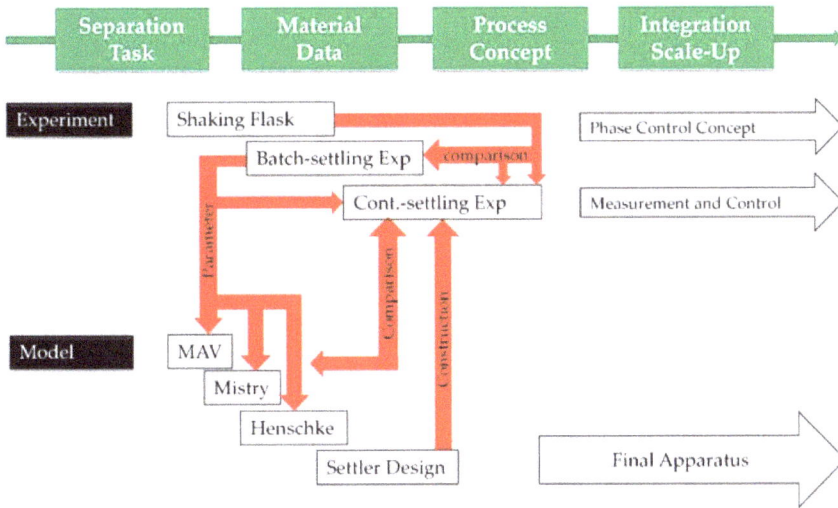

Figure 17. Process development strategy.

The batch-settling experiments in particular enable the identification of the separation times that are necessary to obtain the desired sedimentation profile inside the settler. Knowledge about the sedimentation rates can then be used for more sophisticated calculation of settler dimensions. For 3 h continuous processing of up to 12,000 L cultivation broth, a settler with a length of 3.6 m and a volume of approximately 1.4 m^3 is proposed (Figure 16, H540).

The investigated operating point is characterized by a yield of between 80 and close to 100%, as well as up to 2 log step reductions in cell number. For small system volumes (2–10 L), the calculated settler dimensions are consistent with laboratory tests as well as with pilot-scale volumes of up to 160 L cultivation broth. In order to achieve the above-described yield, the correct system composition (SP1) must be observed. In the case of deviations, this may lead to a reversed direction of phase inversion. Since the heavy phase must be dispersed for a reduction of cells, this is of upmost importance. The focus of further research is shown in an illustrated concept (Figure 17). Further work packages should optimize process integration (i.e., automation, control, and control strategy) as well as the robustness of the process. The applicability for alternative ATPS, i.e., polymer-citrate, as well as for other products, i.e., virus-like particles or different proteins, should also be investigated to evaluate the robustness of the process and the reliability of the development concept.

Acknowledgments: The authors thank the ITVP lab team, especially Frank Steinhäuser and Volker Strohmeyer, for their effort and support. Special thanks are also addressed to Niclas Genthe for excellent laboratory work. This work was kindly supported and funded by Boehringer Ingelheim Global Technology Management.

Author Contributions: Axel Schmidt conceived and designed the experiment as well as wrote the paper. Axel Schmidt performed the described experiments. Michael Richter, Frederik Rudulph, and Jochen Strube substantively revised the work and contributed the materials. Jochen Strube is responsible for conception and supervision.

Conflicts of Interest: The authors declare no conflict of interest.

References

1. Singh, N.; Arunkumar, A.; Chollangi, S.; Tan, Z.G.; Borys, M.; Li, Z.J. Clarification technologies for monoclonal antibody manufacturing processes: Current state and future perspectives. *Biotechnol. Bioeng.* **2016**, *113*, 698–716. [CrossRef] [PubMed]

2.   Luechau, F.; Ling, T.C.; Lyddiatt, A. A descriptive model and methods for up-scaled process routes for interfacial partition of bioparticles in aqueous two-phase systems. *Biochem. Eng. J.* **2010**, *50*, 122–130. [CrossRef]

3.   Albertsson, P.-Å. Particle fractionation in liquid two-phase systems; the composition of some phase systems and the behaviour of some model particles in them; application to the isolation of cell walls from microorganisms. *Biochim. Biophys. Acta* **1958**, *27*, 378–395. [CrossRef]

4.   Albertsson, P.-Å. *Partition of Cell Particles and Macromolecules: Separation and Purification of Biomolecules, Cell Organelles, Membranes, and Cells in Aqueous Polymer Two-Phase Systems and Their Use in Biochemical Analysis and Biotechnology*, 3rd ed.; Wiley: New York, NY, USA, 1986.

5.   Jauregi, P.; Hoeben, M.A.; van der Lans, R.G.J.M.; Kwant, G.; van der Wielen, L.A.M. Recovery of small bioparticles by interfacial partitioning. *Biotechnol. Bioeng.* **2002**, *78*, 355–364. [CrossRef] [PubMed]

6.   Rito-Palomares, M.; Lyddiatt, A. Practical implementation of aqueous two-phase processes for protein recovery from yeast. *J. Chem. Technol. Biotechnol.* **2000**, *75*, 632–638. [CrossRef]

7.   Dennison, C.; Lovrien, R. Three phase partitioning: Concentration and purification of proteins. *Protein Expr. Purif.* **1997**, *11*, 149–161. [CrossRef] [PubMed]

8.   Lee, S.Y.; Khoiroh, I.; Ling, T.C.; Show, P.L. Aqueous Two-Phase Flotation for the Recovery of Biomolecules. *Sep. Purif. Rev.* **2015**, *45*, 81–92. [CrossRef]

9.   Sommerfeld, S.; Strube, J. Challenges in biotechnology production—Generic processes and process optimization for monoclonal antibodies. *Chem. Eng. Process.* **2005**, *44*, 1123–1137. [CrossRef]

10.   Azevedo, A.M.; Rosa, P.A.J.; Ferreira, I.F.; Aires-Barros, M.R. Chromatography-free recovery of biopharmaceuticals through aqueous two-phase processing. *Trends Biotechnol.* **2009**, *27*, 240–247. [CrossRef] [PubMed]

11.   Zobel, S.; Helling, C.; Ditz, R.; Strube, J. Design and Operation of Continuous Countercurrent Chromatography in Biotechnological Production. *Ind. Eng. Chem. Res.* **2014**, *53*, 9169–9185. [CrossRef]

12.   Strube, J.; Grote, F.; Josch, J.P.; Ditz, R. Process Development and Design of Downstream Processes. *Chem. Ing. Tech.* **2011**, *83*, 1044–1065. [CrossRef]

13.   Wiesel, A.; Schmidt-Traub, H.; Lenz, J.; Strube, J. Modelling gradient elution of bioactive multicomponent systems in non-linear ion-exchange chromatography. *J. Chromatogr. A* **2003**, *1006*, 101–120. [CrossRef]

14.   Schulte, M.; Britsch, L.; Strube, J. Continuous preparative liquid chromatography in the downstrem processing of biotechnological products. *Acta Biotechnol.* **2000**, *20*, 3–15. [CrossRef]

15.   Gronemeyer, P.; Ditz, R.; Strube, J. Trends in upstream and downstream process development for antibody manufacturing. *Bioengineering* **2014**, *1*, 188–212. [CrossRef] [PubMed]

16.   Nam, K.-H.; Chang, W.-J.; Hong, H.; Lim, S.-M.; Kim, D.-I.; Koo, Y.-M. Continuous-flow fractionation of animal cells in microfluidic device using aqueous two-phase extraction. *Biomed. Microdevices* **2005**, *7*, 189–195. [CrossRef] [PubMed]

17.   Pinto, R.C.V.; Medronho, R.A.; Castilho, L.R. Separation of CHO cells using hydrocyclones. *Cytotechnology* **2008**, *56*, 57–67. [CrossRef] [PubMed]

18.   Andrews, B.A.; Asenjo, J.A. Protein partitioning equilibrium between the aqueous poly(ethylene glycol) and salt phases and the solid protein phase in poly(ethylene glycol)-salt two-phase systems. *J. Chromatogr. B Biomed. Sci. Appl.* **1996**, *685*, 15–20. [CrossRef]

19.   Rosa, P.A.J.; Azevedo, A.M.; Sommerfeld, S.; Mutter, M.; Aires-Barros, M.R.; Bäcker, W. Application of aqueous two-phase systems to antibody purification: A multi-stage approach. *J. Biotechnol.* **2009**, *139*, 306–313. [CrossRef] [PubMed]

20.   Andrews, B.A.; Asenjo, J.A. Theoretical and Experimental Evaluation of Hydrophobicity of Proteins to Predict their Partitioning Behavior in Aqueous Two Phase Systems: A Review. *Sep. Sci. Technol.* **2010**, *45*, 2165–2170. [CrossRef]

21.   Asenjo, J.A.; Andrews, B.A. Aqueous two-phase systems for protein separation: A perspective. *J. Chromatogr. A* **2011**, *1218*, 8826–8835. [CrossRef] [PubMed]

22.   Polson, A.; Potgieter, G.M.; Largier, J.F.; Mears, G.E.F.; Joubert, F.J. The fractionation of protein mixtures by linear polymers of high molecular weight. *Biochim. Biophys. Acta* **1964**, *82*, 463–475. [CrossRef]

23.   Eggersgluess, J.; Wellsandt, T.; Strube, J. Integration of aqueous two-phase extraction into downstream processing. *Chem. Eng. Technol.* **2014**, *37*, 1686–1696. [CrossRef]

24. Grilo, A.L.; Raquel Aires-Barros, M.; Azevedo, A.M. Partitioning in Aqueous Two-Phase Systems: Fundamentals, Applications and Trends. *Sep. Purif. Rev.* **2015**, *45*, 68–80. [CrossRef]

25. Iqbal, M.; Tao, Y.; Xie, S.; Zhu, Y.; Chen, D.; Wang, X.; Huang, L.; Peng, D.; Sattar, A.; Shabbir, M.A.B.; et al. Aqueous two-phase system (ATPS): An overview and advances in its applications. *Biol. Proced. Online* **2016**, *18*, 18. [CrossRef] [PubMed]

26. De Barros, D.P.C.; Campos, S.R.R.; Azevedo, A.M.; Baptista, A.M.; Aires-Barros, M.R. Predicting protein partition coefficients in aqueous two phase system. *J. Chromatogr. A* **2016**, 50–58. [CrossRef] [PubMed]

27. Raja, S.; Murty, V.R.; Thivaharan, V.; Rajasekar, V.; Ramesh, V. Aqueous Two Phase Systems for the Recovery of Biomolecules—A Review. *SCIT* **2011**, *1*, 7–16. [CrossRef]

28. Rito-Palomares, M. Practical application of aqueous two-phase partition to process development for the recovery of biological products. *J. Chromatogr. B Anal. Technol. Biomed. Life Sci.* **2004**, *807*, 3–11. [CrossRef] [PubMed]

29. Gronemeyer, P.; Ditz, R.; Strube, J. DoE based integration approach of upstream and downstream processing regarding HCP and ATPE as harvest operation. *Biochem. Eng. J.* **2016**, 158–166. [CrossRef]

30. Espitia-Saloma, E.; Vázquez-Villegas, P.; Aguilar, O.; Rito-Palomares, M. Continuous aqueous two-phase systems devices for the recovery of biological products. *Food Bioprod. Process.* **2014**, *92*, 101–112. [CrossRef]

31. Salamanca, M.H.; Merchuk, J.C.; Andrews, B.A.; Asenjo, J.A. On the kinetics of phase separation in aqueous two-phase systems. *J. Chromatogr. B Biomed. Sci. Appl.* **1998**, *711*, 319–329. [CrossRef]

32. Albertsson, P.-Å.; Tjerneld, F. [1] Phase diagrams. *Methods Enzymol.* **1994**, *228*, 3–13.

33. Gu, T. Liquid-liquid partitioning methods for bioseparations. *Sep. Sci. Technol.* **2000**, *2*, 329–364.

34. Kula, M.-R. Extraction and Purification of Enzymes Using Aqueous Two-Phase Systems. *Appl. Biochem. Bioeng. Enzyme Technol.* **1979**. [CrossRef]

35. Tidhar, M.; Merchuk, J.C.; Sembira, A.N.; Wolf, D. Characteristics of a motionless mixer for dispersion of immiscible fluids—II. Phase inversion of liquid-liquid systems. *Chem. Eng. Sci.* **1986**, *41*, 457–462. [CrossRef]

36. Merchuk, J.C.; Andrews, B.A.; Asenjo, J.A. Aqueous two-phase systems for protein separation. *J. Chromatogr. B Biomed. Sci. Appl.* **1998**, *711*, 285–293. [CrossRef]

37. Mistry, S.L.; Kaul, A.; Merchuk, J.C.; Asenjo, J.A. Mathematical modelling and computer simulation of aqueous two-phase continuous protein extraction. *J. Chromatogr. A* **1996**, *741*, 151–163. [CrossRef]

38. Henschke, M. Dimensionierung Liegender Flüssig-Flüssig-Abscheider Anhand Diskontinuierlicher Absetzversuche/Martin Henschke. Ph.D. Thesis, RWTH Aachen, Aachen, Germany, 1994.

antibodies

MDPI

Article

Process Analytical Approach towards Quality Controlled Process Automation for the Downstream of Protein Mixtures by Inline Concentration Measurements Based on Ultraviolet/Visible Light (UV/VIS) Spectral Analysis

Steffen Zobel-Roos [1], Mourad Mouellef [1], Christian Siemers [2] and Jochen Strube [1,]*

[1] Institute for Separation and Process Technology, Clausthal University of Technology, Leibnizstraße 15, 38678 Clausthal-Zellerfeld, Germany; zobel-roos@itv.tu-clausthal.de (S.Z.-R.); mourad.mouellef@tu-clausthal.de (M.M.)

[2] Institute for Process Control, Clausthal University of Technology, Arnold-Sommerfeld-Straße 1, 38678 Clausthal-Zellerfeld, Germany; Christian.siemers@tu-clausthal.de

[*] Correspondence: strube@itv.tu-clausthal.de; Tel.: +49-5323-72-2355

Received: 31 October 2017; Accepted: 6 December 2017; Published: 12 December 2017

Abstract: Downstream of pharmaceutical proteins, such as monoclonal antibodies, is mainly done by chromatography, where concentration determination of coeluting components presents a major problem. Inline concentration measurements (ICM) by Ultraviolet/Visible light (UV/VIS)-spectral data analysis provide a label-free and noninvasive approach to significantly speed up the analysis and process time. Here, two different approaches are presented. For a test mixture of three proteins, a fast and easily calibrated method based on the non-negative least-squares algorithm is shown, which reduces the calibration effort compared to a partial least-squares approach. The accuracy of ICM for analytical separations of three proteins on an ion exchange column is over 99%, compared to less than 85% for classical peak area evaluation. The power of the partial least squares algorithm (PLS) is shown by measuring the concentrations of Immunoglobulin G (IgG) monomer and dimer under a worst-case scenario of completely overlapping peaks. Here, the faster SIMPLS algorithm is used in comparison to the nonlinear iterative partial least squares (NIPALS) algorithm. Both approaches provide concentrations as well as purities in real-time, enabling live-pooling decisions based on product quality. This is one important step towards advanced process automation of chromatographic processes. Analysis time is less than 100 ms and only one program is used for all the necessary communications and calculations.

Keywords: inline concentration measurements; UV/VIS spectral analysis; PAT for monoclonal antibodies; live pooling; peak deconvolution

---

1. Introduction

Inline concentration measurements (ICM) of individual components in a mixture are critical for almost every unit operation in the field of chemical and biotechnological processes. Although most processes are designed to match specific concentration and purity criteria, the actual values still have to be monitored inline or offline [1].

This becomes even more important for batch operations, like chromatography, where concentration and purity vary over time. For such processes, the inline measurement of concentrations and purities become a game winning objective, but often also a major challenge.

In the field of analytical chromatography, concentration or purity quantifications are usually based on peak areas. Therefore, great effort is put into the chromatographic separation to achieve a baseline separation, if possible. Nevertheless, this is not always achieved. Very closely related components are often especially difficult to separate. The deconvolution of overlapping peaks allows for better results as well as shorter and less expensive separations.

For preparative chromatography, complete baseline separation is not desirable; on the contrary, it conflicts with process efficiency and economy. The separation process is optimized for a maximum resin utilization and productivity. The product fractionation is often controlled via timed cut points, where the exact moment for the fraction start and end points are derived from earlier experiments and are therefore not directly related to the current chromatographic run. In this case, critical product attributes like the concentration and purity have to be measured offline after the run. Anomalies in elution behavior often result in shifts of the chromatogram leading to large variations from the expected design chromatogram. Thus, the time-based cut points do not match the purity criteria and might lead to a batch failure or require reprocessing. This problem becomes even worse for continuous chromatography processes. Start- and end-pooling criteria based on real-time, online detection of volume and Ultraviolet/Visible light (UV/VIS) absorbance have proven to be very successful in delivering consistent yield and purity. Online concentration and purity identification as presented here will help to shift the process design from timed to data-based fractionation or column switching and therefore give a greater certainty to match purity and yield targets, thus reducing the risk of batch failure. Also, the safety margins for data based fractionation can be smaller, leading to higher yield and productivity.

Furthermore, the knowledge of concentrations of all components online throughout the chromatographic run will help during process design. At the moment, the chromatographic run has to be fractionated into a lot of small volumes and analyzed offline to identify the actual concentration profiles. This is cost intensive as well as time consuming and, in addition, carries a significant risk of product degeneration of sensitive proteins because of long analysis times. The same problem and solution respectively apply for model parameter determination and model validation for a simulation-based design process [2,3].

Hence, a lot of research has been undertaken to achieve a deconvolution of chromatographic peaks. In general, the approaches can be split into inline and offline methods.

For offline or at-line peak deconvolution in particular two different approaches can be found. A classical at-line method is the division of the complete run into small fractions. These fractions are then investigated by a suitable analytical method afflicted with the method specific failure [4–8]. Although analysis time might be rather short, online sampling always results in a rather large response time. Especially the use of gas or liquid chromatographic analysis often has to deal with overlapping peaks itself due to selectivity variances [9].

Another offline or delayed online approach is a full mathematical analysis of the chromatogram. This analysis assumes that every peak will follow a mathematical function, e.g., Gaussian or modified Gaussian, and estimates the function parameters. This assumes a Gaussian or modified Gaussian shape might be approximately right for analytical chromatography, but is rather uncommon for preparative separations. The identification of peaks occurs by first, second or higher order derivatives of the chromatographic signal [10–12].

To identify a component and determine the parameters for its Gaussian function, the peak must be processed to a certain point. In most cases, at least the first inflection point has to be reached. More often, the peak maximum and second inflection point have to be detected already. Hence, this peak deconvolution method cannot be performed in real-time.

For preparative separations, an improvement can be achieved by the use of process modelling. Having a valid process model implemented during process design, the real behavior and elution of the components can be monitored [13–15].

Spectroscopy is among the most common methods of detection. Hence, major progress was achieved in terms of a multicomponent analysis using ultraviolet (UV), visible (VIS) or infrared spectral data. Mid-infrared spectroscopy (MIR) was used for host-cell protein quantification [16],

UV/VIS spectroscopy for determining the protein and nucleic acid content of viruses [17] as well as for a variety of proteins in a multicomponent mixture [18–20].

According to the Beer–Lambert law, the UV-extinction E_λ of a component at a given wavelength λ is the product of the concentration c, the path length d and a component specific coefficient called extinction coefficient ε_λ as shown in Equation (1) [21,22]:

$$lg\left(\frac{I_{0,\lambda}}{I_{t,\lambda}}\right) = E_{\lambda,i} = \varepsilon_{\lambda,i}\cdot c_i\cdot d. \tag{1}$$

Technically, this law applies to highly diluted mixtures only. Nevertheless, the deviations are often negligible.

The UV/VIS extinction over the wavelengths, viz. the UV/VIS spectrum, is unique for almost every component. Thus, the sum spectrum of a mixture can be disassembled into the single component spectra. It can be found that the absorbance of a mixture of n components sums up from the single component extinctions according to Equation (2) [23,24]:

$$E_\lambda = \sum_{i=1}^{n} \varepsilon_{\lambda,i}\cdot c_i\cdot d. \tag{2}$$

A diode array-based UV/VIS measurement provides as many extinction values as there are diodes in the detector, often 256 or 1024. Each diode represents one specific wavelength sector. Hence, Equation (2) can be formulated for each diode leading to a large set of linear expressions. With prior knowledge of the extinction coefficients of each component $\varepsilon_{\lambda,i}$, this set of linear equations can be solved for the concentrations c_i. This can be done with several mathematical methods like the least-squares or non-negative least-squares (NNLS) [25,26] algorithm. In theory, the experimental effort to measure the extinction coefficients $\varepsilon_{\lambda,i}$, is low. For each single component one injection with known concentration is sufficient. There is no need to calibrate a large set of mixtures with different compositions and concentrations. However, it might be necessary to measure the extinction coefficients $\varepsilon_{\lambda,i}$ for different concentrations of the single components. It can be shown that for higher concentrations, peaks appear in the UV/VIS spectrum that are not visible at lower concentrations. This is due to detector sensitivity only.

The first example in this paper shows the application of the NNLS-algorithm by measuring the concentrations of three proteins from an analytical ion exchange chromatography.

Another approach for UV/VIS-diode array detector (DAD) based concentration measurements was introduced by Brestrich et al. [18,27,28]. Partial least-squares regression is used to create a statistical model. The PLS regression compresses a set of even highly collinear predictor data X into a set of latent variables T. With these orthogonal latent variables observations can be fitted to depended variables Y. In this case, X is the UV/VIS spectra and Y represent the concentrations. For a better understanding of PLS, see [29–31].

In both the work of Brestrich et al. [27] and this work, the SIMPLS-algorithm [30] is used. This does not solve the set of linear equations spanned by Equation (2), but creates a statistic model. This model has to be trained by a set of experiments. Contrary to the non-negative least-squares approach, it is not sufficient to only calibrate for single components. The experimental design has to account for different compositions and different concentrations of each component present in the mixture that should be analyzed later. Thus, the amount of experiments increases dramatically with the number of components. Again, the detector sensitivity has a major impact, as does the detector type, age and utilization status.

The second example in this paper shows the application of the SIMPLS-algorithm for inline and real-time monomer and dimer concentration measurements of monoclonal antibody IgG.

Although the used algorithm is the same, this work differs from other work by a simpler setup. Instead of different programs for each task, only one self-written program is used to do the data acquisition, all calculations, data storage and communication with the programmable logic controller (PLC). Therefore, there is no software bottleneck allowing for very fast measurements. The PLC controls the pumps and valves of a continuous chromatography prototype that was not used in this work. This work is rather a major milestone to achieving a fully automated and self-optimizing system for the prototype, enabling life pooling, purity-based column switching and advanced quality control.

2. Materials and Methods

2.1. Model Proteins, Buffers and Columns

All experiments with proteins were carried out in 20 mM NaPi buffer at pH 6.0. For ion exchange chromatographic separations, this buffer was used as equilibration buffer A. For elution buffer B, 1 M NaCl was added. For analytical size exclusion chromatography (SEC), a buffer containing 100 mM sodium sulfate and 100 mM NaPi was used at pH 6.6. All salts were obtained from Merck KGaA, Darmstadt, Germany.

For the three component mixture experiments Chymotrypsinogen A, Lysozyme (AppliChem GmbH, Darmstadt, Germany) and Cytochrome C (Merck KGaA, Darmstadt, Germany) were used.

IgG was obtained from our own cell culture with an industrial Chinese Hamster Ovary (CHO) cell line and purified by protein A chromatography prior to the experiments.

Ion exchange separations were performed on prepacked strong cation exchange columns with Fractogel® EMD SO$_3^-$ (M) (5–50, 1 mL, Atoll GmbH, Weingarten, Germany).

Analytical size exclusion columns Yarra® SEC-3000 (3 μm, 300 × 4.6 mm) were obtained from Phenomenex® Inc., Torrance, CA, USA.

Protein A chromatography was performed with PA ID Poros® Protein A Sensor Cartridges (Applied Biosystems, Waltham, MA, USA).

2.2. Devices and Instruments

The experimental setup consisted of a standard VWR-Hitachi LaChrom Elite® HPLC system with a quaternary gradient pump L-2130, Autosampler L-2200 and diode array detector L-2455 (VWR International, Radnor, PA, USA). The later was not used for the ICM but for comparative measurements. For the ICM measurements, a Smartline DAD 2600 with 10 mm, 10 μL flow cell from Knauer Wissenschaftliche Geräte GmbH, Berlin, Germany was used. Experimental validations were based on SEC analysis after online fractionation with a Foxy Jr.® from Teledyn Isco, Lincoln, NE, USA.

2.3. Inline Concentration Measurements

The inline concentration measurements are based on UV/VIS spectra measured with a diode array detector (Smartline DAD 2600 from Knauer Wissenschaftliche Geräte GmbH, Berlin, Germany) with 256 diodes and a wavelength range of 190 to 510 nm. The selectable bandwidth is 4 to 25 nm. The highest sampling rate is 10 Hz, which is 100 ms [32]. Data collection and analyses were performed by a conventional Windows desktop computer, which uses a standard EIA RS-232 serial port for the communication with the detector. Since the communication between different applications and programs tends to be a major bottleneck, a self-written program was used. An object-oriented, concurrent and class-based programming language [33] was used (Java). Java runs on its own Java Virtual Machine (JVM), which is available for a variety of platforms and computer architectures. Hence, programming in Java follows the "write once, run anywhere" idea. This makes a program written in Java easily applicable on different machines [33]. The ICM application handles the communication between the DAD and the computer, processes the data, displays the results in different charts and stores the data in "comma separated values" (*.csv) files. The user can choose from several algorithms to implement. Amongst those are the least squares, non-negative least squares [25,26] and SIMPLS (a partial least squares variant) [30], the simplex and powell algorithm and some more. It should be noted that these are not only different algorithm, but different approaches. The non-negative least squares algorithm is used to solve the equation system described earlier (Equation (2)). However, PLS provides a statistical model that is related to the spectral data. Which algorithm should be used depends on the complexity of the given sample. As a starting point, the single-component spectra of each component should be compared. In this work, an application for the non-negative least squares and one for the SIMPLS is shown.

2.4. ICM Examples

2.4.1. Protein Test Mixture on Ion Exchange Column under Analytical Conditions with NNLS

To show the power of ICM for analytical separations, a separation of three proteins on an ion exchange column was performed. Fractogel® EMD SO$_3{}^-$ (M) was used in prepacked 1-mL columns (5–50, Atoll GmbH, Weingarten, Germany).

The binding buffer A was 20 mM NaPi buffer with a pH of 6.0. For elution, buffer B 1 M NaCl was added. The chromatographic run started with 100% buffer A and proceeded with a 10 CV gradient to 100% buffer B. A flow rate of 1 mL/min was used. This method is meant not to achieve baseline separation, but produce widely overlapping peaks.

The test mixture contained Chymotrypsinogen A, Cytochrome C and Lysozyme. 10 mg of each protein were dissolved separately in 1-mL binding buffer, leading to three samples with a concentration of 10 g/L each. To identify the differences in UV/VIS spectra, 20 µL of each sample was injected onto the column and measured with the DAD detector at a certain peak height corresponding to 0.01 g/L.

The single component spectra are shown in Figure 1. It can be seen that Chymotrypsinogene A and Lysozyme have similar spectra. Cytochrome C, however, is relatively unique. Nevertheless, the differences are big enough not to need PLS.

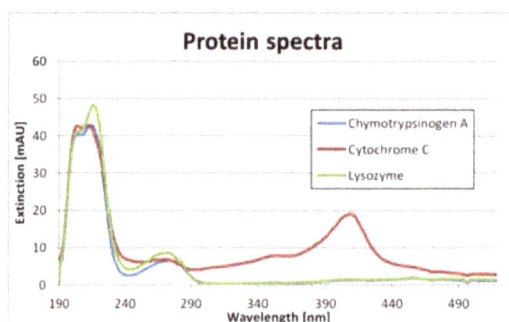

Figure 1. Single-component spectra of Chymotrypsinogene A, Cytochrome C and Lysozyme.

The inline concentration measurements were done by solving the set of linear equations (Equation (2)) for the concentrations with the non-negative least-squares algorithm. Hence, the extinction coefficients $\varepsilon_{\lambda,i}$ for the three proteins were needed.

The extinction coefficients $\varepsilon_{\lambda,i}$ were measured separately for each component. The proteins were dissolved in buffer A separately and injected onto the ion exchange column as described previously. The resulting chromatogram at 280 nm wavelength was converted from extinction to concentration course according to Equation (3):

$$c_i(t) = F \cdot E_{\lambda=280}(t).$$ (3)

The conversion factor F was obtained with Equation (4):

$$F = \frac{m_{i,inj}}{\dot{V} \cdot \int_0^t E_\lambda(t)dt}.$$ (4)

Now, the concentration is known for every time point of the chromatogram. Thus, the extinction coefficients $\varepsilon_{\lambda,i}$, are the only unknown variable left in Equation (2). In this work, Equation (2) was solved for the coefficients at a time point corresponding to 0.1 g/L protein concentration. This calibration has to be done for each protein.

One problem of this approach is the superposition of the protein spectrum with any other spectrum of co eluting components. This might result from UV/VIS active compounds in the chromatographic buffer or from leachables/extractables of the column itself. To overcome this problem, a blank run can be done before the actual measurements to perform a baseline correction.

For some components, one might find that different concentrations lead to different spectra. Physically speaking, this is not possible. However, for rising concentrations, peaks in the spectrum might appear that were not present at lower concentrations. These peaks were simply not visible at lower concentrations due to low detector sensitivity. In this case, it is best to analyze the peak at different concentrations, hence obtaining extinction coefficient values at different concentrations. This should be equal except for these cases where detector sensitivity comes into account. For those coefficients, a spline interpolation over the concentration is done.

The actual separation for testing the peak deconvolution was performed with the same column under the same conditions. The test mixture contained all three proteins.

Quantitative analysis is usually performed based on peak areas. The peak area is a function of the amount of protein injected. To produce the exact same peak areas in the three-component run compared to the single component runs, the same amount of each protein must be injected. Thus, the sample was produced by mixing 0.5 mL of each of the protein solutions used previously. Therefore, this sample has exactly one third of the concentration of each protein compared to the single component samples. Hence, 60 μL were injected to the column.

The peak areas of the single component runs are easily calculated. The corresponding areas from the mixture experiment were obtained by three different methods. First by integrating the concentration curve obtained with the inline concentration measurements. Second, by the perpendicular method. As a typical chromatographic approach, the extinction curve is integrated. The areas between two peaks were divided by a vertical line at the local minima. These areas have to be compared to a calibration curve. The third method is a typical offline peak deconvolution method. The real peak shape is approximated by assuming Gaussian or modified Gaussian behavior. These peaks can then again be integrated and compared to a calibration curve.

2.4.2. Immunoglobulin G (IgG) Monomer and Dime with SIMPLS

A mixture of IgG monomer and dimer was used to show the potential for pharmaceutical production. Both molecules have almost identical UV/VIS spectra. To distinguish between both is therefore relatively hard. Nevertheless, from a product quality point of view, it is very important to know the concentrations of IgG oligomers.

Since the differences in UV/VIS spectra are low, the SIMPLS algorithm was used. This must be calibrated with a training set of data containing different IgG monomer and dimer concentrations.

The monoclonal antibody IgG came from our own fed batch fermentation of an industrial cell culture line. The cell culture was clarified with centrifugation (3000 g) and filtration (0.2 μm syringe filter, VWR International, Radnor, PA, USA). Afterwards, purification was done with protein A chromatography (PA ID Sensor Cartridges, Applied Biosystems, Waltham, MA, USA). The protein A product peak was then loaded to a size exclusion column (Yarra® SEC-3000, 3 μm, 300 × 4.6 mm, Phenomenex Inc., Torrance, CA, USA). Here, only the upper 50% of the dimer and the monomer peak were fractionated. Hence, the overlapping region of monomer and dimer was not used. After protein A and size exclusion chromatography, no other protein besides IgG and no other contaminates are present. We did not find a fixed equilibrium between the monomer and dimer concentration. However, conversions from one form into the other occur, so that there was no pure dimer or monomer. The concentration in the SEC monomer fraction after some conversion time (overnight) was roughly

0.5 mg/mL monomer and 0.02 mg/mL dimer. The dimer fraction contained 0.2 mg/mL monomer and 0.1 mg/mL dimer.

Out of these two fractions, 11 mixtures were created to calibrate the statistic model. Therefore, different volumetric mix ratios were prepared, starting with 100 vol.% monomer and 0 vol.% dimer. The ratios decreased respectively increased with a 10 vol.% step size to 0 vol.% monomer and 100 vol.% dimer. From eleven mixtures, nine were used as a training set and the other three as validation sets, namely 80 vol.% monomer, 50 vol.% monomer and 20 vol.% monomer.

For each sample, analytical size exclusion chromatography was performed again to get the actual monomer and dimer concentrations.

The experiments itself were performed by injecting 0.1 mL into a small stirred tank of 2 mL total volume. The inlet stream consisted of chromatographic buffer A with a volumetric flow of 1 mL/min. The outlet stream, also 1 mL/min, was directly connected to the Smartline DAD 2600. After the DAD, the stream was fractionated with a Foxy Jr.® (Teledyn Isco, Lincoln, NE, USA) fraction collection module. 0.5 min fractions were taken and this 0.5 mL fractions were analyzed offline with the size exclusion chromatography already mentioned.

This setup represents a worst-case scenario, since no column or separation takes place. With moderately overlapping peaks, one might enhance your result by applying some mathematical assumptions; this cannot be done here. Furthermore, this experiment simulates conditions one would find in several other, non-chromatographic unit operations within the downstream of monoclonal antibodies, like filtration. This shows the applicability for inline quality control for example in the last filtration step.

3. Results and Discussion

The presented inline concentration measurement method shows wide, almost general applicability whenever UV/VIS active component mixtures are involved. In this work, the ability to enhance the accuracy of analytical measurements, and to make real time pooling decisions based on real time data could be shown. For the latter case, the calculation time is of major interest. Since only one program is used for data collection and processing, no time delay could be found throughout all experiments. The UV/VIS diode array detector was run with a sampling rate of 100 ms. The data transfer and all calculations were finished prior to the measurement of the next data point. Thus, one ICM concentration measurement lasts less than 100 ms.

It should be noted that every method or calibration comes with a systematic error itself. In this case, impurities present in the assumed to be pure calibration mixture could not be detected later on. This applies for example for low level impurities in the protein standard or in the IgG fractions after size exclusion chromatography.

3.1. Protein Test Mixture on Ion Exchange Column under Analytical Conditions with NNLS

One chromatogram and the corresponding inline concentration measurements are displayed in Figure 2. The reference extinction is 280 nm. It is important to understand that the protein concentrations (light blue, red, green) do not have to match the reference extinction plot (dark blue). The proportions would match only if all three proteins had the same extinction coefficient at 280 nm which is obviously not the case.

Prior to the injection of the three component mixture, the exact same amount of each component was injected separately as described earlier. The comparisons between the single component injections with the ICM peak of the same component in the three component run are shown in Figure 3. It can be seen that the Chymotrypsinogene-A peak in the mixture is pushed a little bit to an earlier elution. The mean residence time shifts from 6.07 min to 5.96 min. Although, the dilution front of this peak is sharper. This is in good agreement with the displacement theory of chromatography. The Cytochrome-C peak has mainly the same shape in both experiments, but tends to co-elute with both neighboring components slightly. The Lysozyme peak is in both cases more or less the same.

For both later proteins, there are only minor changes in the residence time. All values are displayed in Table 1.

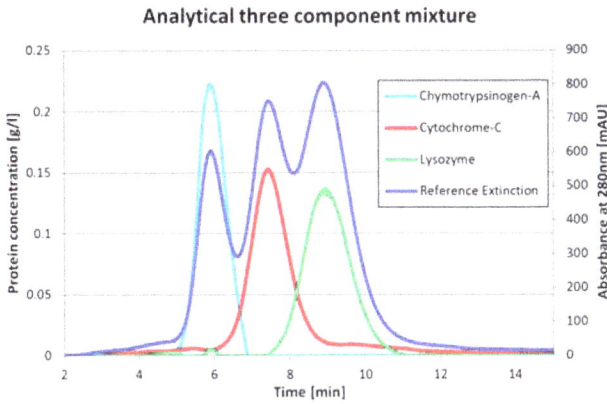

Figure 2. Inline concentration measurement of a three component protein mixture.

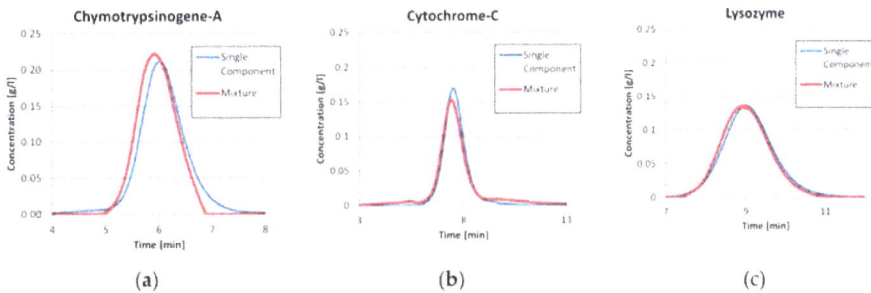

Figure 3. Comparison between the single component injection (blue line) and the deconvoluted peak (red line) of the same component in the mixture for the three proteins Chymotrypsinogene-A (a); Cytochrome-C (b) and Lysozyme (c).

Table 1. Deviation from single component to mixture measurement.

Protein	Mean Residence Time	
	Single Component (min)	Mixture (min)
Chymotrypsinogene-A	6.07	5.96
Cytochrome-C	7.62	7.63
Lysozyme	9.03	9.01

The chromatograms in Figure 3 allow for an easy comparison of the total amount of protein in both experiments. Since the ICM method measures concentrations not extinction, the deviations for this method are calculated comparing the total protein amount measured for single component injections to the three component system. This is calculated from feed concentration and injection volume. The mean deviation for the ICM method used for the chromatogram in Figure 2 is 0.02% with a standard deviation of 0.58% compared to $-16.99\% \pm 7.01\%$ for the perpendicular, $17.58\% \pm 15.82\%$ for the Gaussian and $21.27\% \pm 15.54\%$ for the modified Gaussian approach. The results for each component are listed in Table 2.

Table 2. Deviation from single component to mixture measurement.

Protein	ICM	Perpendicular	Gaussian	Modified Gaussian
	(%)	(%)	(%)	(%)
Chymotrypsinogen-A	−0.63	−7.09	9.92	12.71
Cytochrom-C	0.50	−22.37	3.21	8.02
Lysozyme	0.19	−21.52	39.62	43.09
Mean Value	0.02	−16.99	17.58	21.27
Standard deviation	0.48	7.01	15.82	15.54

It should be noted that the ICM measurements might be sensitive to the gradient. Some gradients, more specific some modifier, are UV/VIS active itself. Even if the gradient does not become visible in the chromatogram at a specific wavelength, there might be an effect to the measurement. When the calibration is done with chromatographic runs of the single components as described, most of this problem is already solved since the gradient is already present in the calibration measurements. Otherwise, a blank run should be made before the measurements. The UV data measured during the actual run are than corrected by subtracting the blank run UV/VIS data.

Figure 4 shows the course of the different peak deconvolution methods. The peak in the middle is left out for clarity. It can be seen that the perpendicular area determination is always neglecting some area. On the other hand, both Gaussian functions (grey and light grey) overestimate the real behavior. It can be seen that the mathematical fits show the first and third peak to be overlapping, which is not the case in reality. The ICM method represents the real behavior best. Because it is not based on some model assumptions, the ICM method further has the potential to also identify tag along effects.

Comparison of different deconvolution methods

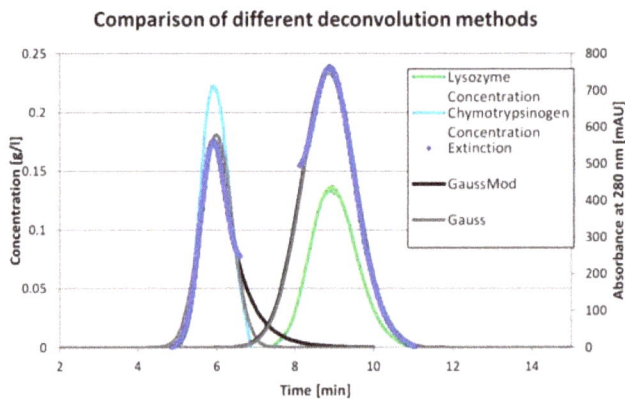

Figure 4. Comparison of the different peak deconvolution methods.

The mathematical methods can only be applied after the run or with huge delay, since the peak needs to be fully developed or at least needs to be developed past the maximum. The ICM measurements presented above were performed inline and with 100 ms between each data point. Since the self-written program is capable of communication with PLC, the data can be used to do live pooling or process optimization during a production run.

The results above were produced with extinction coefficients known beforehand. Similar results could be obtained without knowing the coefficients beforehand. An analysis of the complete run can calculate the extinction coefficients and deconvolute the overlapping peaks by assuming pure components at the beginning and end of the first or last peak respectively. The coefficients for the second component can be calculated by knowing the first and second component coefficients and the total spectra. This would deconvolute the peaks, but not give the exact concentration, since the link between concentration, extinction coefficients and the spectra is missing. In combination with known extinction coefficients for known components, this approach can be used to detect unknown components and give their concentration as a pseudo concentration linked to a known component.

3.2. Immunoglobulin G (IgG) Monomer and Dimer with SIMPLS

As mentioned before, eleven different ratios of IgG monomer and dimer were measured from which nine were used for the SIMPLS calibration and three for the validation. The corresponding UV/VIS spectra for the mixture with the highest and lowest monomer content are shown in Figure 5a. The spectra are standardized (z-score) to emphasize the differences. One statistic model was created to calculate both IgG monomer and dimer concentration at the same time as opposed to creating two models, one for each component. The Variable Importance on Projections (VIP) can be found in Figure 5b. Both Figure 5a,b show, that there are only minor differences in UV/VIS absorption for IgG monomer and dimer. The most important region is from 200 to 290 nm. This predication is supported by Figure 6a, which shows the loadings for each latent variable over wavelength. Again, the highest values are in the region from 200 to 290 nm. The region beyond 390 nm might be dominated by noise. Figure 6b shows the importance of each latent variable for the concentration results (Y-matrix). As expected, the first latent variables are the most important. But especially for the dimer concentration, variables 5 to 7 also show significant contribution.

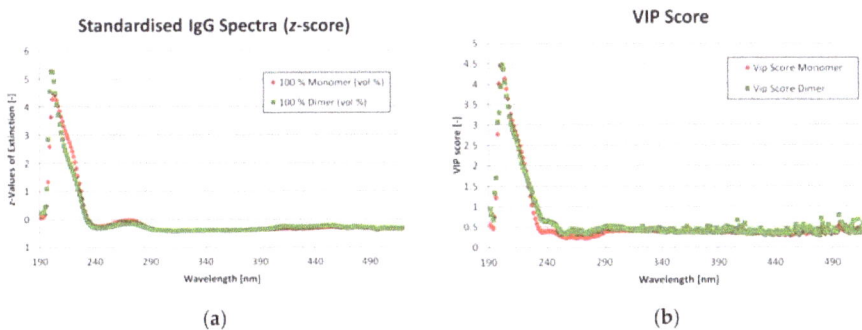

(a)   (b)

Figure 5. (**a**) Ultraviolet/Visible light (UV/VIS) spectra of the trainings data with the highest and lowest Immunoglobulin G (IgG) monomer content. The red dots represent the mixture with 100 vol.% monomer, the green dots with 100 vol.% dimer, respectively. However, this does not mean pure monomer or dimer since a conversion from one into the other takes place; (**b**) Variable Importance on Projections (VIP) scores for monomer (red) and dimer (green).

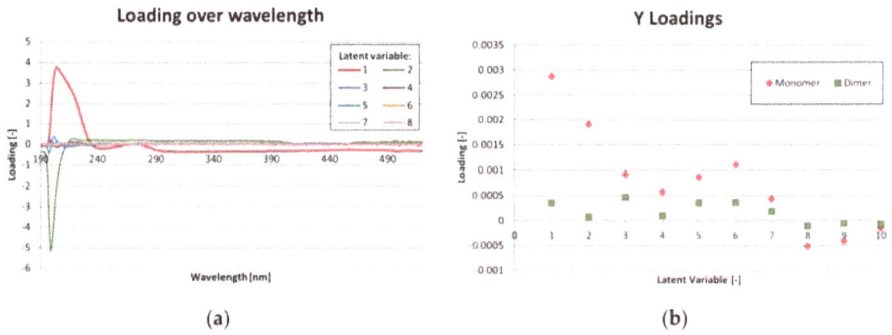

Figure 6. (**a**) Loading over wavelength diagram for the first eight latent variables. The higher the absolute value, the more important is this wavelength for the latent variable; (**b**) loadings of the Y results over latent variables. The higher the absolute value, the higher is the importance of this latent variable for the concentration calculation of monomer or dimer respectively.

As expected, the percentage of explained variance in the response increases with the number of latent variables used in the algorithm. For three latent variables, the percentage of explanation is 80%, for six variables it is over 93%. It maxes out at 98% with eight or more latent variables. Thus, eight latent variables where used for the validation measurements. The root mean square error of calibration (RSMEC) for eight latent variables is 0.14 mg/L for the IgG monomer and 0.025 mg/L for IgG dimer.

Figures 7a and 4b show the results for two of the validation measurements. These experiments represent the injection of the sample with 80 vol.% monomer/20 vol.% dimer (Figure 4a) and 20 vol.% monomer/80 vol.% dimer (Figure 4b), respectively. Again, these numbers represent the volumetric mixing ratio of the SEC fractions as described earlier. Due to conversion of either monomer to dimer or vice versa the mixing ratio does not match the concentration ratio.

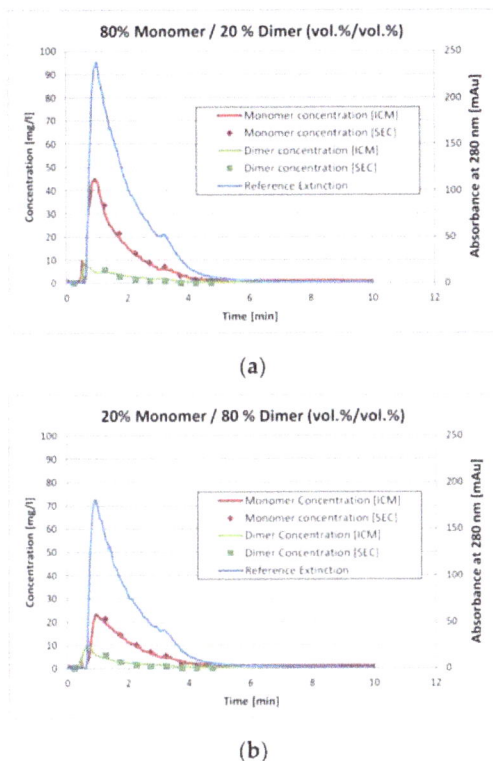

Figure 7. Concentration measurements over time of IgG monomer and dimer after injecting into a continuous stirred-tank reactor: (**a**) Volumetric mixing ratio of 80% monomer and 20% dimer; (**b**) Volumetric mixing ratio of 20% monomer and 80% dimer.

The solid lines represent the ICM measurements whereas the squares with the same color represent the corresponding SEC analysis. The later are done by fractionation the stirred tank outlet with an interval of 0.5 min. The ICM measurements however are continuous. To compare the results, the ICM values for the 0.5 min interval were not averaged. Instead, the ICM concentration value for the corresponding time at the middle of the fractionation interval was used.

The comparison for all validation experiments is shown in Figure 8:

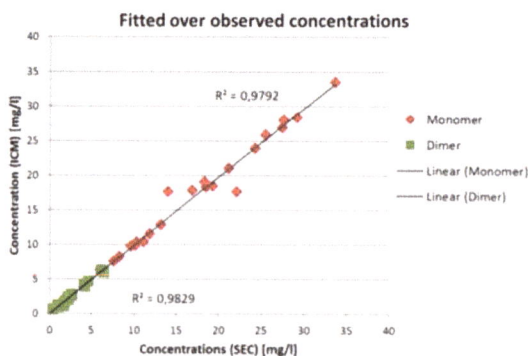

Figure 8. Concentrations based on fractionation and size exclusion chromatography (SEC) analyses compared to concentrations measured with inline concentration measurements (ICM).

The results in Figures 7 and 8 show that the ICM measurements are in good agreement with the SEC measurements. The overall coefficient of determination R^2 is 0.98 for both IgG monomer and dimer. The root mean square error of prediction (RMSEP) is 0.13 mg/L for the IgG monomer and 0.019 mg/L for IgG dimer.

The ICM measurements provide the concentrations and therefore composition of the outlet stream. This can be used to calculate the purity of the product stream in almost real time (100 ms). Purity calculation based on ICM measurements compared to the corresponding offline SEC gives a mean derivation of 0.15%. Within the 24 data points, the minimum deviation was 0.01% and the maximum deviation was 6%.

Since the SIMPLS-algorithm reduces the UV/VIS spectra to a statistic model, it is possible that it only finds a certain fixed distribution between the two components. This would presumably give the same multiplier between both concentrations. Figure 4a,b however show that this is not the case. The extinction values for the 20 vol.% monomer run are roughly 25% lower than for the 80 vol.% monomer run. The IgG monomer concentration however is more than 50% lower and the dimer concentration is 15% higher. Hence, it is reasonable to assume that the algorithm determines the concentrations correctly instead of just splitting the extinction with a fixed proportion.

Since the test mixture was prepared with IgG samples after Protein A and SEC, the purity of the IgG was relatively high. For in-process measurements, for example after protein A, more components and impurities have to be taken into account. Depending on the impurities, the distinction between the IgG monomer and dimer might become more challenging. However, it is hypothesized that at least the distinction between IgG in total and other impurities should be possible.

4. Conclusions

The proposed inline concentration measurement based on UV/VIS spectral data is a new and comparatively easy method for the real time quantification of UV/VIS active components. This indicates a huge potential for analytical as well as preparative and production-scale chromatography as well as for other unit operations. It is a key-enabling methodology for process development by aid of process modelling in order to determine model parameters efficiently—which is currently the major obstacle preventing a general use of simulation instead of empirics in industry for piloting and operation of almost any unit operation in regulated industry under Process Analytical Technology (PAT) and QbD design methods.

The inherent problem of overlapping peaks in analytical chromatography is reduced significantly by applying ICM. The accuracy of the quantitative determination of proteins could be shown to be improved to more than 99% compared to under 85% for conventional, state-of-the-art techniques.

The analysis time is less than 100 ms, so that the rate-limiting step is the data acquisition done by the detector itself. This enables a variety of process control options like live pooling and online optimization of the chromatographic process.

In the case of preparative separations, even for very similar components the purity can be measured at approximately 99.8% accuracy. This is an enormous advantage, since the real-time concentrations and purities are completely unknown for current process analytics. This leads to new possibilities for process control strategies. For example, batch chromatography could perform real-time pooling decisions based on the ICM data, leading to less batch failure and a higher product quality. Since product-related purity data are measured inline, there is a huge potential for cost and time savings by reducing offline analytics. The derived method overcomes obstacles in industrial application resulting from the dependency of extinction coefficients on detector type, age and signal strength by adopting reliable self-adjusting methods based on optimization algorithms, which provide mathematically proven quantitative reliability.

New process control options of multicolumn chromatography processes like Simulated Moving Bed (SMB) and their derivatives, e.g., integrated counter current chromatography (see [34]) based on the ICM-approach will be presented in the near future.

Acknowledgments: The authors thank the ITVP lab team especially Frank Steinhäuser and Volker Strohmeyer for their effort and support. Special thanks are also addressed to Fabian Mestmäcker, Dominik Stein and Thorsten Roth for excellent laboratory work.

Author Contributions: Steffen Zobel-Roos conceived and designed the experiments as well as wrote the paper. Mourad Mouellef wrote the program code and evaluated the experimental results. Christian Siemers substantively revised the work and contributed know-how in process automation. Jochen Strube is responsible for conception and supervision.

Conflicts of Interest: The authors declare no conflicts of interest.

References

1. Degerman, M.; Westerberg, K.; Nilsson, B. Determining Critical Process Parameters and Process Robustness in Preparative Chromatography—A Model-Based Approach. *Chem. Eng. Technol.* **2009**, *32*, 903–911. [CrossRef]
2. Borrmann, C.; Helling, C.; Lohrmann, M.; Sommerfeld, S.; Strube, J. Phenomena and Modeling of Hydrophobic Interaction Chromatography. *Sep. Sci. Technol.* **2011**, *46*, 1289–1305. [CrossRef]
3. Helling, C.; Strube, J. Modeling and Experimental Model Parameter Determination with Quality by Design for Bioprocesses. In *Biopharmaceutical Production Technology*; Subramanian, G., Ed.; Wiley-VCH: Weinheim, Germany, 2012; pp. 409–443.
4. Fahrner, R.L.; Blank, G.S. Real-time control of antibody loading during protein an affinity chromatography using an on-line assay. *J. Chromatogr. A* **1999**, *849*, 191–196. [CrossRef]
5. Kaltenbrunner, O.; Lu, Y.; Sharma, A.; Lawson, K.; Tressel, T. Risk-benefit evaluation of on-line high-performance liquid chromatography analysis for pooling decisions in large-scale chromatography. *J. Chromatogr. A* **2012**, *1241*, 37–45. [CrossRef] [PubMed]
6. Proll, G.; Kumpf, M.; Mehlmann, M.; Tschmelak, J.; Griffith, H.; Abuknesha, R.; Gauglitz, G. Monitoring an antibody affinity chromatography with a label-free optical biosensor technique. *J. Immunol. Methods* **2004**, *292*, 35–42. [CrossRef] [PubMed]
7. Rathore, A.S.; Wood, R.; Sharma, A.; Dermawan, S. Case study and application of process analytical technology (PAT) towards bioprocessing: II. Use of ultra-performance liquid chromatography (UPLC) for making real-time pooling decisions for process chromatography. *Biotechnol. Bioeng.* **2008**, *101*, 1366–1374. [CrossRef] [PubMed]
8. Rathore, A.S.; Parr, L.; Dermawan, S.; Lawson, K.; Lu, Y. Large scale demonstration of a process analytical technology application in bioprocessing: Use of on-line high performance liquid chromatography for making real time pooling decisions for process chromatography. *Biotechnol. Prog.* **2010**, *26*, 448–457. [CrossRef] [PubMed]

9. Gey, M. *Instrumentelle Analytik und Bioanalytik. Biosubstanzen, Trennmethoden, Strukturanalytik, Applikationen*; Springer: Berlin/Heidelberg, Germany, 2008.

10. Kong, H.; Ye, F.; Lu, X.; Guo, L.; Tian, J.; Xu, G. Deconvolution of overlapped peaks based on the exponentially modified Gaussian model in comprehensive two-dimensional gas chromatography. *J. Chromatogr. A* **2005**, *1086*, 160–164. [CrossRef] [PubMed]

11. Vivó-Truyols, G.; Torres-Lapasió, J.R.; van Nederkassel, A.M.; Vander Heyden, Y.; Massart, D.L. Automatic program for peak detection and deconvolution of multi-overlapped chromatographic signals part I: Peak detection. *J. Chromatogr. A* **2005**, *1096*, 133–145. [CrossRef] [PubMed]

12. Vivó-Truyols, G.; Torres-Lapasió, J.R.; van Nederkassel, A.M.; Vander Heyden, Y.; Massart, D.L. Automatic program for peak detection and deconvolution of multi-overlapped chromatographic signals part II: Peak model and deconvolution algorithms. *J. Chromatogr. A* **2005**, *1096*, 146–155. [CrossRef] [PubMed]

13. Helling, C.; Dams, T.; Gerwat, B.; Belousov, A.; Strube, J. Physical characterization of column chromatography: Stringent control over equipment performance in biopharmaceutical production. *Trends Chromatogr.* **2013**, *8*, 55–71.

14. Osberghaus, A.; Drechsel, K.; Hansen, S.; Hepbildikler, S.K.; Nath, S.; Haindl, M.; von Lieres, E.; Hubbuch, J. Model-integrated process development demonstrated on the optimization of a robotic cation exchange step. *Chem. Eng. Sci.* **2012**, *76*, 129–139. [CrossRef]

15. Osberghaus, A.; Hepbildikler, S.; Nath, S.; Haindl, M.; von Lieres, E.; Hubbuch, J. Optimizing a chromatographic three component separation: A comparison of mechanistic and empiric modeling approaches. *J. Chromatogr. A* **2012**, *1237*, 86–95. [CrossRef] [PubMed]

16. Capito, F.; Skudas, R.; Kolmar, H.; Stanislawski, B. Host cell protein quantification by Fourier transform mid infrared spectroscopy (FT-MIR). *Biotechnol. Bioeng.* **2013**, *110*, 252–259. [CrossRef] [PubMed]

17. Porterfield, J.Z.; Zlotnick, A. A simple and general method for determining the protein and nucleic acid content of viruses by UV absorbance. *Virology* **2010**, *407*, 281–288. [CrossRef] [PubMed]

18. Dismer, F.; Hansen, S.; Oelmeier, S.A.; Hubbuch, J. Accurate retention time determination of co-eluting proteins in analytical chromatography by means of spectral data. *Biotechnol. Bioeng.* **2013**, *110*, 683–693. [CrossRef] [PubMed]

19. Hansen, S.K.; Skibsted, E.; Staby, A.; Hubbuch, J. A label-free methodology for selective protein quantification by means of absorption measurements. *Biotechnol. Bioeng.* **2011**, *108*, 2661–2669. [CrossRef] [PubMed]

20. Hansen, S.K.; Jamali, B.; Hubbuch, J. Selective high throughput protein quantification based on UV absorption spectra. *Biotechnol. Bioeng.* **2013**, *110*, 448–460. [CrossRef] [PubMed]

21. Otto, M. *Analytische Chemie*; WILEY-VCH: Weinheim, Germany, 2011.

22. Wedler, G. *Lehrbuch der Physikalischen Chemie*; VCH Verlagsgesellschaft: Weinheim, Germany, 1985.

23. Atkins, P.W.; de Paula, J. *Physikalische Chemie*; Wiley-VCH: Weinheim, Germany, 2012.

24. Sawyer, D.T.; Heinemann, W.R.; Beebe, J.M. *Chemistry Experiments for Instrumental Methods*; Wiley: New York, NY, USA, 1984.

25. Bro, R.; de Jong, S. A fast non-negativity-constrained least squares algorithm. *J. Chemom.* **1997**, *11*, 393–401. [CrossRef]

26. Lawson, C.L.; Hanson, R.J. *Solving Least Squares Problems*; Society for Industrial and Applied Mathematics: Philadelphia, PA, USA, 1995.

27. Brestrich, N.; Briskot, T.; Osberghaus, A.; Hubbuch, J. A tool for selective inline quantification of co-eluting proteins in chromatography using spectral analysis and partial least squares regression. *Biotechnol. Bioeng.* **2014**, *111*, 1365–1373. [CrossRef] [PubMed]

28. Brestrich, N.; Sanden, A.; Kraft, A.; McCann, K.; Bertolini, J.; Hubbuch, J. Advances in inline quantification of co-eluting proteins in chromatography: Process-data-based model calibration and application towards real-life separation issues. *Biotechnol. Bioeng.* **2015**, *112*, 1406–1416. [CrossRef] [PubMed]

29. Andersson, M. A comparison of nine PLS1 algorithms. *J. Chemom.* **2009**, *23*, 518–529. [CrossRef]

30. De Jong, S. SIMPLS: An alternative approach to partial least squares regression. *Chemom. Intell. Lab. Syst.* **1993**, *18*, 251–263. [CrossRef]

31. Wold, S.; Sjöström, M.; Eriksson, L. PLS-regression: A basic tool of chemometrics. *Chemom. Intell. Lab. Syst.* **2001**, *58*, 109–130. [CrossRef]

32. KNAUER. *Smartline UV Detector 2600 Manual/Handbuch*; KNAUER: Berlin, Germany, 2007.

33. Gosling, J. *The Java Language Specification, Java SE 7 Edition*, 4th ed.; Addison-Wesley: Upper Saddle River, NJ, USA, 2013.
34. Zobel, S.; Helling, C.; Ditz, R.; Strube, J. Design and Operation of Continuous Countercurrent Chromatography in Biotechnological Production. *Ind. Eng. Chem. Res.* **2014**, *53*, 9169–9185. [CrossRef]

Article

![antibodies logo] *antibodies*

MDPI

Characterization of Monoclonal Antibody–Protein Antigen Complexes Using Small-Angle Scattering and Molecular Modeling

Maria Monica Castellanos [1,2], James A. Snyder [1], Melody Lee [1], Srinivas Chakravarthy [3], Nicholas J. Clark [4], Arnold McAuley [4] and Joseph E. Curtis [1,*]

[1] NIST Center for Neutron Research, National Institute of Standards and Technology, 100 Bureau Drive, Mail Stop 6102, Gaithersburg, MD 20899, USA; castellanosm@ibbr.umd.edu (M.M.C.); j.a.snyderjr@att.net (J.A.S.); melodylee299@gmail.com (M.L.)

[2] Institute for Bioscience and Biotechnology Research, 9600 Gudelsky Drive, Rockville, MD 20850, USA

[3] Biophysics Collaborative Access Team-Sector 18ID, Illinois Institute of Technology, Advanced Photon Source, Argonne National Laboratory, Lemont, IL 60439, USA; schakrav11@gmail.com

[4] Department of Drug Product Development, Amgen Incorporated, One Amgen Center Drive, Thousand Oaks, CA 91230, USA; nclark01@amgen.com (N.J.C.); arnoldm@amgen.com (A.M.)

* Correspondence: joseph.curtis@nist.gov; Tel.: +1-301-975-3959

Received: 16 November 2017; Accepted: 8 December 2017; Published: 15 December 2017

Abstract: The determination of monoclonal antibody interactions with protein antigens in solution can lead to important insights guiding physical characterization and molecular engineering of therapeutic targets. We used small-angle scattering (SAS) combined with size-exclusion multi-angle light scattering high-performance liquid chromatography to obtain monodisperse samples with defined stoichiometry to study an anti-streptavidin monoclonal antibody interacting with tetrameric streptavidin. Ensembles of structures with both monodentate and bidentate antibody–antigen complexes were generated using molecular docking protocols and molecular simulations. By comparing theoretical SAS profiles to the experimental data it was determined that the primary component(s) were compact monodentate and/or bidentate complexes. SAS profiles of extended monodentate complexes were not consistent with the experimental data. These results highlight the capability for determining the shape of monoclonal antibody–antigen complexes in solution using SAS data and physics-based molecular modeling.

Keywords: antibody; antigen; small-angle scattering; docking; modeling; simulation; Monte Carlo

1. Introduction

Understanding protein–protein interactions is a primary goal of structural biology, which can have direct impact on the manufacturing and therapeutic applications of monoclonal antibodies. Antibody–antigen interactions are specific and often have favorable equilibrium properties. With two potential antigen binding sites per molecule, antibodies can provide insights into protein interactions that are not possible with typical proteins with a single defined binding site per molecule. For decades, the majority of experimental data elucidating antibody–antigen interactions has come from studying such complexes in crystals using X-ray diffraction [1,2] and in solution using nuclear magnetic resonance spectroscopy [3,4]. Knowledge of the specific atomic interactions that define these interactions can be used to engineer new antibody molecules to elucidate the nature of the association to add value to improved candidate antibodies in terms of their efficacy and their ability to be manufactured, stored, and administered. In addition, knowledge of antibody–antigen structures in the case where the antigen has the capability to bind multiple copies of the same monoclonal antibody could have important biochemical impacts on the basic biology of the immune response and in-vivo response in therapeutic settings.

Small-angle scattering (SAS) using X-rays (SAXS) or neutrons (SANS) is a valuable method to obtain low-resolution shape information from a large variety of soft-matter systems including proteins. While the number of unique constraints in SAS data is limited, the use of atomistic models to interpret SAS data can offer useful insight as the atomic interactions and molecular topology are valid, physics-based constraints on the models. SAS is often used to provide shape information for problems in structural biology [5–7]. With selective or random deuteration of hydrogens one can use SANS via contrast matching to elucidate the shape of independent elements of a multidomain complex [8–10].

The sample requirements to measure the SAS of proteins are generally similar to other biophysical characterization methods, with concentrations of 0.5 g/L and higher for measurements in solution [7]. SANS can also be used to study proteins in amorphous and solid phases as there is no upper limit to the protein concentration that can be studied. However, the nature of the scattering can change with increasing concentration due to intermolecular correlations, and thus knowledge of the shape of the proteins can be lost as a result of contributions due to time-averaged spatial ordering of proteins in the sample [11]. Many SAS studies of monoclonal antibodies have been carried out to explore various aspects of antibody function, physical chemistry, and manufacturing [12–26]. Many of these studies report that single structures exist in solution by modeling the data using heuristic methods that lead to overfitting an underdetermined problem. It is understood that antibodies are flexible molecules in solution [27–29] and very few atomic structures of complete antibodies have been determined by X-ray crystallography [22]. The three complete structures of antibodies reported to date consist of two IgG1 structures [30,31] and a single IgG2 structure [32]. These studies considered the antibody structure to be dynamic and the single set of coordinates reported in each case should be considered "snapshots" of likely configurations sampled in solution. Ensembles of structures to model the SAS data of a monoclonal antibody form a more accurate representation of antibody structures in solution [18], and this has been validated by recent studies using atomic force microscopy and individual particle electron tomography [33,34]. SAXS has been used to determine the relative position of domains and characterize the epitope of an antigen–Fab complex [35], but SAS has been under-utilized to characterize antibody–antigen complexes.

Molecular dynamics, Monte Carlo simulation, and physics-based docking protocols to determine protein–protein interactions are viable tools to predict and model experimental data [36]. There are many simulation methods that one can use to enhance sampling and model the physics of the interactions, often using various approximations. Yet it remains a daunting challenge to accurately predict protein–protein interactions for systems of increasing size, and this is complicated further for flexible or disordered molecules. The sampling of antibody structures in solution can be a intractable task using molecular dynamics simulation, as antibodies have ~20,000 atoms and require hundreds of thousands of water atoms to accurately represent a model system to natively explore conformational space. While advances have been made in the simulation community using specialty processors such as Anton 2 [37] and graphical processing units [38–40], generally, systems the size of a single antibody are currently near the upper limit of what one might hope to simulate in a reasonable amount of time if one has access to the specialized computational hardware. That said, molecular simulations using all-atom [18,25,41–44], coarse-grain [45,46], and colloidal models [23,26,47–50] can provide insights into antibody structure and dynamics to probe basic biological, physical, and manufacturing properties [23–26,42,51]. The complexity and tractability of the problem becomes more challenging when one considers the interaction of protein antigens with an ensemble of flexible antibody configurations. There are a variety of tools available to predict antibody–antigen interactions, such as Rosetta, Modeller, and Haddock, among others [52–56]. A recent re-assessment of antibody binding site conformation prediction has provided insight into the viability of computational methods [57].

We have used SAXS and molecular modeling to characterize the stoichiometry and structures of anti-streptavidin IgG2 monoclonal antibody (ASA-IgG2) tetrameric streptavidin (tSA) complexes.

A series of starting models of varied compositions and binding arrangements were then subjected to molecular simulation to provide ensembles of structures to compare to SAXS profiles. Ensembles were generated using backbone torsion-angle Monte Carlo (TAMC) sampling to provide tens of thousands of unique ASA-IgG2–tSA complexes that allowed for a thorough representation of the physical space that ASA-IgG2 and bound tSA could occupy, thus enabling the evaluation of structural models by comparison of theoretical SAXS profiles to the experimental data. This in turn improves the viability of models derived from SAXS (or SANS) data, that inherently contains few constraints. By enhancing the resolution of the experimental SAXS profiles using size-exclusion chromatography (SEC) in line with SAXS detection, together in combination with molecular modeling, it will be shown that specific models of ASA-IgG2–tSA complexes in solution are consistent with the experimental SAXS data.

2. Results

2.1. Binding Affinity Measurements

To assess the specificity of ASA-IgG2 to monomeric streptavidin (mSA) versus tSA, a series of binding affinity measurements were performed using surface-plasmon resonance (SPR) with a Biacore 3000 instrument. Figure 1 shows the measured response after covalently bonding ASA-IgG2 to the surface of the sensor chip and flowing solutions of either mSA and tSA at various concentrations over the chip. The experiment clearly reveals that ASA-IgG2 has no affinity for mSA up to a concentration of 1 μM (M = mol/L). On the contrary, changes in the response were detected after flowing tSA over the immobilized ASA-IgG2. Using a simple 1:1 interaction model, a K_D estimate of 40 nM was obtained for tSA and ASA-IgG2.

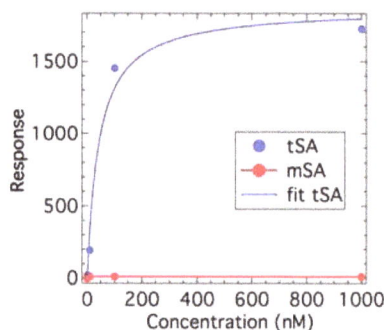

Figure 1. Measured responses for monomeric streptavidin (mSA) and tetrameric streptavidin (tSA) with immobilized anti-streptavidin IgG2 monoclonal antibody (ASA-IgG2).

In addition, fast protein liquid chromatography (FPLC) measurements were performed on mixtures of ASA-IgG2 with mSA and tSA, respectively. Figure 2 presents FPLC measurements of ASA-IgG2 in buffer compared to a mixture of ASA-IgG2 with each type of streptavidin. Both chromatograms show the results for ASA-IgG2 only and mixtures of ASA-IgG2 with streptavidin. For the ASA-IgG2 samples, a major peak was observed for the monomer, although about 2% of dimer was observed in the chromatograms. For the mixtures of tSA with ASA-IgG2, two major peaks were observed, representing tSA and the complex of tSA and ASA-IgG2. In the case of mSA and ASA-IgG2, no species eluted before ASA-IgG2, and the two major peaks observed correspond to ASA-IgG2 and mSA. Therefore, no complex of mSA with ASA-IgG2 was detected.

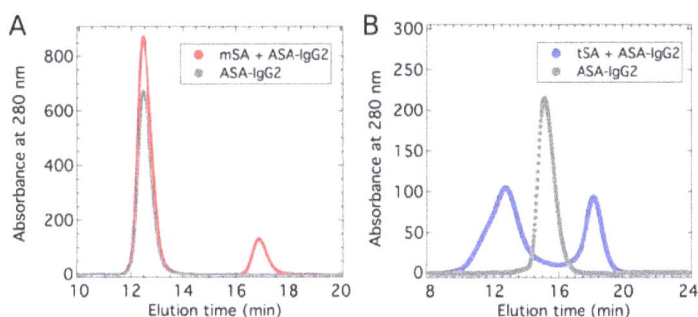

Figure 2. Fast protein liquid chromatography (FPLC) chromatograms of samples with ASA-IgG2 and streptavidin. (**A**) mSA and ASA-IgG2 with a molar ratio of 4:1 (flow rate 0.75 mL/min). (**B**) tSA and ASA-IgG2 with a molar ratio of 5:2 (flow rate 0.50 mL/min). Different concentrations of the free species were used in each chromatogram.

Since no binding was observed between mSA and ASA-IgG2, no further studies were performed with mSA. The following results refer to tSA and the complex formed with ASA-IgG2.

To further characterize the complex of tSA and ASA-IgG2, size-exclusion high-pressure liquid chromatography measurements were coupled with multi-angle light scattering (MALS), known as SEC-MALS, to study various concentration ratios of antigens to antibodies. Figure 3A displays the chromatograms representing the free and bound species after mixing tSA and ASA-IgG2 at different molar ratios. In these chromatograms, the free ASA-IgG2 eluted at about 10 min, whereas tSA eluted after 12 min. The antigen–antibody complex eluted first, which can be seen as a main peak at 8 min with an overlapping left peak or shoulder, depending on the molar ratio. These results suggest that there is not a single complex stoichiometry between tSA and ASA-IgG2. Depending on which free species was in excess and the molar ratios, different amounts of the complex species were formed. When the stoichiometry of the species was close to 1:1, the maximum amount of the complex was formed as most of the ASA-IgG2 and all the tSA were consumed. Regardless, the molecular weight of the main peak in the complex was constant at the center of the peak and during the remainder of the elution, as seen in Figure 3B. Therefore, the most abundant species of the antigen–antibody complex was monodisperse in molecular weight and can be studied by collecting fractions of the main complex peak. Note that the left shoulder in the complex peak became more pronounced with higher molar ratios of tSA:ASA-IgG2 and when free ASA-IgG2 species were depleted.

Figure 4 shows the percentage of complex formed after mixing tSA and ASA-IgG2 at different molar ratios. The maximum percentage of the complex was formed when the molar ratio of the species was close to one. Moreover, when the antigen and antibody were mixed at the same molar concentration, very low amounts of the free species were present, suggesting that most of the antigens and antibodies were used to form the complex up to the point that one of the species was depleted. Thus, the equilibrium is favored toward the formation of the complex, in agreement with the low K_D estimated from SPR. These results also suggest that molar ratios close to one should be used for the SAXS measurements to obtain higher concentrations of the bound species.

From the SEC-MALS measurements, the molecular weights of the free and bound species were obtained. Table 1 shows the molecular weights of tSA, ASA-IgG2, and the main complex formed. The values for the free species are in agreement with the known values for these molecules. The molecular weight obtained for the main complex is 447 ± 12 kDa. This is an average of eight measurements. The only stoichiometry that results in the experimentally calculated value is a combination of two molecules of tSA and two of ASA-IgG2. Using the values of Table 1, the molecular weight of two molecules of tSA and two of ASA-IgG2 is 447 ± 7 kDa, in agreement with the experimental value for the main complex. For the remainder of the discussion of the results, unless specified, the main complex refers to two ASA-IgG2 molecules associated with two tSA molecules.

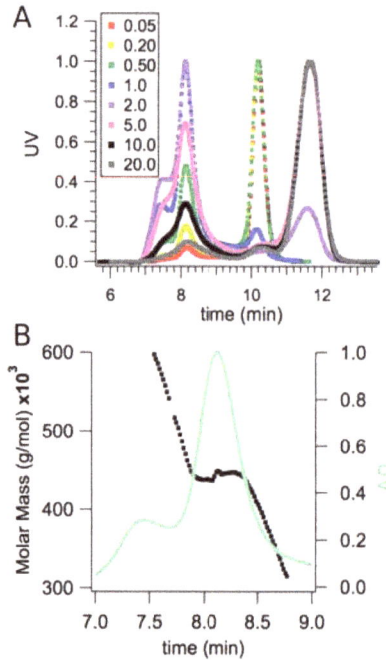

Figure 3. Size-exclusion chromatograms (SECs) of ASA-IgG2 at pH 6.5. (**A**) Different ratios of molar concentration of tSA:ASA-IgG2. (**B**) Molecular weight of the ASA-IgG2–tSA complex peak.

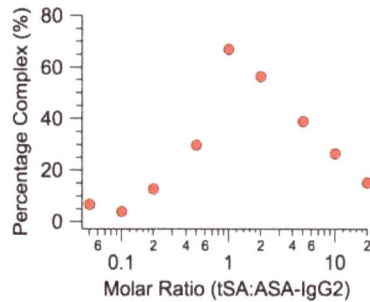

Figure 4. Percentages of the main ASA-IgG2–tSA complex formed at pH 6.5 after mixing ASA-IgG2 and tSA at different molar ratios.

Table 1. Molecular weight of the free and bound species for the main ASA-IgG2–tSA complex

Species	Molecular Weight (kDa)	Number of Measurements
tSA	63.3 ± 0.8	3
ASA-IgG2	160 ± 3	5
main complex	447 ± 12	8

The formation of the main complex as well as peak shape and separation in the chromatogram were evaluated in acidic pH. However, ASA-IgG2 was unstable at pH 3. Figure 5 shows the SEC chromatogram combined with the MALS data close to neutral and acidic pH. As shown in Figure 2, ASA-IgG2 was mostly monomeric at pH 6.5, with only 2% aggregates. In the case of pH 3.0, the main peak that eluted at 9.5 min represents the ASA-IgG2 monomer, whereas the first elution peaks account for dimers and aggregates of higher molecular weight. Based on the area of the peaks, only 67% of the sample was monomeric at pH 3.0. The peak that eluted before the monomer corresponds to the dimer, and the first peak, which overlaps with the dimer peak, corresponds to aggregates with a wide range of molecular weights. Although acidic pH is not suitable for this system because of aggregation of ASA-IgG2, mixtures of the antigen and the antibody were studied with SEC. Regardless of aggregation, the main complex did not form at acidic pH (data not shown), as the peak area was directly proportional to the prepared concentration of antibody and antigen and neither were associated in a complex. Note that data shown in Figures 1–5 correspond to single measurements and do not represent mean values from multiple measurements.

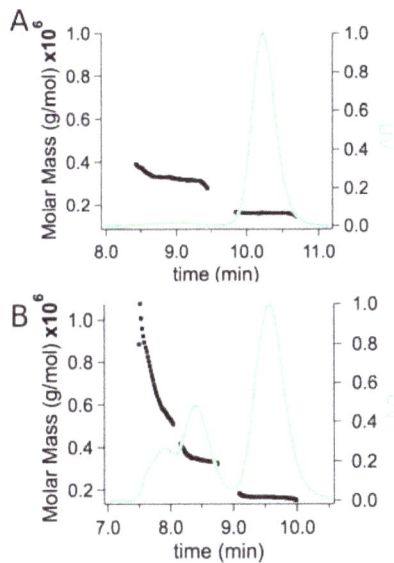

Figure 5. Size-exclusion chromatography coupled with multi-angle light scattering (SEC-MALS) data of ASA-IgG2 at (**A**) pH 6.5 and (**B**) pH 3.0. Light green represents the UV absorption data. Black marks represent the molecular weight of the eluted species.

2.2. Small-Angle X-ray Scattering

Based on the binding affinity results, SAXS measurements were performed in antigen–antibody mixtures with similar molar ratios at pH 6.5. Figure 6 presents the SAXS data of the antigen–antibody complex using different separation techniques. Although peak fractionation and separation of the main complex peak were clearly needed for SAXS measurements, scattering measurements were performed in bulk (without fractionation), after collecting fractions of the main complex peak (fractionation), and coupling SEC with SAXS (SEC-SAXS). The bulk measurement consisted of a mixture of tSA and ASA-IgG2 at a molar ratio of 1:1, in which the main complex corresponds to 67% of the sample according to the SEC. The SAXS profile for this sample shows a linear slope in the intermediate Q region (0.01–0.1 Å^{-1}), which is characteristic of polydisperse systems. In addition, the profile displays an increase in intensity at low Q, indicating the presence of aggregates and large species. The SAXS profile after SEC-fractionation of the main complex peak shows significantly less polydispersity and lower amounts of aggregates. This is confirmed by the SEC-MALS analysis, which shows that 86% of the fractionated sample corresponds to the main complex. Finally, the SEC-SAXS displays the profile of the main complex right after eluting from the SEC column. Although all profiles show similar features, the profile obtained with SEC-SAXS shows the lowest polydispersity and aggregation.

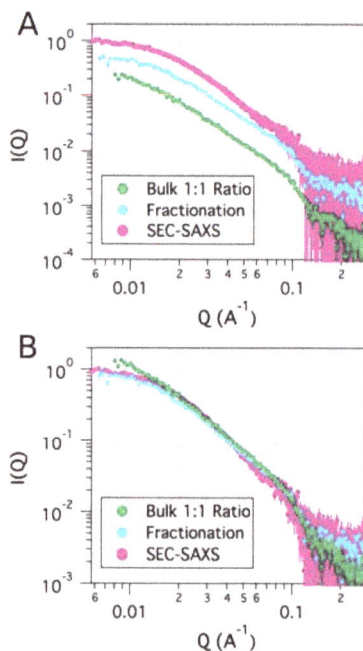

Figure 6. Small-angle X-ray scattering (SAXS) data of the ASA-IgG2–tSA complex at pH 6.5 using different separation methods. (**A**) Profiles are arbitrarily shifted for better visualization. (**B**) Scaled profiles. Error bars correspond to ±1 propagated standard error.

Figure 7 shows the SAXS profiles for tSA, ASA-IgG2, and the main complex from the SEC-SAXS measurement. tSA and the ASA-IgG2 had the expected curvature and features in the intermediate Q region for a globular protein and an antibody, respectively [44]. Figure 7B presents the scaled profiles of these species, which show higher intensities at low Q for the complex, followed by the ASA-IgG2 and tSA. The low Q intensity is proportional to molecular weight, in agreement with the SEC-MALS results. In addition, the radius of gyration (R_g) for each sample can be calculated using Guinier analysis. Table 2 displays the results of the Guinier analysis using the SAXS profiles in Figures 6 and 7. As expected, the main complex is the species with the largest size. However, depending on the method used to separate the main complex, some differences are observed. Using fractionation or no separation for the main complex results in larger radii of gyration due to the presence of other higher molecular weight species (see Figure 3). Nonetheless, the method of fractionation provides results that are comparable to those of SEC-SAXS, with a difference in size of 4 Å.

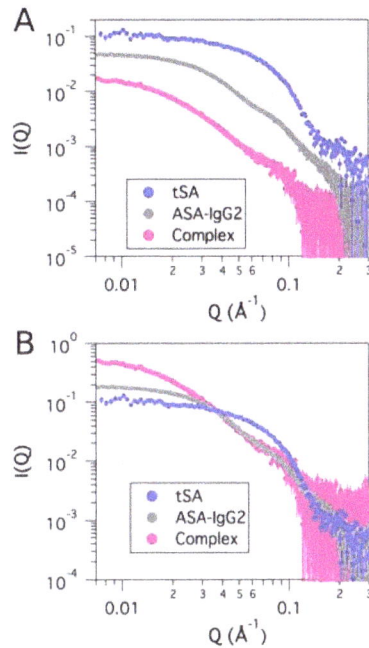

Figure 7. SAXS data of the ASA-IgG2–tSA complex and the free species at pH 6.5. (**A**) Profiles are arbitrarily shifted for better visualization. (**B**) Scaled profiles. Error bars correspond to the ±1 propagated standard error.

Table 2. Guinier Analysis of the SAXS profiles for the free and bound species.

Sample	Method	Radius of Gyration (Å)	$Q_{min}R_g$	$Q_{max}R_g$	r^2
tSA	Bulk	27.4	0.52	1.3	0.98
ASA-IgG2	SEC	49.2	0.44	1.0	0.99
ASA-IgG2–tSA complex	SEC	84.8	0.56	1.3	0.95
ASA-IgG2–tSA complex	Fractionation	88.9	0.69	1.3	0.91
ASA-IgG2–tSA complex	Bulk	123	1.0	1.2	0.62

Figure 8 shows the pair distribution function and Kratky plot for tSA, ASA-IgG2, and the main complex using different separation methods. tSA and ASA-IgG2 show the expected profiles for a globular protein and an antibody, respectively [44]. For the complex, a shoulder is observed in the pair distribution function at about 50 Å and a maximum at 88 Å. The first shoulder matches the maximum of the pair distribution function for tSA and ASA-IgG2, suggesting that it corresponds to distribution of distances in each of the antibody domains and subdomains of the components. On the contrary, the maximum at 88 Å is shifted by 10 Å to larger distances compared to the ASA-IgG2 distribution, which suggests that the peak maximum corresponds to distances between ASA-IgG2 domains and the antigen-binding fragment (Fab) with tSA. A Kratky plot is useful to qualitatively assess the folded or globular state of proteins. ASA-IgG2 has a Kratky plot that is asymmetric, as shown in Figure 8B, indicating some degree of non-globular shape most likely due to inherent flexibility that has been noted for other flexible monoclonal antibodies [18,44]. However, due to the low signal-to-noise in the samples with ASA-IgG2–tSA complexes, it is difficult to judge the degree of flexibility of the complex compared to antibodies alone or to globular proteins. For tSA, a bell-shaped curve characteristic of globular proteins was obtained.

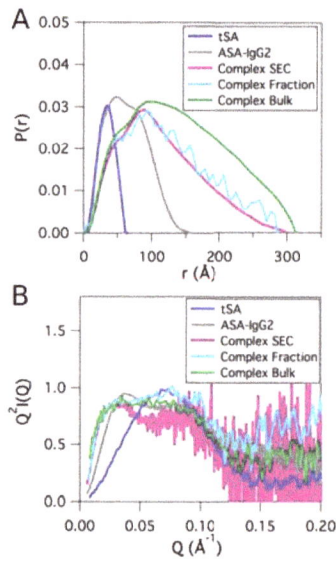

Figure 8. (**A**) Pair distribution function and (**B**) Kratky plot for the ASA-IgG2–tSA complex and its components at pH 6.5.

2.3. Model Building

The generation of atomistic models for use to model the SAXS data was carried out in two steps as described in Materials and Methods, and is summarized here. First, a series of building blocks were created in order to systematically create variant models of the various species involved. These structures included ASA-IgG2, tSA, the ASA-IgG2 Fab–tSA complex derived from docking, ASA-IgG2 (Fab)2-tSA derived by symmetrizing the ASA-IgG2 Fab–tSA complex, and the ASA-IgG2 fragment crystallizable region (Fc) domain as shown in Figure 9. As the stoichiometry derived from SEC-SAXS indicates that complexes with two ASA-IgG2 and two tSA molecules are the major components in the main fraction, the building blocks were used to create models of monodentate (single shared tSA) and bidentate (doubly-shared tSA) models, as shown in Figure 10.

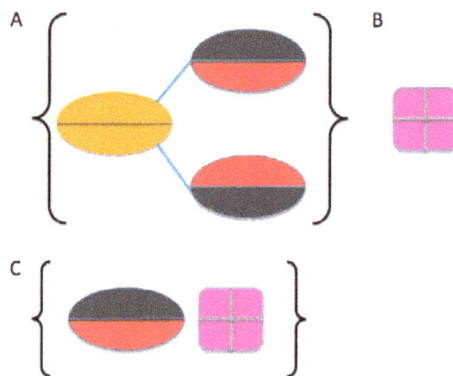

Figure 9. *Cont.*

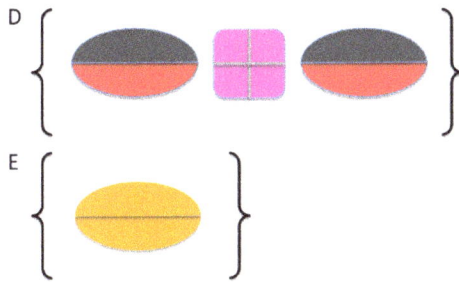

Figure 9. Schematic representation of protein structures used to create models. Fc: orange; Fab light chains: red; Fab heavy chains: grey; linkers between Fc and Fab domains: blue; tSA: mulberry. Brackets indicate that ensembles of structures were created. (**A**) ASA-IgG2; (**B**) tSA; (**C**) Fab–tSA from the docking protocol; (**D**) the Fab–tSA–Fab (Fab2–tSA) complex created by symmetrizing structures from (**C,E**) Fc. Coordinates for Fc and Fab were from the original ASA-IgG2 model [18]. Fc: fragment crystallizable region; Fab: antigen-binding region.

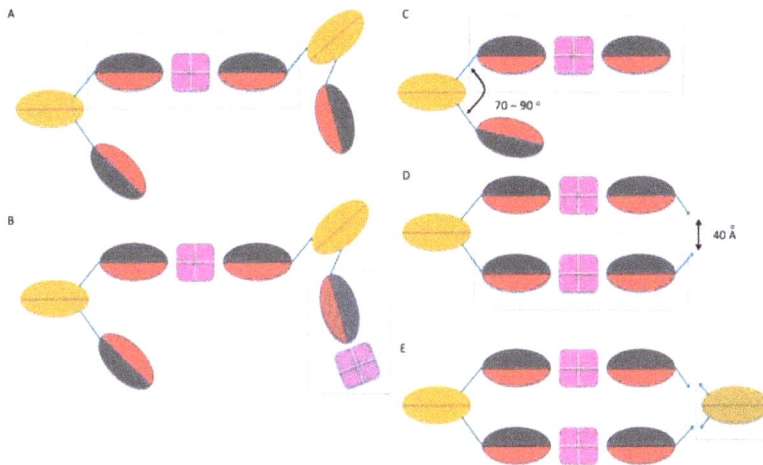

Figure 10. Schematic of model building process for monodentate (**A,B**) and bidentate (**C–E**) ASA-IgG2-tSA complexes. (**A**) Fc and a secondary Fab structure were aligned to ensembles of Fab2–tSA (depicted in blue box) to create preliminary Fab2–tSA 2ASA-IgG2 structures. (**B**) Ensembles of Fab–tSA were added to structures built in (**A**). Note that the predicted Fab–tSA docked configurations were maintained. (**C**) Fc and secondary Fab structures were aligned to ensembles of Fab2–tSA (depicted in blue boxes) to create preliminary Fab2–tSA single ASA-IgG2 structures. Only structures with Fab–Fc–Fab angles between 70 and 90 degrees were considered. (**D**) Fc and secondary Fab structures were aligned to ensembles of Fab2–tSA (depicted in blue boxes) to create preliminary Fab2–tSA single ASA-IgG2–two Fab structures. Only structures with distances between terminal C-alpha atoms of heavy chain residues 211 of less than 40 Å were considered. (**E**) Fc structures were added to structures from (**D**) that were subsequently energy-minimized and equilibrated using molecular dynamics simulation.

2.4. Comparison of Models to SAXS Data

In order to thoroughly model the SEC-SAXS data, a comprehensive comparison of ensembles of models representing potential molecular species in solution was carried out. While many of these models do not have molecular weights as found in the SEC-MALS data, it is informative to compare the theoretical SAXS profiles to the experimental SEC-SAXS data. As shown in Figure 11 models of tSA, ASA-IgG2, ASA-IgG2–tSA and ASA-IgG2–(tSA)2 are not in agreement with the experimental data as $\chi^2 > 150$ for all models for each ensemble, as shown in Figure 11A–D. Note that the blue volumetric densities represent the physical space occupied for each simulation and the single structure depicted within each density plot represents the single best structure from that ensemble determined by comparison of the theoretical SAXS profile to the experimental data. The ensemble of structures of two ASA-IgG2 plus one tSA molecule contained configurations that were by themselves consistent with the SAXS profiles (see Figure 11E) but have to be ruled out based on the measured molecular weight by SEC-MALS.

Evaluation of monodentate (Figure 11F) and bidentate (Figure 11G) models indicates that both contain structures that are in agreement with the SEC-SAXS data. The comparison of the scattering profiles from the structural models to SEC-SAXS profiles indicates that larger extended structures are not consistent with the experimental data and that the most likely set of configurations are compact monodentate or bidentate structures.

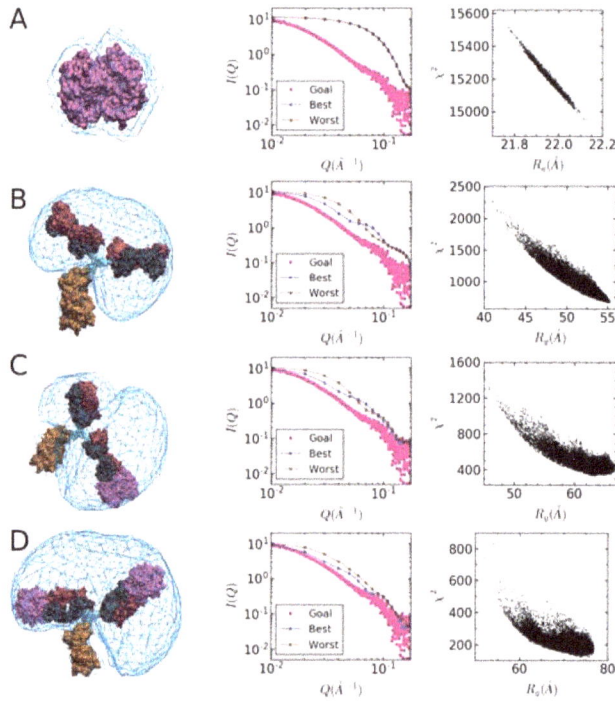

Figure 11. *Cont.*

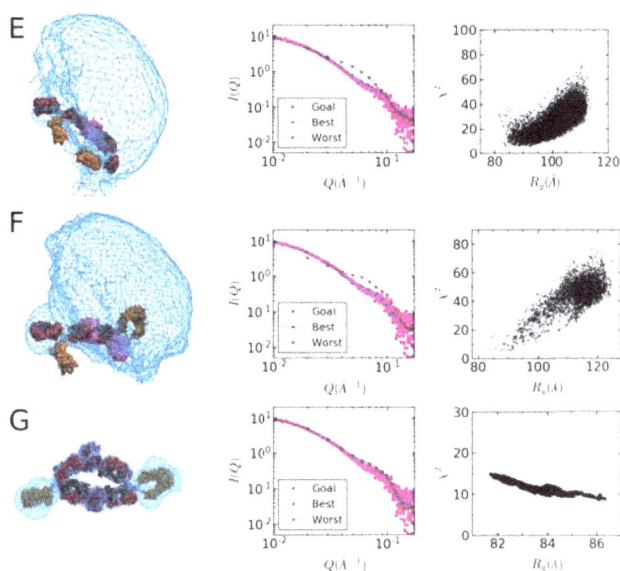

Figure 11. Structures and comparisons of the experimental and simulated scattering profiles. (**A**) tSA; (**B**) ASA-IgG2; (**C**) One ASA-IgG2 and one tSA molecule; (**D**) One ASA-IgG2 and two tSA molecules; (**E**) Two ASA-IgG2 and one tSA molecule; (**F**) Two ASA-IgG2 and two tSA molecules in monodentate configuration; (**G**) Two ASA-IgG2 and two tSA molecules in bidentate configuration. Blue mesh represents the configurational space samples by the ensembles referenced to one Fc. "Goal" represents the experimental data; "Best" represents the best match to the experimental data; and "Worst" represents the worst match to the experimental data. The plots in the right column represent the goodness-of-fit of the models to the experimental data in terms of reduced χ^2 as a function of the size of the models via radius of gyration (R_g).

3. Discussion

We have applied SEC-MALS, and SEC-SAXS with molecular docking and modeling to derive models of the complex formed by a monoclonal antibody and an antigen with the propensity to bind two Fab domains simultaneously. As expected for antibody–antigen complexes, the obtained K_D was in the nanomolar range for tSA and ASA-IgG2. The engineered mutations in mSA significantly affected the binding affinity with the ASA-IgG2 and no binding was observed up to micromolar concentrations of the antigen in its monomeric form. Combining SEC with a light scattering detector provided the stoichiometry of the main complex formed by the antibody–antigen complex and the suitable ratios of the free species for the SAXS measurements.

The use of SEC-SAXS improved the homogeneity and monodispersity of the samples, which enabled the evaluation of the structure of the complex using high resolution atomistic models. By using ensembles of viable models obtained from simulation, we were able to explore the specificity of SEC-SAXS to evaluate atomistic models consistent with the solution scattering data. Knowledge of the molecular weight of the species provided discriminating information to rule out models of the two ASA-IgG2 molecules and single tSA molecule in the case where the theoretical SAXS profiles were in agreement with the experimental data.

Finally, we found that compact models of monodentate and bidentate complexes were consistent with the SEC-SAXS data, although further discrimination of whether the monodentate and bidentate complexes are in equilibrium is not available without further experimental constraints. Thus, the use of SEC-SAXS and detailed ensemble modeling is a viable method to characterize antibody–antigen complexes and could have impact for those cases where stoichiometry and/or symmetry allows a discrimination of scattering profiles for putative models. Furthermore, the use of modern docking protocols to create physics-based structures to calculate SAS data in order to compare to SEC-SAXS data is a valuable orthogonal constraint that can be useful in modeling such data.

4. Materials and Methods

4.1. Sample Preparation

ASA-IgG2 was provided by Amgen in frozen solutions of 10 mM sodium acetate buffer, 10 mM acetic acid, 9% *w/v* sucrose, with pH 5.2 at concentrations of 30 mg/mL. The tSA (product number S4762 Sigma-Aldrich, St. Louis, MO, USA) was received as a lyophilized powder and stored at −80 °C. The mSA (product number 1385, Kerafast, Boston, MA, USA) is an engineered protein with high affinity to biotin that prevents multivalent interactions. mSA was provided in a 50-mM Tris (pH 7.5) with 150 mM of NaCl buffer and stored at 4 °C for up to two weeks. Frozen samples were thawed overnight at 4 °C one day before usage.

A phosphate buffer solution was prepared using sodium phosphate dibasic anhydrous (product number MSX0720-1, EMD Millipore, Burlington, MA, USA) and potassium phosphate monobasic (product number 231-913-4, Sigma-Aldrich, St. Louis, MO, USA) in Millipore SuperQ water, adjusting the pH to 6.5. Solutions were buffer exchanged using Slide-A-Lyzer dialysis cassettes (product number 66330, Thermo Scientific, Grand Island, NY, USA) with a 3.5-K molecular weight cutoff for mSA. For ASA-IgG2 and tSA samples, solutions were buffer-exchanged using Float-a-Lyzer dialysis devices (product number G235031, SpectrumLabs, Rancho Dominguez, CA, USA) with a 8–10 K molecular weight cutoff. The samples were immersed for at least 10 hours in fresh buffer up to three times while stirring to reach more than 99.9% of the final desired buffer composition. Solutions at pH 3.0 were prepared using a buffer with sodium phosphate dibasic anhydrous (product number MSX0720-1, EMD Millipore, Burlington, MA, USA) and phosphoric acid (product number 79617, Sigma-Aldrich, St. Louis, MO, USA) in Millipore SuperQ adjusted to pH 3.0.

Sample concentration was performed using Amicon Ultra-0.5 centrifugal filters (product number UFC501096, EMD Millipore, Burlington, MA, USA) with a 10-kDa molecular weight cutoff in a swinging bucket centrifuge (product number 75003181, Thermo Scientific, Grand Island, NY, USA) at 4000 relative centrifugal force until reaching the desired concentration. Final protein concentration was measured with a Nanodrop 2000 spectrometer (ND-2000, Thermo Scientific, Grand Island, NY, USA) using percent extinction coefficients ($\varepsilon_{percent}$) of 24.1, 31.7, and 16.0 for the mSA, tSA, and ASA-IgG2, respectively.

4.2. Binding Measurements

Binding affinity was assessed using the Biacore 3000. A protein A sensor chip (product number 29127558, GE Healthcare, Pittsburg, PA, USA) was used with a pre-immobilized recombinant protein A that has high affinity for the Fc region of antibodies. A solution of ASA-IgG2 at a concentration of 30 μg/mL was flowed through the cell until reaching a steady-state response, that was used as a baseline. Solutions containing antigen were then injected at concentrations from 1 to 1000 nM for 60 s. Samples and buffer were filtered using a 0.2-μm filter and degassed prior to the measurements. Size exclusion chromatography was performed using a Superdex200 10/300 GL column with an ÄKTA Purifier system (GE Healthcare, Pittsburg, PA, USA). All measurements were performed at 25 °C. This system was used for fractionation, collecting 1 mL of volume(s), and the resulting fractions were concentrated to 2 mg/mL for the SAXS measurements as described above.

High-performance liquid chromatography coupled with multi-angle light scattering was performed using a Thermo Scientific/Dionex U3000. Samples were injected onto a TSKgel G3000SWxl column (product number 08541, Tosoh Bioscience, South San Franciso, CA, USA) and absorbance was measured at 280 nm. After equilibrating the system and the column with the mobile phase, the flow rate was set to 0.8 mL/min for the measurements. Phosphate buffer at pH 6.5 was used as the mobile phase. Light scattering was performed with a DAWN HELEOS II detector (Wyatt Technology, Santa Barbara, CA, USA) at a wavelength of 664 nm. An Optilab T-rEX instrument (Wyatt Technology, Santa Barbara, CA, USA) was used for differential refractive index detection. Data were analyzed with the ASTRA®V software (Wyatt Technology, Santa Barbara, CA, USA).

4.3. Small-Angle X-ray Scattering Measurements

SAXS measurements were performed using an in-house Rigaku X-ray source and the SAXSLab Ganesha platform at the Institute for Bioscience and Biotechnology Research. Samples were loaded into a 96-well plate and sealed with tape to prevent solvent evaporation. Then, 20 μL of each sample was loaded into a 1.3-mm capillary by an automated robot. Sample to detector distance was varied from 0.7 to 1.7 m and a wavelength of 1.5418 Å was used to cover the range $0.005 \text{ Å}^{-1} < Q < 0.45 \text{ Å}^{-1}$. Scattered photons were detected with a two-dimensional Pilatus 300 K detector (Dectris, Baden-Dättwil, Switzerland). Data reduction was performed using RAW [58]. This setup was used for all samples, with the exception of those measurements for the complex that were coupled with SEC. Pair distribution functions and Guinier fits were calculated using RAW [58].

SAXS was also performed at BioCAT (beamline 18ID at the Advanced Photon Source, Chicago, IL, USA) with in-line size exclusion chromatography (SEC-SAXS) to separate the species of interest from other species and contaminants, thus ensuring optimal sample quality. Samples were loaded onto a Superdex-200 Increase 10/300 GL column (GE Healthcare, Pittsburg, PA, USA), which was run at 0.75 mL/min, and the eluate, after passing through the UV monitor, was directed through a SAXS flow cell. The SAXS flow cell had a 1.5-mm quartz capillary with 10-μm walls. Scattering intensity was recorded using a Pilatus3 1M detector (Dectris, CH) that was placed ∼3.5 m from the sample resulting in a q-range of 0.0057 to 0.36 Å^{-1}. 0.5 s exposures were acquired every 2 s during elution and data was reduced by the beam line specific pipeline that uses the ATSAS program suite [59]. Exposures corresponding to the regions flanking the elution peak were averaged to generate a buffer file which was subtracted from all the exposures. The buffer-subtracted exposures corresponding to the elution peak were used for subsequent analysis. This setup was used for the antigen–antibody complex and the antibody.

The profiles for ASA-IgG2 were consistent between the two SAXS instruments and methods. No radiation damage was observed when using the in-house source or the synchrotron facility. This was confirmed by collecting short exposures as the sample eluted and comparing the radius of gyration and low-Q scattering profile of the initial and final exposures. The free species were measured at concentrations of 1 and 2 mg/mL. No inter-particle interaction effects were observed at the highest concentration used. The complex sample was obtained by mixing the free-species using a molar ratio of 1:1 and separated using SEC-SAXS and fractionation as described in the results.

Certain commercial equipment, instruments, materials, suppliers, or software are identified in this paper to foster understanding. Such identification does not imply recommendation or endorsement by the National Institute of Standards and Technology, nor does it imply that the materials or equipment identified are necessarily the best available for the purpose.

4.4. Molecular Modeling

A sequence homology model of the ASA-IgG2 with 19,668 atoms, from an earlier study [18], was energy-minimized for 2500 steps, and subject to 1 ns dynamics as described previously. Note that the disulfide bonds in this model correspond to the IgG2-A form [60] characterized by structurally independent Fab domains and hinge region. The resulting structure was used as a starting structure for torsion-angle Monte Carlo (TAMC) studies. The program SASSIE [61] was used to generate 47,319 non-overlapping configurations by sampling backbone angles Φ or Ψ of the amino acid residues 212–214 of the upper hinge region of the heavy chain. Details of the TAMC method applied to ASA-IgG2 are described elsewhere [18].

A model of the tSA protein was created using the coordinates from the crystal structure (PDB ID: 1SWB) [62] obtained from the protein data bank [63] where all missing atoms were added using the program PSFGEN distributed as part of NAMD [64]. The complete model was energy-minimized for 2500 steps with the molecular dynamics program NAMD [64] using the CHARMM22 force field [65]. Subsequently, the structure was immersed in a previously equilibrated 200 Å cubic box of water (using the TIP3P water model [66]) and overlapping waters were removed and a neutralizing number of ions were added. The system was equilibrated at 300 K and 1 bar followed by a production run for 2 ns in the isothermal-isobaric (NPT) ensemble and compared to the SEC-SAXS data shown in Figure 11A.

For clarity, in the remainder of this section, Fab refers to ASA-IgG2 Fab domain and Fc refers to ASA-2 IgG2 Fc domain. Starting with coordinates taken from the TAMC ensemble for the ASA-IgG2 and the final tSA structure from the solvated molecular dynamics simulation, a starting structure for docking was constructed. This structure consisted of a Fab subunit and the tSA protein initially positioned near the Fab paratope region. The Fab subunit consisted of the light chain and residues 1–211 of the heavy chain. The RosettaDock [67] full protocol was used to create 150,000 docking decoys by generating random reorientations of both the Fab and tSA proteins independently. Optimization of side chain conformations of both partners were performed prior to docking using the RosettaDock pre-packing protocol. Subsequent post-analysis involved ranking the decoys using clustering. The set of decoys were sorted by score and the top 10,000 decoys were selected for clustering analysis. The Calibur program [68] was used for clustering with the Rosetta option selected for finding the clustering threshold. The largest cluster was selected, that contained 383 structures.

Assembly of the complexes shown in Figures 9 and 10 involved alignment using the heavy chain of Fab. Structural alignment was carried out using the Align module in SASSIE. For these alignments, care was taken to retain, specifically, the heavy chain which was originally interfaced with tSA from the docking protocol. Moreover, the region of the heavy chains used for alignment was selected to exclude residues which comprised or were near the interfacial region between the tSA and Fab binding site defined by the predictions from docking. Following alignment, atom distances between the pair of aligned molecules were monitored and if any of these distances were less than 8.0 Å, the aligned structure was rejected.

Fab–tSA–Fab complexes (Fab2–tSA) comprising the ensemble shown in Figure 9D were derived by symmetric replication of tSA from the ensemble of Figure 9C. The Fc and secondary Fab were then added to the complex by alignment of the heavy chains of both the full-length ASA-IgG2 structure from the TAMC ensemble (Figure 9A) and the Fab2–streptavidin (SA) complex to give the full-length Fab2–tSA 2 ASA-IgG2 complex shown in Figure 10A. The monodentate ASA-IgG2–tSA complex (Figure 9B) was then constructed by aligning Fab heavy chains of Fab2–tSA structures from the ensemble of Figure 9C and the Fab2–SA 2 ASA-IgG2 complex (Figure 10A). The resulting structures were energy-minimized and subjected to 10 ps Born implicit solvent molecular dynamics (MD). These structures were then used to generate approximately 50,000 accepted configurations in a TAMC simulation. Each TAMC configuration was energy minimized for 2500 steps, and then subjected to 10 ps MD to relax the structural ensemble to compare to SEC-SAXS data shown in Figure 11F.

To construct the bidentate complex, a subset of the TAMC ensemble of full-length ASA-IgG2 with Fab–Fc–Fab angles between 70 and 90 degrees was first collected to provide structures poised to represent more likely candidates for the general bidendate configuration. Subsequently, Fab heavy chains of Fab2–tSA structures from the ensemble of Figure 9C were aligned with heavy chains of both Fab subunits of the selected subset of ASA-IgG2. Only structures with distances between C-alpha atoms of residue 211 of less than 40 Å were retained. Fc structures were then added. The resulting structures were energy-minimized for 2500 steps, and then subjected to 10 ps MD to relax the structural ensemble. A single structure from the ensemble was used for a 10 ns generalized Born MD simulation to provide a trajectory to compare to the SEC-SAXS data shown in Figure 11G.

Theoretical SAS profiles were calculated using the SasCalc [69] module within SASSIE. These profiles were then filtered by calculating reduced χ^2 values for each of the SAS profiles relative to a SAS profile calculated for the experimental SAXS data. Nineteen grid points of momentum transfer, Q, between 0 and 0.19 Å^{-1} were used. All the images of protein structures were generated using visual molecular dynamics [70].

Acknowledgments: The authors thank Daniel Scott (NIST, IBBR), James Hoopes (UMD, IBBR), and Marco Blanco (NIST, IBBR) for providing assistance with the Biacore, the AKTA Purifier and the SEC-MALS instruments, respectively. The authors acknowledge professor Scott Walsh (UMD, IBBR) for providing access to the Biacore instrument. MMC acknowledges financial support from the NIST biomanufacturing initiative. Support for the Summer High School Intern Program for ML was provided by the Center for High Resolution Neutron Scattering, a partnership between the National Institute of Standards and Technology and the National Science Foundation under Agreement No. DMR-1508249. This work used CCP-SAS software developed through a joint EPSRC (EP/K039121/1) and NSF (CHE-1265821) grant. P41 GM103622 Thomas C. Irving PI and use of the Pilatus 3 1M detector were provided by grant 1S10OD018090-01 from NIGMS.

Author Contributions: M.M.C., N.J.C., A.M. and J.E.C. conceived and designed the experiments; M.M.C., M.L., and S.C. performed experiments; J.A.S. and J.E.C. created atomistic models and performed simulations; J.A.S., J.E.C., N.J.C., A.M. and M.M.C. analyzed the data and compared the experiment to theoretical predictions; N.J.C. and A.M. provided and characterized ASA-IgG2 and streptavidin samples prior to SAXS measurements; and M.M.C. and J.E.C. wrote the paper.

Conflicts of Interest: The authors declare no conflict of interest.

Abbreviations

The following abbreviations are used in this manuscript:

SAXS	small-angle X-ray scattering
SANS	small-angle neutron scattering
SAS	small-angle scattering
SEC	size-exclusion chromatography
SEC-SAXS	size-exclusion chromatography coupled with small-angle X-ray scattering
SEC-MALS	size-exclusion chromatography coupled with multi-angle light scattering
SPR	surface plasmon resonance
MD	molecular dynamics
MC	Monte Carlo
Fab	antigen-binding region
Fc	fragment crystallizable region
IgG	immunoglobulin
SA	streptavidin

References

1. Egli, M. Diffraction Techniques in Structural Biology. In *Current Protocols in Nucleic Acid Chemistry*; John Wiley & Sons, Inc.: Hoboken, NJ, USA, 2001; Volume 65, pp. 7.13.1–7.13.41.
2. Shi, Y. A Glimpse of Structural Biology through X-ray Crystallography. *Cell* **2014**, *159*, 995–1014.
3. Markwick, P.R.L.; Malliavin, T.; Nilges, M. Structural Biology by NMR: Structure, Dynamics, and Interactions. *PLoS Comput. Biol.* **2008**, *4*, 1–7.
4. Marion, D. An Introduction to Biological NMR Spectroscopy. *Mol. Cell. Proteom.* **2013**, *12*, 3006–3025.
5. Stuhrmann, H.B.; Miller, A. Small-angle scattering of biological structures. *J. Appl. Crystallogr.* **1978**, *11*, 325–345.
6. Svergun, D.I.; Koch, M.H.J. Small-angle scattering studies of biological macromolecules in solution. *Rep. Prog. Phys.* **2003**, *66*, 1735–1782.
7. Jacques, D.A.; Trewhella, J. Small-angle scattering for structural biology—Expanding the frontier while avoiding the pitfalls. *Protein Sci.* **2010**, *19*, 642–657.
8. Stuhrmann, H.B. Small-angle scattering and its interplay with crystallography, contrast variation in SAXS and SANS. *Acta Crystallogr. Sect. A* **2008**, *64*, 181–191.
9. Whitten, A.E.; Trewhella, J. Small-Angle Scattering and Neutron Contrast Variation for Studying Bio-Molecular Complexes. In *Micro and Nano Technologies in Bioanalysis: Methods and Protocols*; Foote, S.R., Lee, W.J., Eds.; Humana Press: Totowa, NJ, USA, 2009; pp. 307–323.

10. Krueger, S. Designing and Performing Biologicial Solution Small-Angle Neutron Scattering Contrast Variation Experiments on Multi-component Assemblies. In *Biological Small Angle Scattering: Techniques, Strategies and Tips*; Chaudhuri, B., Muñoz, I.G., Urban, V., Qian, S., Eds.; Springer: New York, NY, USA, 2017.
11. Castellanos, M.M.; McAuley, A.; Curtis, J.E. Investigating Structure and Dynamics of Proteins in Amorphous Phases Using Neutron Scattering. *Comput. Struct. Biotechnol. J.* **2017**, *15*, 117–130.
12. Perkins, S.J.; Bonner, A. Structure determinations of human and chimaeric antibodies by solution scattering and constrained molecular modelling. *Biochem. Soc. Trans.* **2008**, *36*, 37–42,
13. Perkins, S.J.; Okemefuna, A.I.; Nan, R.; Li, K.; Bonner, A. Constrained solution scattering modelling of human antibodies and complement proteins reveals novel biological insights. *J. R. Soc. Interface* **2009**, *6*, S679–S696.
14. Abe, Y.; Gor, J.; Bracewell, D.G.; Perkins, S.J.; Dalby, P.A. Masking of the Fc region in human IgG4 by constrained X-ray scattering modelling: Implications for antibody function and therapy. *Biochem. J.* **2010**, *432*, 101–114.
15. Ashish; Solanki, A.K.; Boone, C.D.; Krueger, J.K. Global structure of HIV-1 neutralizing antibody IgG1 b12 is asymmetric. *Biochem. Biophys. Res. Commun.* **2010**, *391*, 947–951.
16. Mosbæk, C.R.; Konarev, P.V.; Svergun, D.I.; Rischel, C.; Vestergaard, B. High concentration formulation studies of an IgG2 antibody using small angle X-ray scattering. *Pharm. Res.* **2012**, *29*, 2225–2235.
17. Lilyestrom, W.G.; Shire, S.J.; Scherer, T.M. Influence of the cosolute environment on IgG solution structure analyzed by small-angle X-ray scattering. *J. Phys. Chem. B* **2012**, *116*, 9611–9618.
18. Clark, N.J.; Zhang, H.; Krueger, S.; Lee, H.J.; Ketchem, R.R.; Kerwin, B.; Kanapuram, S.R.; Treuheit, M.J.; McAuley, A.; Curtis, J.E. Small-Angle Neutron Scattering Study of a Monoclonal Antibody Using Free-Energy Constraints. *J. Phys. Chem. B* **2013**, *117*, 14029–14038, doi:10.1021/jp408710r.
19. Castellanos, M.M.; Pathak, J.A.; Leach, W.; Bishop, S.M.; Colby, R.H. Explaining the non-Newtonian Character of Aggregating Monoclonal Antibody Solutions Using Small-Angle Neutron Scattering. *Biophys. J.* **2014**, *107*, 469–476.
20. Tian, X.; Langkilde, A.E.; Thorolfsson, M.; Rasmussen, H.B.; Vestergaard, B. Small-Angle X-ray Scattering Screening Complements Conventional Biophysical Analysis: Comparative Structural and Biophysical Analysis of Monoclonal Antibodies IgG1, IgG2, and IgG4. *J. Pharm. Sci.* **2017**, *103*, 1701–1710.
21. Tian, X.; Vestergaard, B.; Thorolfsson, M.; Yang, Z.; Rasmussen, H.B.; Langkilde, A.E. In-depth analysis of subclass-specific conformational preferences of IgG antibodies. *IUCrJ* **2015**, *2*, 9–18.
22. Rayner, L.E.; Hui, G.K.; Gor, J.; Heenan, R.K.; Dalby, P.A.; Perkins, S.J. The Solution Structures of Two Human IgG1 Antibodies Show Conformational Stability and Accommodate Their C1q and FcγR Ligands. *J. Biol. Chem.* **2015**, *290*, 8420–8438.
23. Yearley, E.; Zarraga, I.; Shire, S.; Scherer, T.; Gokarn, Y.; Wagner, N.; Liu, Y. Small-Angle Neutron Scattering Characterization of Monoclonal Antibody Conformations and Interactions at High Concentrations. *Biophys. J.* **2013**, *105*, 720–731.
24. Yearley, E.J.; Godfrin, P.D.; Perevozchikova, T.; Zhang, H.; Falus, P.; Porcar, L.; Nagao, M.; Curtis, J.E.; Gawande, P.; Taing, R.; et al. Observation of Small Cluster Formation in Concentrated Monoclonal Antibody Solutions and Its Implications to Solution Viscosity. *Biophys. J.* **2014**, *106*, 1763–1770.
25. Hui, G.K.; Wright, D.W.; Vennard, O.L.; Rayner, L.E.; Pang, M.; Yeo, S.C.; Gor, J.; Molyneux, K.; Barratt, J.; Perkins, S.J. The solution structures of native and patient monomeric human IgA1 reveal asymmetric extended structures: Implications for function and IgAN disease. *Biochem. J.* **2015**, *471*, 167–185.
26. Castellanos, M.M.; Clark, N.J.; Watson, M.C.; Krueger, S.; McAuley, A.; Curtis, J.E. Role of Molecular Flexibility and Colloidal Descriptions of Proteins in Crowded Environments from Small-Angle Scattering. *J. Phys. Chem. B* **2016**, *120*, 12511–12518, doi:10.1021/acs.jpcb.6b10637.
27. Yguerabide, J.; Epstein, H.F.; Stryer, L. Segmental flexibility in an antibody molecule. *J. Mol. Biol.* **1970**, *51*, 573–590.
28. McCammon, J.A.; Karplus, M. Internal motions of antibody molecules. *Nature* **1977**, *268*, 765–766.
29. Hanson, D.C.; Yguerabide, J.; Schumaker, V.N. Segmental flexibility of immunoglobulin G antibody molecules in solution: A new interpretation. *Biochemistry* **1981**, *20*, 6842–6852, doi:10.1021/bi00527a016.
30. Harris, L.J.; Skaletsky, E.; McPherson, A. Crystallographic structure of an intact IgG1 monoclonal antibody. *J. Mol. Biol.* **1998**, *275*, 861–872.

31. Saphire, E.O.; Parren, P.W.H.I.; Barbas, Carlos, F.I.; Burton, D.R.; Wilson, I.A. Crystallization and preliminary structure determination of an intact human immunoglobulin, b12: An antibody that broadly neutralizes primary isolates of HIV-1. *Acta Crystallogr. Sect. D* **2001**, *57*, 168–171.

32. Harris, L.J.; Larson, S.B.; Hasel, K.W.; Day, J.; Greenwood, A.; McPherson, A. The three-dimensional structure of an intact monoclonal antibody for canine lymphoma. *Nature* **1992**, *360*, 369–372.

33. Chaves, R.C.; Teulon, J.M.; Odorico, M.; Parot, P.; Chen, S.W.W.; Pellequer, J.L. Conformational dynamics of individual antibodies using computational docking and AFM. *J. Mol. Recognit.* **2013**, *26*, 596–604.

34. Zhang, X.; Zhang, L.; Tong, H.; Peng, B.; Rames, M.J.; Zhang, S.; Ren, G. 3D Structural Fluctuation of IgG1 Antibody Revealed by Individual Particle Electron Tomography. *Sci. Rep.* **2015**, *5*, doi:10.1038/srep09803.

35. Wilhelm, P.; Friguet, B.; Djavadi-Ohaniance, L.; Pilz, I.; Goldberg, M.E. Epitope localization in antigen-monoclonal-antibody complexes by small-angle X-ray scattering. *Eur. J. Biochem.* **1987**, *164*, 103–109.

36. Keskin, O.; Tuncbag, N.; Gursoy, A. Predicting Protein–Protein Interactions from the Molecular to the Proteome Level. *Chem. Rev.* **2016**, *116*, 4884–4909, doi:10.1021/acs.chemrev.5b00683.

37. Shaw, D.E.; Grossman, J.P.; Bank, J.A.; Batson, B.; Butts, J.A.; Chao, J.C.; Deneroff, M.M.; Dror, R.O.; Even, A.; Fenton, C.H.; et al. Anton 2: Raising the Bar for Performance and Programmability in a Special-Purpose Molecular Dynamics Supercomputer. In Proceedings of the SC14: International Conference for High Performance Computing, Networking, Storage and Analysis, New Orleans, LA, USA, 16–21 November 2014; pp. 41–53.

38. Friedrichs, M.S.; Eastman, P.; Vaidyanathan, V.; Houston, M.; Legrand, S.; Beberg, A.L.; Ensign, D.L.; Bruns, C.M.; Pande, V.S. Accelerating Molecular Dynamic Simulation on Graphics Processing Units. *J. Comput. Chem.* **2009**, *30*, 864–872.

39. Salomon-Ferrer, R.; Götz, A.W.; Poole, D.; Le Grand, S.; Walker, R.C. Routine Microsecond Molecular Dynamics Simulations with AMBER on GPUs. 2. Explicit Solvent Particle Mesh Ewald. *J. Chem. Theory Comput.* **2013**, *9*, 3878–3888, doi:10.1021/ct400314y.

40. Stone, J.E.; Phillips, J.C.; Freddolino, P.L.; Hardy, D.J.; Trabuco, L.G.; Schulten, K. Accelerating molecular modeling applications with graphics processors. *J. Comput. Chem.* **2007**, *28*, 2618–2640.

41. Brandt, J.P.; Patapoff, T.W.; Aragon, S.R. Construction, {MD} Simulation, and Hydrodynamic Validation of an All-Atom Model of a Monoclonal IgG Antibody. *Biophys. J.* **2010**, *99*, 905–913.

42. Fortunato, M.E.; Colina, C.M. Effects of Galactosylation in Immunoglobulin G from All-Atom Molecular Dynamics Simulations. *J. Phys. Chem. B* **2014**, *118*, 9844–9851, doi:10.1021/jp504243e.

43. Lapelosa, M.; Patapoff, T.W.; Zarraga, I.E. Molecular Simulations of the Pairwise Interaction of Monoclonal Antibodies. *J. Phys. Chem. B* **2014**, *118*, 13132–13141, doi:10.1021/jp508729z.

44. Castellanos, M.M.; Howell, S.; Gallagher, D.T.; Curtis, J.E. Characterization of the NISTmAb Reference Material using Small-Angle Scattering and Molecular Simulation Part I: Dilute Solution Structures. *Anal. Bioanal. Chem.* **2017**, doi:10.1021/acs.jpcb.6b10637.

45. Chaudhri, A.; Zarraga, I.E.; Kamerzell, T.J.; Brandt, J.P.; Patapoff, T.W.; Shire, S.J.; Voth, G.A. Coarse-Grained Modeling of the Self-Association of Therapeutic Monoclonal Antibodies. *J. Phys. Chem. B* **2012**, *116*, 8045–8057, doi:10.1021/jp301140u.

46. Franco-Gonzalez, J.F.; Ramos, J.; Cruz, V.L.; Martinez-Salazar, J. Exploring the dynamics and interaction of a full ErbB2 receptor and Trastuzumab-Fab antibody in a lipid bilayer model using Martini coarse-grained force field. *J. Comput.-Aided Mol. Des.* **2014**, *28*, 1093–1107.

47. Zhou, J.; Tsao, H.K.; Sheng, Y.J.; Jiang, S. Monte Carlo simulations of antibody adsorption and orientation on charged surfaces. *J. Chem. Phys.* **2004**, *121*, 1050–1057, doi:10.1063/1.1757434.

48. Calero-Rubio, C.; Saluja, A.; Roberts, C.J. Coarse-Grained Antibody Models for "Weak" Protein–Protein Interactions from Low to High Concentrations. *J. Phys. Chem. B* **2016**, *120*, 6592–6605, doi:10.1021/acs.jpcb.6b04907.

49. De Michele, C.; De Los Rios, P.; Foffi, G.; Piazza, F. Simulation and Theory of Antibody Binding to Crowded Antigen-Covered Surfaces. *PLoS Comput. Biol.* **2016**, *12*, e1004752.

50. Corbett, D.; Hebditch, M.; Keeling, R.; Ke, P.; Ekizoglou, S.; Sarangapani, P.; Pathak, J.; Van Der Walle, C.F.; Uddin, S.; Baldock, C.; et al. Coarse-Grained Modeling of Antibodies from Small-Angle Scattering Profiles. *J. Phys. Chem. B* **2017**, *121*, 8276–8290, doi:10.1021/acs.jpcb.7b04621.

51. Arzenšek, D.; Kuzman, D.; Podgornik, R. Hofmeister Effects in Monoclonal Antibody Solution Interactions. *J. Phys. Chem. B* **2015**, *119*, 10375–10389, doi:10.1021/acs.jpcb.5b02459.

52. Sivasubramanian, A.; Sircar, A.; Chaudhury, S.; Gray, J.J. Toward high-resolution homology modeling of antibody F(v) regions and application to antibody-antigen docking. *Proteins* **2009**, *74*, 497–514.
53. Weitzner, B.D.; Jeliazkov, J.R.; Lyskov, S.; Marze, N.; Kuroda, D.; Frick, R.; Adolf-Bryfogle, J.; Biswas, N.; Dunbrack, R.L., Jr.; Gray, J.J. Modeling and docking of antibody structures with Rosetta. *Nat. Protoc.* **2017**, *12*, 401–416.
54. Šali, A.; Blundell, T.L. Comparative Protein Modelling by Satisfaction of Spatial Restraints. *J. Mol. Biol.* **1993**, *234*, 779–815.
55. Dominguez, C.; Boelens, R.; Bonvin, A.M.J.J. HADDOCK: A Protein-Protein Docking Approach Based on Biochemical or Biophysical Information. *J. Am. Chem. Soc.* **2003**, *125*, 1731–1737, doi:10.1021/ja026939x.
56. Van Zundert, G.; Rodrigues, J.; Trellet, M.; Schmitz, C.; Kastritis, P.; Karaca, E.; Melquiond, A.; van Dijk, M.; de Vries, S.; Bonvin, A. The HADDOCK2.2 Web Server: User-Friendly Integrative Modeling of Biomolecular Complexes. *J. Mol. Biol.* **2016**, *428*, 720–725.
57. Almagro, J.C.; Teplyakov, A.; Luo, J.; Sweet, R.W.; Kodangattil, S.; Hernandez-Guzman, F.; Gilliland, G.L. Second antibody modeling assessment (AMA-II). *Proteins Struct. Funct. Bioinform.* **2014**, *82*, 1553–1562.
58. Nielsen, S.S.; Toft, K.N.; Snakenborg, D.; Jeppesen, M.G.; Jacobsen, J.K.; Vestergaard, B.; Kutter, J.P.; Arleth, L. *BioXTAS RAW*, a software program for high-throughput automated small-angle X-ray scattering data reduction and preliminary analysis. *J. Appl. Crystallogr.* **2009**, *42*, 959–964.
59. Petoukhov, M.V.; Franke, D.; Shkumatov, A.V.; Tria, G.; Kikhney, A.G.; Gajda, M.; Gorba, C.; Mertens, H.D.T.; Konarev, P.V.; Svergun, D.I. New developments in the *ATSAS* program package for small-angle scattering data analysis. *J. Appl. Crystallogr.* **2012**, *45*, 342–350.
60. Wypych, J.; Li, M.; Guo, A.; Zhang, Z.; Martinez, T.; Allen, M.J.; Fodor, S.; Kelner, D.N.; Flynn, G.C.; Liu, Y.D.; et al. Human IgG2 Antibodies Display Disulfide-mediated Structural Isoforms. *J. Biol. Chem.* **2008**, *283*, 16194–16205.
61. Curtis, J.E.; Raghunandan, S.; Nanda, H.; Krueger, S. SASSIE: A program to study intrinsically disordered biological molecules and macromolecular ensembles using experimental scattering restraints. *Comput. Phys. Commun.* **2012**, *183*, 382–389.
62. Freitag, S.; Le Trong, I.; Klumb, L.; Stayton, P.S.; Stenkamp, R.E. Structural studies of the streptavidin binding loop. *Protein Sci.* **1997**, *6*, 1157–1166.
63. Berman, H.M.; Westbrook, J.; Feng, Z.; Gilliland, G.; Bhat, T.N.; Weissig, H.; Shindyalov, I.N.; Bourne, P.E. The Protein Data Bank. *Nucleic Acids Res.* **2000**, *28*, 235–242.
64. Phillips, J.C.; Braun, R.; Wang, W.; Gumbart, J.; Tajkhorshid, E.; Villa, E.; Chipot, C.; Skeel, R.D.; Kalé, L.; Schulten, K. Scalable molecular dynamics with NAMD. *J. Comput. Chem.* **2005**, *26*, 1781–1802.
65. Brooks, B.R.; Bruccoleri, R.E.; Olafson, B.D.; States, D.J.; Swaminathan, S.; Karplus, M. CHARMM: A program for macromolecular energy, minimization, and dynamics calculations. *J. Comput. Chem.* **1983**, *4*, 187–217.
66. Jorgensen, W.L.; Chandrasekhar, J.; Madura, J.D.; Impey, R.W.; Klein, M.L. Comparison of simple potential functions for simulating liquid water. *J. Chem. Phys.* **1983**, *79*, 926–935.
67. Gray, J.J.; Moughon, S.; Wang, C.; Schueler-Furman, O.; Kuhlman, B.; Rohl, C.A.; Baker, D. Protein–Protein Docking with Simultaneous Optimization of Rigid-body Displacement and Side-chain Conformations. *J. Mol. Biol.* **2003**, *331*, 281–299.
68. Li, S.C.; Ng, Y.K. Calibur: A tool for clustering large numbers of protein decoys. *BMC Bioinform.* **2010**, *11*, 25.
69. Watson, M.C.; Curtis, J.E. Rapid and accurate calculation of small-angle scattering profiles using the golden ratio. *J. Appl. Crystallogr.* **2013**, *46*, 1171–1177.
70. Humphrey, W.; Dalke, A.; Schulten, K. VMD: Visual molecular dynamics. *J. Mol. Graph.* **1996**, *14*, 33–38.

antibodies

MDPI

Article

Controlling the Glycosylation Profile in mAbs Using Time-Dependent Media Supplementation

Devesh Radhakrishnan [1], Anne S. Robinson [2] and Babatunde A. Ogunnaike [1,*]

[1] Department of Chemical and Biomolecular Engineering, University of Delaware, Newark, DE 19716, USA; devesh@udel.edu

[2] Department of Chemical and Biomolecular Engineering, Tulane University, New Orleans, LA 70118, USA; asr@tulane.edu

* Correspondence: ogunnaike@udel.edu; Tel.: +1-302-831-4504

Received: 16 October 2017; Accepted: 15 December 2017; Published: 21 December 2017

Abstract: In order to meet desired drug product quality targets, the glycosylation profile of biotherapeutics such as monoclonal antibodies (mAbs) must be maintained consistently during manufacturing. Achieving consistent glycan distribution profiles requires identifying factors that influence glycosylation, and manipulating them appropriately via well-designed control strategies. Now, the cell culture media supplement, $MnCl_2$, is known to alter the glycosylation profile in mAbs generally, but its effect, particularly when introduced at different stages during cell growth, has yet to be investigated and quantified. In this study, we evaluate the effect of time-dependent addition of $MnCl_2$ on the glycan profile quantitatively, using factorial design experiments. Our results show that $MnCl_2$ addition during the lag and exponential phases affects the glycan profile significantly more than stationary phase supplementation does. Also, using a novel computational technique, we identify various combinations of glycan species that are affected by this dynamic media supplementation scheme, and quantify the effects mathematically. Our experiments demonstrate the importance of taking into consideration the time of addition of these trace supplements, not just their concentrations, and our computational analysis provides insight into what supplements to add, when, and how much, in order to induce desired changes.

Keywords: cell culture; glycosylation; media supplementation; $MnCl_2$; controllability analysis

---

1. Introduction

The global market for pharmaceuticals is predicted to grow to $1.5 trillion by 2021, with biologics such as monoclonal antibodies (mAbs), hormones, and therapeutic enzymes accounting for nearly 20% of the market share, based on current projections. With US sales rising from $8.29 billion in 2005 to $24.6 billion in 2012 [1,2]—coupled with the increase in regulatory agencies approvals of mAb treatments for different indications ranging from cancer to rheumatoid arthritis and Crohn's disease [3,4], the growing market for mAbs has resulted in active development of these biological products. While several different mammalian cell expression systems can be used to synthesize mAb therapeutics, over half of all currently approved mAbs are produced in Chinese Hamster Ovary (CHO) cell lines. CHO cells are the preferred host for a large number of recombinant mAb therapeutics because of the ease of adapting them to suspension growth, and the availability of powerful gene amplification systems to improve specific productivity [5]. However, more importantly, the post-translational modification machinery in CHO cells produces human-like structures in mAbs, thus ensuring biocompatibility.

One important post-translational modification in mAbs is N-glycosylation, a process by which an oligosaccharyltransferase complex in the endoplasmic reticulum adds a sugar substrate (glycan) to the Asn-X-Ser/Thr motif in the heavy chain of the mAb (where X is any amino acid other than Pro). As the mAb traverses the Golgi complex, the core glycan undergoes a series of non-template driven reactions that are mediated by the localized glycosyltransferase enzymes in the different Golgi compartments [6,7]. The result is a heterogeneous distribution of glycan residues or glycoforms, which affects the immunogenicity, effector functions, and the pharmacokinetic properties of the mAb, and consequently the final drug product quality [8–10]. Thus, there is considerable motivation for manufacturers and regulators to understand, characterize, and, if necessary, modulate the glycoform distribution in mAbs during production, in order to maintain a consistent glycan profile [11–14]. However, manipulating the glycan distribution effectively requires (i) identifying the factors that can influence the glycan distribution; (ii) quantifying the degree to which these factors affect the concentration of the glycoform species; and (iii) using such information to design effective control systems.

Previous studies have demonstrated that protein glycosylation can be influenced by various factors [15], such as pH [16,17], temperature [18–20], dissolved oxygen [21,22], ammonia [23], and media supplements such as nucleotide sugar precursors [24] and manganese chloride ($MnCl_2$) [25–28]. Each study focused on an individual factor, establishing empirical relationships between the individual factor in question and the specific set of glycan species it affects. However, modulating the complete glycan distribution profile requires manipulating multiple input factors simultaneously, and to be effective, such action must be based on a thorough, holistic understanding of how these inputs individually and jointly affect various glycan species. Such process understanding can be generated systematically by using statistical design of experiments whereby input factors are judiciously varied simultaneously to generate data on the main and interaction effects they exert on all the output responses of interest. Such structural information indicates which inputs to manipulate, and by how much, in order to alter the relative concentrations of different glycan species appropriately. In most cases, however, the available inputs are fewer than the glycan species to be controlled, resulting in a system with insufficient degrees of freedom. Consequently, we must first answer a fundamental question: given a limited set of inputs, to what extent can we independently control the concentrations of all the desired glycan species? In other words, is the desired change in the glycan distribution achievable using the available inputs? We address this question using "controllability analysis", by which we can determine quantitatively the extent to which the system is controllable. (Informally, a system is considered completely controllable if it is possible to drive the complete set of outputs from some initial value to any arbitrarily specified final, desired value, by manipulating the available set of inputs.) We have previously introduced, in [29], the concept of output controllability, demonstrated how to use it to assess the controllability of the glycan reaction network using data generated from statistical design of experiments, and in [30] we illustrated practical applicability by using it to identify in an experimental system, the glycan species whose concentrations can be controlled using such media supplements as $MnCl_2$, galactose and NH_4Cl as manipulated variables.

The role of different media supplements in modulating critical quality attributes of the mAb in general, and the glycan distribution profile in particular, has received considerable attention recently [31]. Typically, supplements such as $MnCl_2$, which are known to affect the expression and activity of several glycosyltransferase enzymes, are added to the media at the start of the batch to alter the glycan distribution. However, over the course of the batch run, as the cells continue to grow and produce mAb molecules, it is clear that changes in the cellular availability of supplements will influence not just the antibody productivity but also the activity of the glycosyltransferase enzymes, thus affecting the final glycan distribution. Consequently, we postulate that it is possible to control the glycosylation profile in mAbs by introducing specific media supplements at different stages of cell growth. We postulate further that introducing a chelating agent to the media can alter the effect of $MnCl_2$ addition on the glycan distribution. Consequently, a chelating agent such as EDTA provides an additional degree of freedom for fine-tuning the effect of $MnCl_2$, the media supplement of primary interest. Although EDTA is toxic to cells and can titrate both Mn^{++} and other bivalent ions in the medium that are necessary for cell growth and may be cofactors for other glycosyltransferase enzymes, it has been used in this study primarily to elucidate the effects of a dynamic addition of the two media supplements. (It is important to stress that the proposed time-dependent media supplementation is distinct from current fed-batch strategies where the objective is primarily to meet the nutrient demand in the culture, for the express purpose of enhancing cell growth and productivity—not to ensure that the product quality meets specific targets.) Specifically, we aim to identify the glycan species that can be controlled by adding $MnCl_2$ during lag, exponential, and stationary phases of cell growth, and to quantify the effect of such time-dependent $MnCl_2$ additions on the glycan distribution.

In this work, we use a mixed factorial experimental design to add $MnCl_2$ and EDTA at various stages of cell growth and, via appropriate analysis of the resulting data, we quantify the effect of time-dependent media supplementation on the antibody productivity and the glycosylation profile in mAbs. Subsequently, we use controllability analysis to identify the glycan species whose relative percentages can be controlled effectively by introducing $MnCl_2$ and EDTA to the media at different time points, and quantify the effect of these time-dependent additions. Overall, our results highlight the importance of taking into account the dynamic nature of media supplementation, and also provide concepts that can be exploited to develop new strategies for controlling the glycosylation profile in mAbs.

2. Materials and Methods

2.1. Cell Culture

All experiments were conducted using an IgG1 producing CHO-K1 cell line donated by Genentech, San Francisco, California. The cells were scaled up in a custom CD OptiCHO^{TM} medium formulation (Thermo Fisher Scientific, Waltham, MA, USA) that was supplemented with 4 mM glutamine, 5 g/L glucose and 25 nM MTX. The osmolality was adjusted to 300 mOsm by adding NaCl stock solution. The concentration of $MnCl_2$ in the media was adjusted using a 0.5 M stock solution (Sigma Aldrich, St. Louis, MO, USA). Similarly, a 0.5 M EDTA sterile stock solution was prepared and added to the media as required. The cells were inoculated with an initial seeding density of 0.5×10^6 cells/mL in vented-cap Erlenmeyer shake flasks with a working volume of 50 mL and grown in batch in suspension in an incubator maintained at 37 °C with a 5% CO_2 overlay, with supplements only as indicated by experimental design below. Cell count measurements were taken every two days using a hemocytometer. Metabolite (glutamine, glucose, glutamate and lactate) concentrations, media pH, and osmolality were measured using a Bioprofile 100+ analyzer (Nova Biomedical, Waltham, MA, USA). Antibody titer was measured with an Agilent 1200 HPLC instrument using 1X PBS buffer on a Thermo Scientific^{TM} MAbPac Protein A chromatography column (12 micron particle size, 35 × 4.0 mm I.D., Thermo Fisher Scientific, Waltham, MA, USA).

2.2. Experimental Design

The shake flask experiments were conducted according to a $(2^2, 3^2)$ mixed level experimental design for the following factors: (i) $MnCl_2$ concentration (high and low levels); (ii) EDTA concentration (high and low levels); (iii) time of addition of $MnCl_2$ (high, intermediate, and low levels); and (iv) time of addition of EDTA (high, intermediate and low levels). The concentration of $MnCl_2$ in the basal media corresponds to the low-level condition (−1) for $MnCl_2$, while the high-level condition (+1) corresponds to the final concentration of $MnCl_2$ supplemented media (0.04 mM). High and low levels for $MnCl_2$ were-based upon previous experiments performed using the same cell line [30]. Similarly, the low-level condition (−1) for EDTA corresponds to "no EDTA" added to the media, while the high level (+1) corresponds to 0.08 mM EDTA added to the media. $MnCl_2$ and EDTA are added on day 0 (D0), day 3 (D3), or day 6 (D6) after inoculation, corresponding respectively to the low (−1), intermediate (0), and high (+1) levels. Thus, this full factorial mixed level $(2^2, 3^2)$ experimental design yields a total of 36 different possible shake flask conditions to be tested. However, the 36 conditions are not unique because some of the cases correspond to identical experimental conditions. For instance, 9 of the 36 conditions correspond to $MnCl_2$ and EDTA at low levels (−1), with the time of addition at low (-1), intermediate (0) and high levels ($+1$). The low level condition for $MnCl_2$ represents basal concentrations, while the low level for EDTA represents no EDTA supplementation. Thus, these 9 cases represent identical conditions where the flask has basal levels of $MnCl_2$ with no EDTA supplementation either on D0, D3, or D6. A single flask (F1) was used for all nine cases and was treated as the control flask because of what the conditions represent—basal level of $MnCl_2$ and no EDTA supplementation on any day. It can be shown that there are in fact only 16 unique experimental cases/conditions, as listed in Table 1. Each condition was tested with two biological replicates. The glycan distribution profile was determined using the permethylation assay described below, and the resulting relative glycan percentages data obtained for each condition were analyzed in MINITAB using standard analysis of variance (ANOVA) to obtain the factor effects/coefficients and associated *p*-values.

Table 1. Experimental conditions tested in the mixed level factorial design.

Experimental Condition	$MnCl_2$ Conc. (mM)	EDTA Conc. (mM)	Time of Addition of $MnCl_2$ [1]	Time of Addition of EDTA [1]	Label
1	0.01	0	D0	D0	Control
2	0.01	0.08	D0	D0	ED D0
3	0.04	0.08	D0	D0	Mn D0/ED D0
4	0.04	0.08	D3	D0	Mn D3/ED D0
5	0.04	0.08	D6	D0	Mn D6/ED D0
6	0.01	0.08	D0	D3	ED D3
7	0.04	0.08	D0	D3	Mn D0/ED D3
8	0.04	0.08	D3	D3	Mn D3/ED D3
9	0.04	0.08	D6	D3	Mn D6/ED D3
10	0.01	0.08	D0	D6	ED D6
11	0.04	0.08	D0	D6	Mn D0/ED D6
12	0.04	0.08	D3	D6	Mn D3/ED D6
13	0.04	0.08	D6	D6	Mn D6/ED D6
14	0.04	0	D0	D0	Mn D0
15	0.04	0	D3	D0	Mn D3
16	0.04	0	D6	D0	Mn D6

[1] D0, D3 and D6 refer to Day 0, Day 3 and Day 6 after inoculation, respectively.

2.3. Glycan Permethylation Assay

On day 8 after inoculation, the cells were centrifuged at 3000 rpm for 10 min and the spent media was harvested. The IgG1 antibody was then purified from the spent media using a PhyNexus Benchtop MEA2 system using Protein A chromatography resin packed in a 2 mL PhyTip column (PhyNexus, San Jose, CA, USA). The glycan permethylation assay was then carried out with 100 microgram of the purified antibody using a previously described method [30]. Briefly, the antibody was first digested with trypsin (Promega, Madison, WI, USA) for four hours in an incubator held at 37 °C, followed by enzymatic deglycosylation using PNGase-F (ProZyme, Hayward, CA, USA) for a minimum of 16 h at 37 °C. The free separated glycans were captured on Hypersep Hyper Carb SPE cartridges (Thermo Fisher Scientific, Waltham, MA, USA) and permethylated following the Ciucanu method using methyl iodide and NaOH in the presence of DMSO [32,33]. The permethylated glycans were purified in a liquid-liquid extraction step with chloroform (Sigma Aldrich, St. Louis, MO, USA), dried and resuspended in 80% methanol (Sigma Aldrich, St. Louis, MO, USA). The resuspended glycans were spotted onto a MALDI/TOF plate with a DHB matrix and analyzed using a 4800 MALDI TOF/TOF Analyzer (ABSciex) in positive ion, reflector mode. The data collected using the mass spectrometer was then exported to DataExplorer to obtain the peak heights for the identified glycans (see Supplementary Table S8). The relative glycan distribution in each sample was calculated from the sum of the peak heights for all the identified glycans in that sample.

2.4. Glycosylation Index

For each experimental condition, glycosylation indices were calculated from the relative percentages of individual glycan species [17,34]. For example, the galactosylation index (GI), defined as the percentage of mono- and di-galactosylated species in the total glycan distribution, was determined according to Equation (1):

$$GI = \frac{2 \times G_2 + G_1}{2 \times (G_0 + G_1 + G_2)} \%, \tag{1}$$

where G_0 is the sum of all agalactosylated species, G_1 is the sum of all monogalactosylated species, and G_2 is the sum of all digalactosylated species. Similarly, we calculated the fucosylation index (FI) for each distribution as

$$FI = \frac{F_1}{(F_0 + F_1)} \%, \tag{2}$$

where F_0 and F_1 are the sum of all afucosylated and fucosylated species, respectively.

2.5. Controllability Analysis

Using the technique presented in St.Amand et al. [29], we performed controllability analysis to quantify the effect of time dependent media supplementation on the glycosylation profile in mAbs. Briefly, we note first that estimates of factor coefficients obtained after analyzing the mixed factorial design data correspond to the various "process gains", defined as the change observed in the glycan distribution (output), Δy, in response to a unit change in the input factor with which the coefficient in question is associated. By selecting statistically significant factor coefficients (at the significance level of $\alpha = 0.05$) and setting all non-significant coefficients to zero, we genereated the process gain matrix $\mathbf{K}$ so that:

$$\Delta \mathbf{y} = \mathbf{K} \Delta \mathbf{u}, \tag{3}$$

where $\Delta \mathbf{u}$ represents the change in the input factor. Singular value decomposition of the process gain matrix produces the diagonal singular value matrix, $\mathbf{\Sigma}$, and the unitary matrices, $\mathbf{W}$ and $\mathbf{V}^{\mathsf{T}}$, that are subsequently used to obtain the orthogonal input (μ) and output (η) modes, which, along with the corresponding singular values are used to assess controllability.

3. Results

3.1. Early Addition of EDTA Was Detrimental to Cell Growth and Reduces Antibody Titer

Figures 1 and 2 show the effect that introducing media supplements ($MnCl_2$ or EDTA) at different time points had on viable cell density and final antibody titer, compared to corresponding results obtained from a control flask (F1), which contained $MnCl_2$ at basal media concentrations and no EDTA (see Figure S2 in supplementary information for the normalized glucose concentration in each shake flask).

Compared to the conditions in the control flask, early addition of EDTA on D0 reduced the cell density significantly, which is consistent with hampered cell growth. However, this effect was offset somewhat by introducing $MnCl_2$ in addition to EDTA on D0. The average viable cell density (VCD) measured on day 4 for samples in which both $MnCl_2$ and EDTA were introduced on D0, was 1.31×10^9 cells/L, which was nearly three times as large as the value of the VCD in the flasks where only EDTA was added on D0 ($\sim 0.4 \times 10^9$ cells/L). Introducing $MnCl_2$ on D3 or D6 after the addition of EDTA on D0 did not improve the VCD. Similarly, when EDTA was introduced on D3, the VCD in samples with no additional $MnCl_2$ supplementation dropped sharply. By contrast, when the media was supplemented with $MnCl_2$ on D0 or D3, the VCD was slightly higher than when the media was supplemented with just EDTA on D3. Supplementing the samples with $MnCl_2$ on D6 after the addition of EDTA did not improve the VCD significantly. In all cases, the observed VCD was generally higher than when EDTA was introduced on D0. Addition of EDTA on D6 had no impact on the VCD, regardless of the time of introduction of $MnCl_2$. Similarly, as shown in Figure 1d, early addition of $MnCl_2$ on D0 and D3 reduced the VCD, while addition of $MnCl_2$ on D6 did not alter the VCD. Thus, in summary, early addition of EDTA reduced cell viability in the absence of $MnCl_2$ supplementation, but the addition of $MnCl_2$ by itself did not alter cell viability significantly.

Figure 2 shows the effect of media supplementation on antibody titers. The average mAb titer in the control flask F1 (with no $MnCl_2$ or EDTA supplementation) was 0.13 g/L. The addition of EDTA to the media on D0 in the absence of $MnCl_2$ supplementation decreased the titer by about a fourth, to 0.03 g/L. This decrease in the titer was marginally offset when $MnCl_2$ was introduced on D0, D3, or D6, with earlier $MnCl_2$ supplementation resulting in higher titers than later supplementation. When EDTA was introduced on D3, the resulting titer was 0.10 g/L, three times higher than that observed with EDTA supplementation on D0. Further supplementing the media with $MnCl_2$ on D0, D3, or D6, increased the titer observed with EDTA supplementation on D3 to values comparable to that of the control case. Supplementing the flasks on D6 with EDTA increased titers even further to 0.15 g/L, beyond values obtained in the control case. The titer values were also higher when EDTA supplementation on D6 was combined with $MnCl_2$ supplementation on D0 (0.17 g/L), D3 (0.17 g/L) or D6 (0.16 g/L). Finally, $MnCl_2$ supplementation alone on D0, D3, or D6 increased the titer to an average of 0.14 g/L. Thus, while early EDTA supplementation has an adverse effect on the antibody titer, late EDTA addition improves the final titer. Consequently, we conclude that early addition of EDTA reduces antibody titer, while latter addition of EDTA helps to increase the titer by a marginal amount.

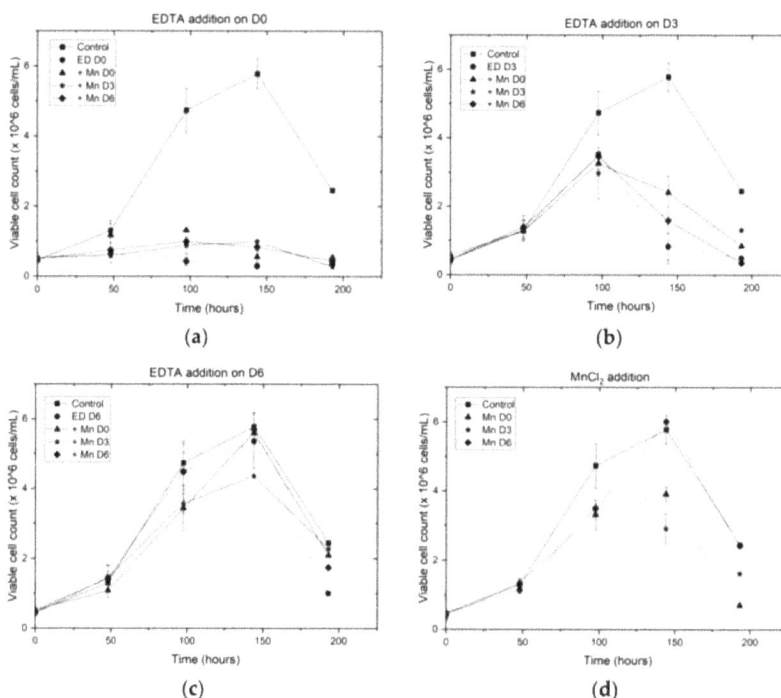

Figure 1. Average viable cell concentration data for CHO-K1 cells when (**a**) EDTA is added by itself on D0 (●), or in the presence of $MnCl_2$ on D0 (▲), D3 (*), and D6 (♦); (**b**) EDTA is added by itself on D3 (●), or in the presence of $MnCl_2$ on D0 (▲), D3 (*), and D6 (♦); (**c**) EDTA is added by itself on D6 (●), or in the presence of $MnCl_2$ on D0 (▲), D3 (*), and D6 (♦); and (**d**) $MnCl_2$ is added on D0 (▲), D3 (*), and D6 (♦).

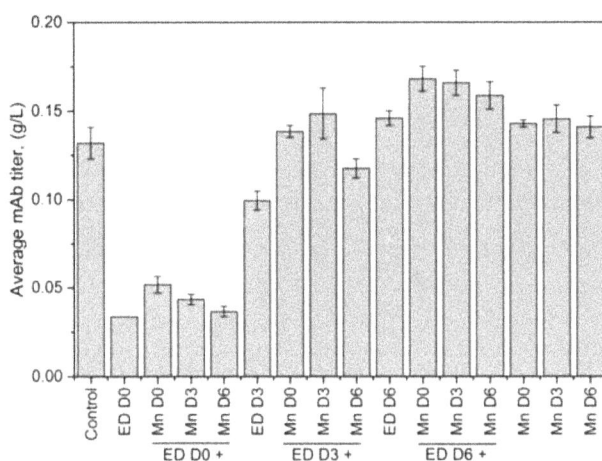

Figure 2. Average antibody titer for the 16 experimental conditions. Mn and ED refer to the media supplements MnCl$_2$ and EDTA, while D0, D3 and D6 refer to Day 0, Day 3 and Day 6 after inoculation, respectively. Error bars represent the range of biological replicates ($n = 2$).

3.2. Early Addition of MnCl$_2$ Alters the Glycan Distribution Significantly

The addition of EDTA and MnCl$_2$ at different time points altered the glycan distribution, with earlier addition of MnCl$_2$ having the more significant effect. Figure 3 shows the effect of media supplementation on select glycan species, with panels 3(a), (b) and (c) showing the changes in the glycan profile as a result of media supplementation with EDTA on D0, D3, and D6, with or without additional MnCl$_2$ supplementation, and 3(d) showing the impact of MnCl$_2$ addition in the absence of EDTA supplementation. (See Supplementary Information Figure S3 for the average relative glycan percentage of all the glycan species.)

Adding EDTA to the cell culture media on D0 (Figure 3a) decreased the amount of biantennary fucosylated species, FA2, by 5.58%, which was offset by a concomitant increase in its galactosylated isoforms, FA2G1 (4.52%) and FA2G2 (1.52%). (See supplementary information Table S1 for a complete list of experimentally observed glycan species and their structures.) The mannosylated species M5 also increased by 1.52%, with a corresponding reduction of 1.79% in the concentration of the biantennary group A2 (which is produced from M5 in the Golgi compartment). Adding MnCl$_2$ on D0, D3, or D6 to media in which EDTA was introduced on D0 resulted in a further increase in the relative concentrations of the galactosylated isoforms FA2G1 and FA2G2, in addition to a decrease in the relative percentage of FA2, A2, and A2G1 species, compared to the distribution in the control flask. However, the increase in FA2G1 and FA2G2 was more pronounced when MnCl$_2$ was introduced on D3 or D6 after adding EDTA. When EDTA was added on D3 (Figure 3b) the relative percentages of the biantennary species FA2, FA2G1, and A2G1 decreased, while the relative percentages of glycan species M5, A1, and FA1 increased. When MnCl$_2$ was added on D0 followed by EDTA supplementation on D3, the concentrations of M5 and FA1 increased, while the concentration of FA2 decreased. Additionally, there was an increase in the relative percentage of A2G1 that was offset by the decrease in the concentration of A1G1. When both EDTA and MnCl$_2$ were added to the media on D3, the change in the glycan distribution mirrors the trend observed when only EDTA was introduced on D3. The main difference occurs when MnCl$_2$ was added to the media after EDTA addition, i.e., on D6. Here, the relative percentages of M5, FA1, FA2G1, and FA2G2 increased while the relative percentages of A2, A2G1, and FA2 decreased, similar to the trends observed with the addition of MnCl$_2$ on D3 or D6 following EDTA supplementation on D0. Although the late addition of EDTA on D6 of the cell culture did not affect the glycan profile significantly (Figure 3c), for those glycan species whose relative percentages change, the trend was similar to that observed with EDTA supplementation on D0 and D3, i.e., an increase in the relative percentage of M5 and A1 accompanied by a decrease in that of FA2 and FA2G1. When the media was first supplemented with MnCl$_2$ on D0 or D3, followed by EDTA supplementation on D6, the relative percentage of FA2 species decreased while the relative percentages of M5, A2 and FA2G1 species increased. A simultaneous addition of both EDTA and MnCl$_2$ toward the end of the batch increased the relative percentage of A2G1 while decreasing the relative amount of its fucosylated isoform FA2G1.

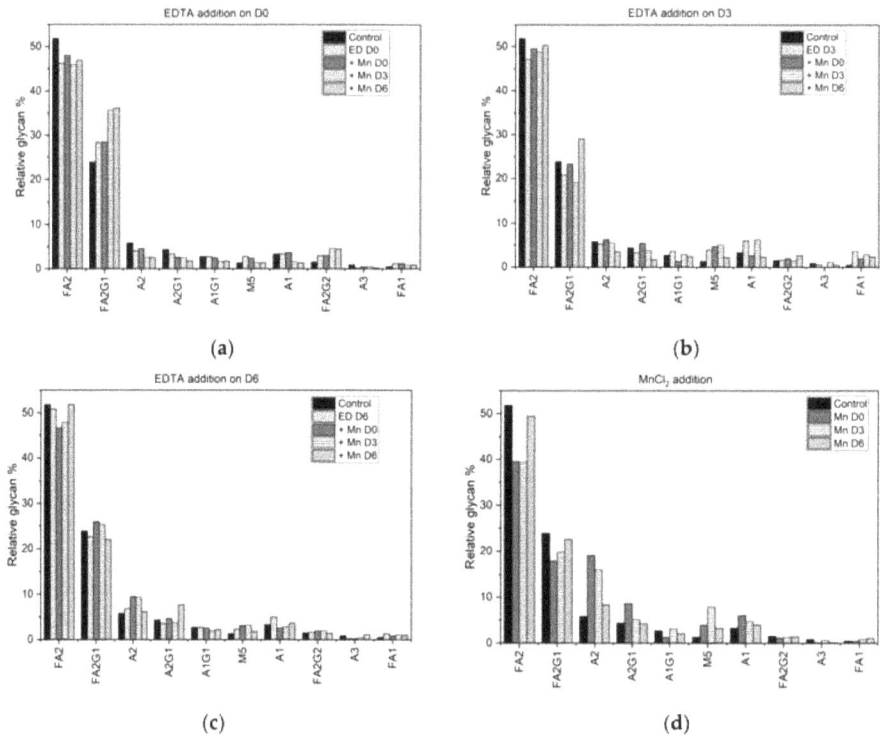

Figure 3. Average relative glycan percentage of select IgG1 glycans produced in CHO-K1 cells when: (a) EDTA is added on D0 with no MnCl$_2$ supplementation or with MnCl$_2$ supplementation on D0, D3, and D6; (b) EDTA added on D3 with no MnCl$_2$ supplementation or with MnCl$_2$ supplementation on D0, D3, and D6; (c) EDTA added on D6 with no MnCl$_2$ supplementation or with MnCl$_2$ supplementation on D0, D3, and D6; and (d) MnCl$_2$ is added on D0, D3, and D6.

The solo effect of MnCl$_2$ on the glycan profile was observed by supplementing the media with MnCl$_2$ in the absence of EDTA on D0, D3 or D6 (Figure 3d). Adding MnCl$_2$ on D0 decreased the relative percentage of the FA2 species from the control value of 51.79% (with a 95% confidence interval range of ±0.94%) to 39.50% (±1.55%). Similarly, the relative percentage of the monogalactosylated species FA2G1 dropped from the control value of 23.87% (±0.94%) to 17.95% (±1.55%). This decrease in the fucosylated species was offset by an increase in the biantennary species A2 (which increased by 13.24%), the mannosylated species M5 (2.62%), and the galactosylated biantennary species A2G1 (4.29%). It is important to note that this trend (observed in the glycan distribution when only MnCl2 is added to the shake flask) is consistent with the trend previously observed in experiments performed using the same cell line [30]. Introducing MnCl$_2$ to the culture medium on D3 resulted in a similar trend, with the decrease in the fucosylated species being offset by a significant increase in the biantennary species A2 (which increased by 10.20%), the mannosylated species M5 (increasing by 6.59%) and only a marginal increase in the A2G1 species (by 0.76%). Adding $MnCl_2$ during the peak exponential phase (D6) produced similar changes in the glycan distribution, but these changes were smaller in comparison to the changes in the glycan distribution due to earlier addition of $MnCl_2$. The concentrations of the fucosylated species FA2 and FA2G1 decreased by 2.36% and 1.28% respectively, with the corresponding increases in the concentrations of A2 and M5 species being 2.52% and 1.94%, respectively.

While such qualitative discussions of the glycan distribution profiles may be instructive in general, for our purposes in this work, a quantitative analysis relating the changes in the experimental conditions to the observed changes in the glycan profile is more informative.

3.3. The Type of Media Supplement and the Time of Addition both Showed Statistically Significant Effects on Glycan Distribution

The concentrations of $MnCl_2$ and EDTA and the time of addition of each one constitute the four factors in the $(2^2, 3^2)$ mixed factorial experimental design used in this study. In general, the full complement of this factorial experimental design consists of $4 \times 9 = 36$ independent experimental conditions, and the resulting model, in principle, consists of 35 main and interaction effects. The statistically significant factor coefficients (at the significance level of $\alpha = 0.05$) estimated using ANOVA, were used to generate a "gain" matrix (see Table S3 in Supplementary Information), whose elements represent "by how much" each output variable (relative glycan distribution) will change in response to a unit change in each input factor, including multi-factor interactions (since multiple combinations of single inputs are considered as valid inputs in this case). However, in this case, the full factorial experiment consists of only 16 unique cases (see Materials and Methods); consequently, eliminating the redundant rows in our gain matrix resulting in a reduced gain matrix with 15 input factors, some of which are multi-factor combinations (see Supplementary Information Table S4). Figure 4 shows a heat map of the elements of the gain matrix, indicating which input factor affects which glycan. The input factors include both main effects such as the concentration of media supplements (EDTA, $MnCl_2$), times of addition (Mn T1, ED T1, etc.) as well as interaction effects, such as those between the times of addition (e.g., Mn T1-ED T1) or between concentration and the times of addition (e.g., EDTA—Mn T2, EDTA—Mn T1—ED T2, etc.). The colors indicate the direction of the effect (green for increase; red for decrease) and the intensity indicates magnitude.

An examination of the heat map indicates that the glycan species FA2, FA2G1, and A2 are affected by most of the factors and their combinatorial interactions. For instance, the concentration of the most abundant glycoform, FA2 (which accounts for nearly 52% of the glycan concentration in the control sample), is affected by the concentration of $MnCl_2$, the two-way interaction of $MnCl_2$ and EDTA, and the late stage addition of $MnCl_2$. In particular, a unit change in the concentration of $MnCl_2$ causes an increase in the average concentration of FA2 (as indicated by the positive coefficient for the $MnCl_2$ effect), while introducing $MnCl_2$ on D3 causes a reduction in the average concentration of FA2. By contrast, the monogalactosylated form FA2G1 is not affected by changes in the media concentration of $MnCl_2$; rather it is influenced by changes in the concentration of EDTA, the early addition of $MnCl_2$, and the two-way interaction of $MnCl_2$ and EDTA. Further, we observe that two of the interaction factors, Mn T1-ED T1 and Mn T2-ED T1, have statistically significant and opposing effects on the concentration of FA2G1. The factor Mn T1-ED T1 represents the two-way interaction of adding MnCl2 on D0 and adding EDTA on D0, whereas Mn T2-ED T1 represents the two-way interaction of adding MnCl2 on D3 when EDTA is introduced on D0. We observe therefore that the concentration of FA2G1 is affected by multiple input factors, including complex interactions between the amount of supplements added, and the times of addition of the supplements. One is thus able to assess the impact of each input factor on the response of all other individual glycan species in similar fashion. As indicated by the heat map, the effects of higher order interactions on most of the glycan species are negligible (if they exist at all) because estimates of the coefficients associated with most interaction effects are not statistically significant.

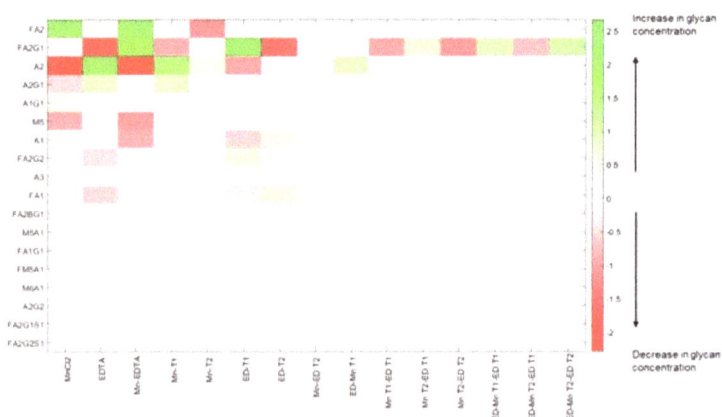

Figure 4. Heat map of significant factor coefficients ($\alpha = 0.05$, see Table S3 for details) obtained from experimental data analysis using ANOVA. The input factors are on the horizontal axis, and individual glycan species on the vertical axis. The color red indicates a decrease in the concentration of a particular glycan in response to a positive change on the input factor in questions, while green indicates an increase. The color intensity represents the magnitude of the significant coefficient, with increasingly darker hues indicating increasingly larger magnitudes and progressively lighter hues indicating commensurately lower magnitudes.

4. Discussion

While the addition of specific media supplements has been studied widely for its effect on the quality attributes of mAbs, such studies have been limited to the introduction of media supplements exclusively at the start of the culture, and the results, when quantitative, have yielded only isolated single factor relationships. The results from the current study show that introducing media supplements at different time points during cell culture does in fact have an effect on cell growth conditions and antibody glycosylation distribution, and the effects can be quantified globally and potentially used to design effective control schemes.

Specifically, we have shown that earlier addition of EDTA is detrimental to cell growth and results in a decrease in antibody titer. When EDTA was added on D0 (at the inoculation stage) the peak viable cell density (VCD) remained close to the seeding density, indicating a hampering of cell growth. This result is consistent with the well-known fact that EDTA is toxic to cells [35]. The decline in the viable cell densities due to EDTA addition during the growth phase can also be attributed to the removal of such trace metals as Ca^{++}, Zn^{++} etc. that are essential for cell survival, from the media (as a result of EDTA's chelating effect). By contrast, when both EDTA and $MnCl_2$ were added on D0, the peak VCD increased two-fold. While this peak VCD is lower than the peak VCD of the control flask (5.78×10^6 cells/mL), the increased viability can be attributed to the presence of excess $MnCl_2$ titrating EDTA, resulting in reduced cytotoxicity. When EDTA was added on D3, the cells were in the middle of the growth phase and the addition of the cytotoxic EDTA hampered further growth, leading to a steep decline in the cell viability beyond D3 (Figure 1b). By contrast, when $MnCl_2$ was added to the media (on D0 or D3) in the presence of EDTA, the cells did not experience a similar reduction in viability. When EDTA was added on D6, the cells were already at the end of the growth phase and hence the introduction of EDTA did not alter the cell viability. However, unlike the other cases where EDTA addition on D0 or D3 resulted in low titers, adding EDTA on D6 along with $MnCl_2$ supplementation on D0, D3 or D6, resulted in end of run (EOR) titers higher than the titer value in the control flask. The increase in antibody titer in the presence of EDTA was observed by others as well [36,37] and is attributed to the inhibition of antibody reduction during cell lysis. Further, analyzing the EOR titer data using ANOVA shows that the factor coefficients for the concentration of EDTA, the concentration of $MnCl_2$, and time of addition of EDTA are all statistically significant (at a significance level of $\alpha = 0.05$). The expected change in the EOR titer in response to a unit change in any of these factors is quantified by the magnitude of that factor coefficient, while the sign of the factor coefficient indicates the direction of change. Thus, for example, a unit positive step change in the concentration of EDTA (with a factor coefficient of -0.017) or time of addition of EDTA (factor coefficient -0.032) results in a decrease in EOR titer; a unit positive step change in the concentration of $MnCl_2$ (factor coefficient 0.008) causes an increase in titer. Quantifying the effect of these input factors on the EOR titer provides a rational basis for selecting which particular supplement to add, and how much of it to add, at a given stage of cell culture in order to maximize product yield. However, any media supplementation strategy must meet not just the desired specifications for final titer but also for product quality, i.e., to be effective, the implemented media supplementation strategy must not alter the glycan distributions significantly.

The EOR titer represents the total amount of antibody accumulated at the end of the batch, while the measured glycan distribution represents the relative amount of each individual glycan isoform that has accumulated over the duration of the batch. Now, the relative amount of individual glycan species is a function of antibody productivity and it changes over the course of the batch. In our case, the addition of different media supplements at different stages of batch cultures affected both viability and antibody titer; consequently, the observed change in the glycan distribution has been induced by dynamic media supplementation and changes in productivity. To develop a mechanistic understanding of the effect of dynamic media supplementation on the glycosylation profile, therefore, we first decouple the effect of antibody productivity on the glycan distribution from the overall change observed at the end of the batch. One such decoupling approach involves estimating the mass fractions of specific glycoforms produced at different stages of cell culture using the expression [38]:

$$f_i = \frac{\left[mAb_i\big|_{t= t_1}\right] - \left[mAb_i\big|_{t= t_2}\right]}{\left[mAb_{Tot}\big|_{t= t_1}\right] - \left[mAb_{Tot}\big|_{t= t_2}\right]}, \tag{4}$$

for f_i, the fraction of mAb glycoform i produced in the time period $[t_1, t_2]$ relative to the total amount of antibody secreted in the same period. However, we cannot use this expression for our purpose because we only measured EOR titer and final glycan distribution, not intermittent antibody titer or glycan concentration. Consequently, we propose an alternative metric-based solely on the final titer and glycan measurements.

To illustrate, consider the glycan distribution in the control flask and in the flask with $MnCl_2$ added on D6. In both flasks, the cell growth profile and antibody productivity will be the same until Day 6, when $MnCl_2$ is introduced to the latter flask. Thus, the amount of ith glycoform fractions accumulated between Day 6 (D6) and the end of the run (EOR) for the two flasks can be written as:

$$f_{Mn6/i} = \frac{\left[mAb_{Mn6/i}\big|_{t= EOR}\right] - \left[mAb_{Mn6/i}\big|_{t= D6}\right]}{\left[mAb_{Mn6/Tot}\big|_{t= EOR}\right] - \left[mAb_{Mn6/Tot}\big|_{t= D6}\right]}, \tag{5}$$

$$f_{control/i} = \frac{\left[mAb_{control/i}\big|_{t= EOR}\right] - \left[mAb_{control/i}\big|_{t= D6}\right]}{\left[mAb_{control/Tot}\big|_{t= EOR}\right] - \left[mAb_{control/Tot}\big|_{t= D6}\right]}, \tag{6}$$

Recognizing that the D6 values in Equations (5) and (6) above are identical, we can eliminate the intermittent time point by simple arithmetic manipulations and obtain the change in the accumulation of the ith glycoform-based solely on EOR values, as:

$$\Delta f_i = \frac{\left[mAb_{Mn6/i}\big|_{t= EOR}\right] - \left[mAb_{control/i}\big|_{t= EOR}\right]}{\left[mAb_{Mn6/Tot}\big|_{t= EOR}\right] - \left[mAb_{control/Tot}\big|_{t= EOR}\right]}, \tag{7}$$

Thus, this fractional difference allows us to group together different experimental conditions with similar antibody titers, making it possible to compare final glycan distributions and hence quantify the effect of individual media supplements on the glycan profile appropriately. Such analyses extended to other experimental conditions produced the comparative fractional difference in the glycoform distribution shown in Figure 5.

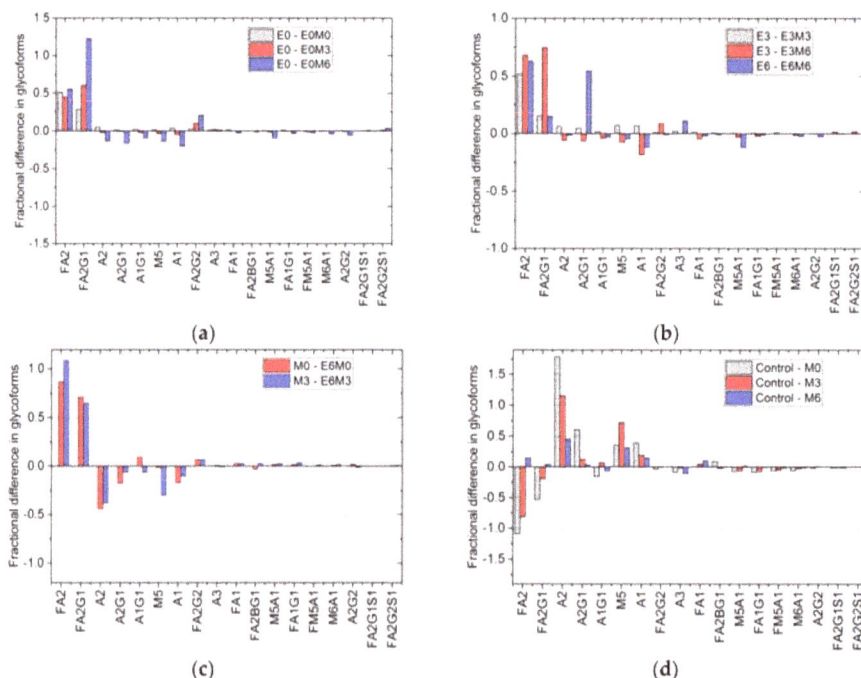

Figure 5. Fractional difference in glycoform distribution due to (**a**) $MnCl_2$ addition on D0 (cross-hatched), D3 (red bar), and D6 (blue bar) after EDTA addition on D0, relative to the baseline glycan distribution in EDTA D0 supplemented flask; (**b**) $MnCl_2$ addition on D3 (cross-hatched), and D6 (red) after EDTA addition on D3 relative to the baseline glycan distribution in EDTA D3 supplemented flask, and $MnCl_2$ addition on D6 after EDTA addition on D6 (blue) relative to the baseline glycan distribution in EDTA D6 supplemented flask; (**c**) EDTA addition on D6 after $MnCl_2$ addition on D0 (red) and D3 (blue) relative to the respective glycan distributions due to $MnCl_2$ addition on D0 and D3; and (**d**) $MnCl_2$ addition on D0 (cross-hatched), D3 (red bar), and D6 (blue bar) relative to the glycan distribution in the control case.

A comparison of the fractional difference in the glycan distribution in flasks where $MnCl_2$ is added to D0 EDTA supplemented flasks relative to the glycan distribution in D0 EDTA supplemented flasks (Figure 5a), shows the following: an increase in the amount of FA2 (by nearly 50% in all cases), FA2G1 (by 29% during D0 supplementation, 60% during D3 supplementation, and 122% during D6 supplementation), and FA2G2 (by 3%, 10%, and 20% respectively), with a relative decrease in A2 and M5 by 2% and 4% when $MnCl_2$ is added on D3, and nearly 14% when $MnCl_2$ is added on D6. A similar trend is observed in the fractional difference in the glycan distribution in flasks where $MnCl_2$ is added on D3 and D6 to D3 EDTA supplemented flasks, and when $MnCl_2$ is added on D6 to D6 EDTA supplemented flasks (Figure 5b). Again, we notice an increase in FA2, FA2G1 and FA2G2, with a decrease in A2, M5, and A2G1 observed only when $MnCl_2$ is added on day 6 after the addition of EDTA on D3. Previous studies have shown that adding $MnCl_2$ produces an upregulation of galactosyltransferase enzymes [30], and subsequently in increased galactosylation [26,27]. Hence, the increase in the amount of FA2G1 and FA2G2 species can be attributed to the effect of late stage manganese addition on the galactosyltransferase enzyme. Figure 5c shows fractional difference when EDTA is added on D6 after $MnCl_2$ has been added to the culture on D0 and D3. These differences are calculated relative to the glycan distribution observed due to the addition of $MnCl_2$ on D3 and D6, respectively. We note that the fractional difference in the fucosylated species FA2 and FA2G1 is positive when EDTA is added after $MnCl_2$ supplementation, indicating that the addition of EDTA increases the concentration of these species relative to their respective concentrations in $MnCl_2$ supplemented cultures. Also, a comparison of the fractional difference in the glycan distribution when $MnCl_2$ was introduced on D0, D3, or D6 relative to the glycan distribution in the control flask (Figure 5d), shows that the relative concentrations of FA2 and FA2G1 species decreased in flasks with D0, while late stage addition of $MnCl_2$ did not have a significant effect on the overall glycan distribution. Taken together, our findings indicate that the latter addition of EDTA reverses the changes in the glycan distribution induced by $MnCl_2$.

Fractional difference analysis helps to identify which glycan species are altered as a result of the addition of specific media supplements, but not why those particular glycan species changed. To identify the kinetic mechanisms underlying the changes observed in the glycan distribution upon adding $MnCl_2$ to the media, we use an existing dynamic mathematical model of glycosylation [39] to simulate the system response under control conditions and when $MnCl_2$ is added on D0, and obtain predictions of both the control and D0 glycan distributions (see Supplementary Information section S4 for details). The simulation results show that changes in the glycan distribution due to the addition of $MnCl_2$ can be directly attributed to the changes in the reaction rates associated with the FucT enzyme.

In addition to fractional difference analysis, we use the glycan indices computed for each experimental condition (Table 2), as a metric for quantitative assessment of the change in the final glycosylation profile caused by the addition of different media supplements. Specifically, a comparison of the individual glycan indices for each condition against the corresponding values under control conditions allows us to establish, objectively, that altering the availability of $MnCl_2$ in the media using a chelating agent reverses the changes in the glycan distribution. While $MnCl_2$ addition on D0 resulted in a decrease in the fucosylation index from 79.9% in the control flask to 60.5%, we see that subsequent introduction of EDTA on either D0, D3, or D6 reversed that trend. Early stage addition of EDTA on D0 increased the fucosylation index to 82.46%, while adding EDTA on D3 and D6 resulted in fucosylation indices of 78.8%, and 76.5% respectively. Similarly, the decrease in the galactosylation index, upon adding $MnCl_2$ on D0, from 17.9% in the control flask to 15.8%, could be reversed by subsequently adding EDTA on D3. Adding $MnCl_2$ to the media on D3 reduced the fucosylation index to 61.9%, but if we then added EDTA on D6, the fucosylation index increased to 77.8%, which is comparable to the value of 79.9% in the control flask. It is important to note however, that the reversal in the glycan indices observed due to the addition of EDTA on D0 and D3 is achieved at the expense of reduced titer and reduced viability, as discussed above. The indication from our results is that changes in the glycan distribution due to $MnCl_2$ addition can be reversed only when EDTA is added to the media after $MnCl_2$ addition. Thus, the effect of $MnCl_2$ supplementation can be reversed, without decreasing productivity, by adding EDTA on D6 to $MnCl_2$ supplemented media, providing a means of ensuring higher productivity without altering glycan distribution.

Table 2. Galactosylation index (GI) and fucosylation index (FI) for each experimental condition.

Experimental Condition [1]	Galactosylation Index (GI)	Fucosylation Index (FI)
Control	17.9	79.9
Mn D0	15.8	60.5
Mn D0/ED D0	20.6	82.5
Mn D0/ED D3	18.0	78.8
Mn D0/ED D6	19.3	76.5
Mn D3	15.7	61.8
Mn D3/ED D0	25.2	88.6
Mn D3/ED D3	15.1	73.9
Mn D3/ED D6	18.0	77.8
Mn D6	16.7	76.3
Mn D6/ED D0	24.8	90.3
Mn D6/ED D3	19.9	86.4
Mn D6/ED D6	17.9	77.3
ED D0	21.6	81.0
ED D3	16.5	75.3
ED D6	16.9	77.7

[1] Mn and ED refer to the media supplements $MnCl_2$ and EDTA, while D0, D3 and D6 refer to Day 0, Day 3 and Day 6 after inoculation, respectively.

Although such observations as these provide useful qualitative information about the system, they cannot be used to develop an automatic control strategy; that requires a quantitative representation (and hence understanding) of the effects of media supplementation on glycan distribution. Such quantitative representation can be obtained via formal analysis of the experimental data using analysis of variance (ANOVA) to generate the process gain matrix, $\mathbf{K}$, as described in Materials and Methods. Singular value decomposition of the gain matrix $\mathbf{K}$ produces a (diagonal) matrix of singular values, Σ, and two unitary matrices, $\mathbf{W}$ and $\mathbf{V}^T$. Together these three matrices provide a particularly insightful representation of the process information encapsulated in the gain matrix, $\mathbf{K}$: Equation (3) is transformed into a series of n individual and independent equations (see [29]) where, in each case, a linear combination of the original process input factors, with weighting coefficients from the matrix $\mathbf{W}$, now constituting an "input mode", μ_i, is connected through the associated singular value σ_i, to the corresponding linear combination of the output glycans, (with weighting coefficients from the matrix $\mathbf{V}$), now constituting an output mode η_i [29]. Furthermore, as a result of this decoupling transformation, the magnitude of each singular value naturally quantifies the extent to which the output mode in question will change in response to a change in the corresponding input mode. Thus, the larger the value of σ_i, the greater will be the change in the corresponding output mode η_i as a result of changes in the input mode μ_i, so that output modes associated with larger values of σ_i will be more "controllable" than modes associated with smaller values of σ_i.

The first ten singular values (σ_1–σ_{10}) for our experimental system are listed in Table 3 in decreasing order of magnitude. As modes associated with singular values of smaller magnitude are less controllable, we limit our analysis only to those modes that are practically controllable; we do this by using a threshold cutoff value, σ^*, arbitrarily selected to be 0.50 in this example, thereby limiting our analysis to the first five singular values. From a process control perspective, modes associated with singular values below this threshold are considered to be of no practical importance since, for all intents and purposes, they are not controllable.

Table 3. Singular values obtained from singular value decomposition of the matrix of significant factor coefficients.

σ_1	6.22
σ_2	3.68
σ_3	2.21
σ_4	0.80
σ_5	0.61
σ_6	0.45
σ_7	0.41
σ_8	0.31
σ_9	0.10
σ_{10}	0.05

Next, since by definition, each input–output mode pair represents the linear combination of output glycan species that can be controlled by manipulating the specific input factors in the corresponding input mode, the coefficients of each output factor in the output modes and of each input factor in the input modes provide further information about the relative influences exerted by each original input factor on each output factor. Specifically, the coefficient of a particular output factor in a particular mode represents the magnitude by which the relative percentages of those particular glycans will change in response to a unit change in the input mode. On the other hand, the coefficient of a particular input factor in the associated input mode corresponds to the relative contribution of that input factor to the unit change the input mode in question. Thus, the inputs with the largest coefficients in an input mode represent the dominant factors and hence the largest contributors to the influence of that mode, while the output glycans with the largest coefficients in an output mode represent those species whose relative percentages will change the most under the influence of the input modes. The input-output mode pairs and their associated coefficients are shown in Figure 6.

Because it is associated with the largest singular value of $\sigma_1 = 6.62$, the first output mode η_1 is the most controllable output mode. The value of σ_1 represents the change in the overall output mode η_1 resulting from a unit change in the input mode μ_1. For this output mode, we note that the dominant glycan species are A2, with a coefficient of 0.64, followed by FA2G1 (with a coefficient of −0.62), FA2 (−0.38), M5 (0.14), and A1 (0.13), indicating that a unit positive change in the input mode μ_1 will result in an increase in A2, M5, and A1, accompanied by a decrease in FA2G1 and FA2, each in the amount indicated by the identified coefficients. The biantennary species A2, with the largest coefficient, is the most controllable glycan in the first mode, followed by FA2G1 and FA2. The indicated coupling between the glycans A2, FA2G1, and FA2 makes sense because an increase in the afucosylated glycoforms occurs at the expense of the fucosylated forms, as our experimental results show. The associated input mode μ_1 is a linear combination of different input factors representing the media supplements $MnCl_2$ and EDTA as well as the times of their addition. Mode μ_1 is primarily dominated by the interaction of MnCl2 and EDTA, and the concentrations of EDTA and $MnCl_2$, with associated coefficients −0.58, 0.44, and −0.39 respectively, indicating that the addition of these two media supplements has opposing individual effects on output mode 1. Early stage addition of EDTA and $MnCl_2$, denoted by the factors EDT1 and MnT1, with associated coefficients −0.38 and 0.25 respectively, also exert important influences on the first output mode. Based on the different elements that comprise the first input mode, this analysis indicates that one can control the glycans in output mode η_1 by adjusting the concentrations of the two supplements at the early stages of the cell culture.

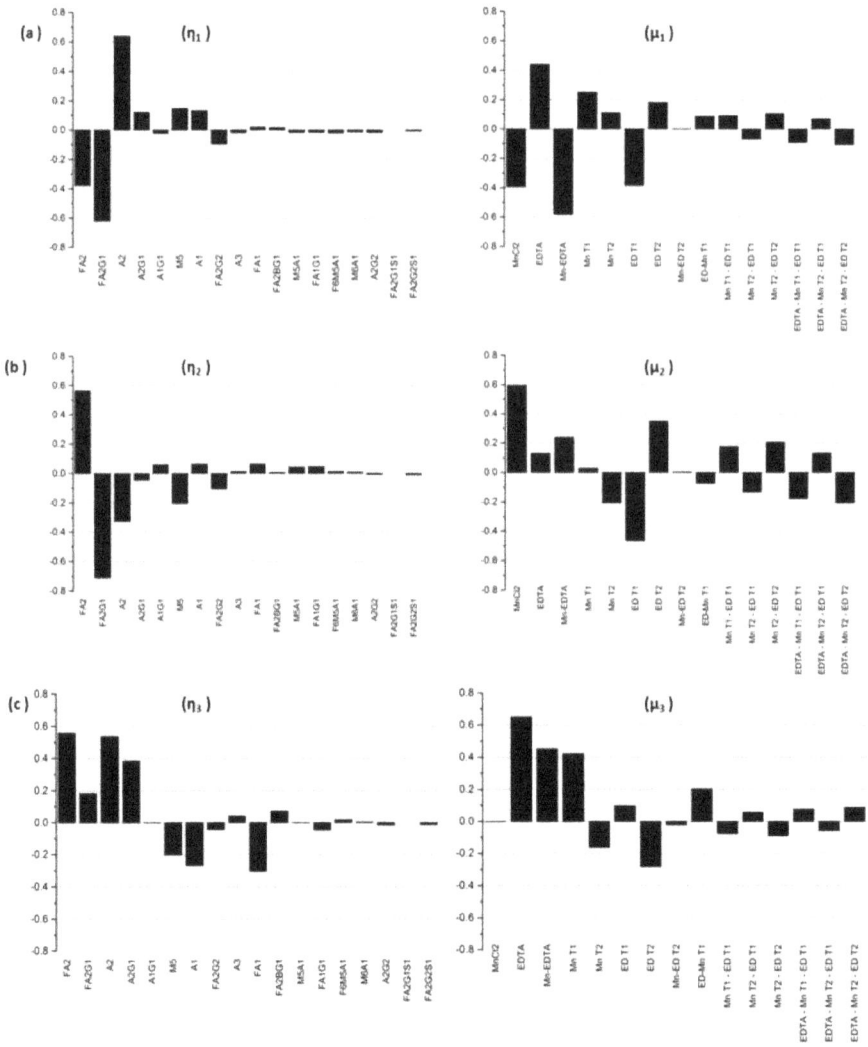

Figure 6. *Cont.*

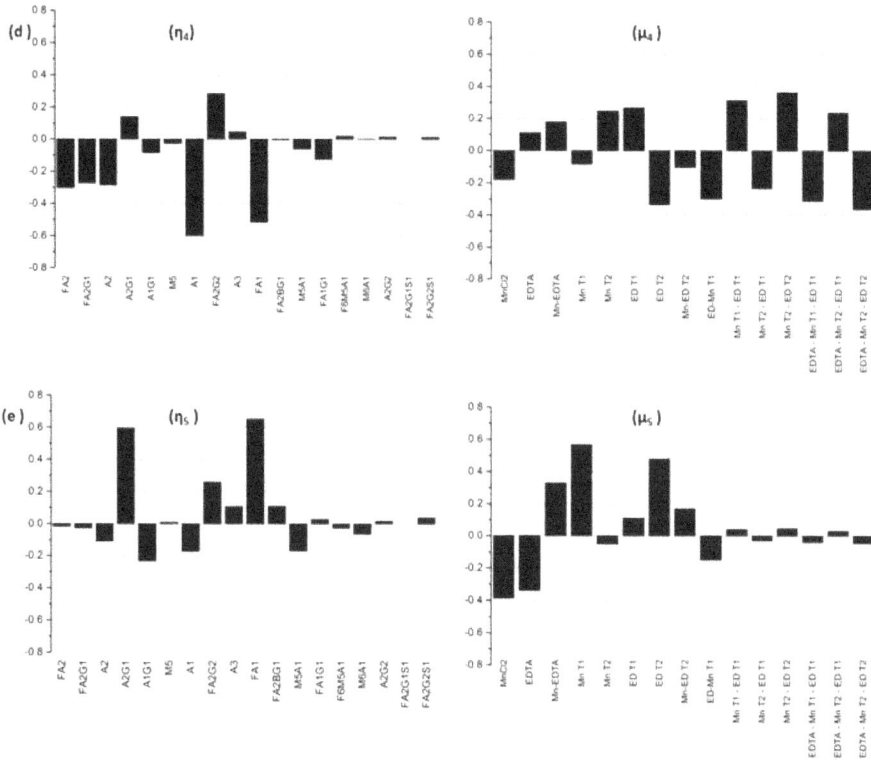

Figure 6. Graphical representation of the coefficients associated with the first five input and output modes (with $\sigma_i \geq \sigma^* = 0.5$) obtained from controllability analysis. (**a**) Output mode η_1 and the corresponding input mode μ_1; (**b**) Output mode η_2 and the corresponding input mode μ_2; (**c**) Output mode η_3 and the corresponding input mode μ_3; (**d**) Output mode η_4 and the corresponding input mode μ_4; (**e**) Output mode η_5 and the corresponding input mode μ_5.

The next controllable output mode η_2 is associated with the singular value $\sigma_2 = 3.68$, and the linear combination of glycans represented by this mode is dominated by the glycan species FA2G1, FA2, A2, M5, and FA2G2. The coefficients associated with these glycans are $-0.71, 0.56, -0.33, -0.20$ and -0.11, respectively, indicating that a unit positive change in the input mode μ_2 will cause a relative increase in FA2 while causing the other glycan species to decrease. The increase in FA2 coupled with the decrease in FA2G1 and FA2G2 indicates that perturbations to the input mode μ_2 affect the galactosylated species particularly. Since the singular values are arranged in decreasing order, the influence of mode μ_2 on output mode η_2 is less than that of μ_1 on mode η_1. The largest coefficients in mode μ_2 are associated with the input factors MnCl$_2$ (with a coefficient of 0.6), the early stage addition of EDTA denoted by factors EDT1 (with a coefficient of -0.47) and EDT2 (0.35), and the interaction effect of MnCl$_2$ and EDTA (0.24).

A unit positive change to the input mode μ_3 causes the following changes in the relative concentrations of the glycan species that comprise output mode η_3 (with $\sigma_3 = 2.21$): A2, A2G1, FA2, and FA2G1 increase, and M5, A1, and FA1 decrease, simultaneously. This indicates that input mode μ_3 can be used to increase the biantennary species, but at the expense of a (perfectly logical) concomitant decrease in the species that are upstream of these biantennary species. The input mode μ_3 is dominated by EDTA (with associated coefficient 0.65), the interaction effect of MnCl$_2$ and EDTA (with coefficient 0.45), and the early stage addition of $MnCl_2$ denoted by the factor MnT1). In each of the three input modes, the coefficients associated with the interaction effect of $MnCl_2$ and EDTA indicate the importance of this combination input factor in altering the concentrations of the glycan species associated with the respective modes.

The fourth and the fifth modes are less controllable, as they are associated with singular values of comparatively smaller magnitudes ($\sigma_4 = 0.80$ and $\sigma_5 = 0.61$). η_4 is dominated by glycans A1, A2, FA1, FA2, FA2G1, and FA2G2, while η_5 is dominated by FA1, A2G1, and FA2G2. The input mode μ_4 is dominated by the interaction effects Mn T2-ED T2 and EDTA*Mn T2-ED T2 representing the interaction between the time of addition of the media supplements. By contrast, the predominant factors in mode μ_5 are MnT1 and EDT2, which represent the addition of MnCl2 on D0 and EDTA on D3 respectively.

It is worth mentioning the following facts about the controllability analysis presented here: (i) even though it is a theoretical analysis, the process gain matrix, **K** (on which the analysis is-based), was obtained entirely from experimental data. The information it contains about how changes in the inputs (amount of supplement added, and time of addition) affect glycan species, come directly from our designed experiments; (ii) Notwithstanding, a follow-up independent experimental validation of the modal analysis predictions (which contributor to which mode is affected in what manner) will be valuable. Such a validation experimental study, which lies outside the intended scope of this current work, is slated for future work. That study's focus will not only be to validate the controllability analysis results; it will also encompass the more consequential design and implementation of a control system to achieve effective control of glycan distribution during a batch.

For now, we note that the controllability analysis provides a data-based quantification of the effect that the addition of specified amounts of particular media supplements and the respective times of addition jointly have on the output glycan distribution at the end of the batch. As discussed above, introducing these media supplements dynamically also results in quantifiable changes in the antibody titer. Thus, the dynamic supplementation strategies discussed here present a challenging problem involving a trade-off between product yield and product quality. To be effective, a control strategy based on these considerations, must therefore be carefully designed to resolve these conflicts appropriately in order to optimize both the titer and the quality simultaneously. These matters will be addressed theoretically and experimentally in a follow up study.

In closing, we note that

(1) the relationships between time-dependent changes in media supplements and the corresponding changes in glycan distribution are (understandably) complex and not necessarily obvious or easily amenable to qualitative thinking; but

(2) controllability analysis via singular value decomposition, and the resulting input-output mode pairs determined specifically for our experimental system, using data from carefully designed experiments, have enabled us to identify which input factors are best manipulated, in order to effect changes in the relative percentage of specific glycan species;

(3) in addition, the coefficients in the equations representing the input and output modes allow us to quantify by how much we expect the glycan species to change in response to specific time-dependent media supplementation actions.

5. Conclusions

There is growing interest in evaluating the role of media supplements, especially $MnCl_2$, in modulating glycoform distributions in mAbs. However, most media supplementation studies (where supplements are added to the media before starting the batch) do not take into consideration the impact of introducing media supplements at different stages of cell growth. In this paper, we have presented a systematic approach for evaluating the effects of time dependent media supplementation on the glycan profile, and provided a methodology for quantifying and analyzing the complex effects. Our results show that while it is important to consider which supplements are to be added to the media in order to alter specific glycan species, when they are to be added is just as important. In addition to this general observation about what to change and when, our results and analysis technique also demonstrate how to quantify "by how much" we can expect specific glycan species to change as a result of the changes in the media supplements.

For instance, we observe that early stage addition of $MnCl_2$ affects fucosylated species and alters the glycan distribution more significantly than a late stage addition. Similarly, early stage addition of EDTA affects not just the antibody titer, but also the relative percentages of biantennary and galactosylated species; late stage addition of EDTA does not alter the glycan distribution significantly. The glycan distribution profile is also affected by the addition of both EDTA and $MnCl_2$ to the media at different time points and it is possible to develop a mechanistic understanding of the effect of individual media components on the glycan distribution by studying the fractional difference in the glycan distribution. However, we note that early stage addition of EDTA is detrimental to cell growth and antibody production, while late stage addition of EDTA improves titer with no appreciable effect on glycan distribution. This study does not recommend using EDTA as a media supplement—instead, our work presents a systematic approach to studying the complex interactions between media supplements when introduced at different time points during batch culture. Additionally, our results demonstrate that changes in the glycan distribution profile due to the addition of $MnCl_2$ are not immutable; they can be reversed by adding EDTA after $MnCl_2$ has been added to the media. We were able to identify the specific combinations of input factors which, when manipulated, result in quantifiable changes in the relative percentage of specific glycan species via controllability analysis using our experimental data. For the specific experimental system, our analyses show that A2, FA2G1 and FA2 are the most controllable glycan species whose concentrations can be changed by early stage supplementation of EDTA and late stage supplementation of $MnCl_2$.

While this work has focused on establishing a rational framework for studying the influence of time-dependent media supplementation on the glycosylation profile, the techniques introduced here can be extended to tackle the complementary problem of designing and implementing appropriate control strategies to achieve desired glycan distribution profiles. Traditionally, the composition of cell culture media is fixed prior to starting the batch. However, we have demonstrated that introducing supplements at different points in time can influence both the productivity and the quality attribute of the antibody. Although we have examined only two specific media components in the current set of experiments, in principle, the systematic approach presented here can be extended easily to other media components such as amino acids, or trace metals. Future work will involve a comprehensive validation study where a control system will be designed on the basis of these results, and implemented on our experimental systems.

Supplementary Materials: The following are available online at http://www.mdpi.com/2073-4468/7/1/1/s1, Table S1: Experimentally observed glycan species and their masses; Table S2: Full factorial experimental design table; Table S3: Gain matrix generated from statistically significant ($p \leq 0.05$) coefficients obtained from ANOVA of the full factorial design experimental data; Table S4: Main and interaction effects and corresponding confounding factors; Table S5: "Reduced" gain matrix (**K**) obtained by eliminating the redundant rows from the gain matrix listed in Table S3; Table S6: Unitary matrix **W** obtained from SVD of gain matrix **K**; Table S7: Diagonal matrix of singular values (Σ) obtained from SVD of reduced gain matrix **K**; Table S8: Unitary matrix $\mathbf{V^T}$ obtained from SVD of reduced gain matrix **K**; Table S9: Reaction rules for generating the glycan reaction network used in simulations of the dynamic glycosylation model; Table S10: Enzyme and co-substrate concentrations used in the simulations of the dynamic glycosylation model; Figure S1: Comparison of experimental data and model fit for the glycan distribution profile obtained from (a) control flask; and (b) when $MnCl_2$ is added on D0; Figure S2: Normalized glucose concentration data for each condition tested; Figure S3: Average relative glycan percentage of IgG1 glycans produced in CHO-K1 cells.

Acknowledgments: The authors thank Life Technologies (Thermo Fisher Scientific) for donating custom media for this work, and Leila Choe at the Delaware Biotechnology Institute (Newark, DE, USA) for help with the mass spectrometry experiments and for extensive discussions on MALDI data analysis. The authors are also grateful for support from the National Science Foundation via grant # NSF-CBET 1034213.

Author Contributions: D.R., A.R. and B.O. conceived the paper; D.R. performed the experiments; D.R., A.R. and B.O. analyzed the data and wrote the manuscript.

Conflicts of Interest: The authors declare no conflict of interest.

References

1. Aggarwal, S. What's fueling the biotech engine? *Nat. Biotechnol.* **2007**, *25*, 1097–1104. [CrossRef] [PubMed]
2. Aggarwal, S. What's fueling the biotech engine-2012 to 2013. *Nat. Biotechnol.* **2014**, *32*, 32–39. [CrossRef] [PubMed]
3. Elvin, J.G.; Couston, R.G.; van der Walle, C.F. Therapeutic antibodies: Market considerations, disease targets and bioprocessing. *Int. J. Pharm.* **2013**, *440*, 83–98. [CrossRef] [PubMed]
4. Reichert, J.M. Marketed therapeutic antibodies compendium. *mAbs* **2012**, *4*, 413–415. [CrossRef] [PubMed]
5. Kim, J.; Kim, Y.-G.; Lee, G. Cho cells in biotechnology for production of recombinant proteins: Current state and further potential. *Appl. Microbiol. Biotechnol.* **2012**, *93*, 917–930. [CrossRef] [PubMed]
6. Kornfeld, R.; Kornfeld, S. Assembly of asparagine-linked oligosaccharides. *Annu. Rev. Biochem.* **1985**, *54*, 631–664. [CrossRef] [PubMed]
7. Cumming, D.A. Pathways and functions of mammalian protein glycosylation. *New Compr. Biochem.* **2003**, *38*, 433–455.
8. Berger, M.; Kaup, M.; Blanchard, V. Protein glycosylation and its impact on biotechnology. In *Genomics and Systems Biology of Mammalian Cell Culture*; Hu, W.S., Zeng, A.-P., Eds.; Springer: Berlin/Heidelberg, Germany, 2012; Volume 127, pp. 165–185.
9. Del Val, I.J.; Kontoravdi, C.; Nagy, J.M. Towards the implementation of quality by design to the production of therapeutic monoclonal antibodies with desired glycosylation patterns. *Biotechnol. Prog.* **2010**, *26*, 1505–1527. [CrossRef] [PubMed]
10. Liu, L. Antibody glycosylation and its impact on the pharmacokinetics and pharmacodynamics of monoclonal antibodies and fc-fusion proteins. *J. Pharm. Sci.* **2015**, *104*, 1866–1884. [CrossRef] [PubMed]
11. Read, E.K.; Park, J.T.; Brorson, K.A. Industry and regulatory experience of the glycosylation of monoclonal antibodies. *Biotechnol. Appl. Biochem.* **2011**, *58*, 213–219. [CrossRef] [PubMed]
12. Beck, A.; Wurch, T.; Bailly, C.; Corvaia, N. Strategies and challenges for the next generation of therapeutic antibodies. *Nat. Rev. Immunol.* **2010**, *10*, 345–352. [CrossRef] [PubMed]
13. Jefferis, R. Glycosylation as a strategy to improve antibody-based therapeutics. *Nat. Rev. Drug Discov.* **2009**, *8*, 226–234. [CrossRef] [PubMed]
14. Wacker, C.; Berger, C.N.; Girard, P.; Meier, R. Glycosylation profiles of therapeutic antibody pharmaceuticals. *Eur. J. Pharm. Biopharm.* **2011**, *79*, 503–507. [CrossRef] [PubMed]
15. Hossler, P.; Khattak, S.F.; Li, Z.J. Optimal and consistent protein glycosylation in mammalian cell culture. *Glycobiology* **2009**, *19*, 936–949. [CrossRef] [PubMed]
16. Yoon, S.K.; Choi, S.L.; Song, J.Y.; Lee, G.M. Effect of culture ph on erythropoietin production by chinese hamster ovary cells grown in suspension at 32.5 and 37.0 degrees C. *Biotechnol. Bioeng.* **2005**, *89*, 345–356. [CrossRef] [PubMed]
17. Ivarsson, M.; Villiger, T.K.; Morbidelli, M.; Soos, M. Evaluating the impact of cell culture process parameters on monoclonal antibody n-glycosylation. *J. Biotechnol.* **2014**, *188*, 88–96. [CrossRef] [PubMed]
18. Sou, S.N.; Sellick, C.; Lee, K.; Mason, A.; Kyriakopoulos, S.; Polizzi, K.M.; Kontoravdi, C. How does mild hypothermia affect monoclonal antibody glycosylation? *Biotechnol. Bioeng.* **2015**, *112*, 1165–1176. [CrossRef] [PubMed]
19. Ahn, W.S.; Jeon, J.J.; Jeong, Y.R.; Lee, S.J.; Yoon, S.K. Effect of culture temperature on erythropoietin production and glycosylation in a perfusion culture of recombinant cho cells. *Biotechnol. Bioeng.* **2008**, *101*, 1234–1244. [CrossRef] [PubMed]
20. Gawlitzek, M.; Estacio, M.; Furch, T.; Kiss, R. Identification of cell culture conditions to control n-glycosylation site-occupancy of recombinant glycoproteins expressed in cho cells. *Biotechnol. Bioeng.* **2009**, *103*, 1164–1175. [CrossRef] [PubMed]
21. Serrato, J.A.; Palomares, L.A.; Meneses-Acosta, A.; Ramirez, O.T. Heterogeneous conditions in dissolved oxygen affect n-glycosylation but not productivity of a monoclonal antibody in hybridoma cultures. *Biotechnol. Bioeng.* **2004**, *88*, 176–188. [CrossRef] [PubMed]
22. Kunkel, J.P.; Jan, D.C.H.; Jamieson, J.C.; Butler, M. Dissolved oxygen concentration in serum-free continuous culture affects n-linked glycosylation of a monoclonal antibody. *J. Biotechnol.* **1998**, *62*, 55–71. [CrossRef]

23. Borys, M.C.; Linzer, D.I.H.; Papoutsakis, E.T. Ammonia affects the glycosylation patterns of recombinant mouse placental lactogen-i by chinese hamster ovary cells in a ph-dependent manner. *Biotechnol. Bioeng.* **1994**, *43*, 505–514. [CrossRef] [PubMed]

24. Wong, N.S.C.; Wati, L.; Nissom, P.M.; Feng, H.T.; Lee, M.M.; Yap, M.G.S. An investigation of intracellular glycosylation activities in cho cells: Effects of nucleotide sugar precursor feeding. *Biotechnol. Bioeng.* **2010**, *107*, 321–336. [CrossRef] [PubMed]

25. Pacis, E.; Yu, M.; Autsen, J.; Bayer, R.; Li, F. Effects of cell culture conditions on antibody n-linked glycosylation—What affects high mannose 5 glycoform. *Biotechnol. Bioeng.* **2011**, *108*, 2348–2358. [CrossRef] [PubMed]

26. Gramer, M.J.; Eckblad, J.J.; Donahue, R.; Brown, J.; Shultz, C.; Vickerman, K.; Priem, P.; van den Bremer, E.T.J.; Gerritsen, J.; van Berkel, P.H.C. Modulation of antibody galactosylation through feeding of uridine, manganese chloride, and galactose. *Biotechnol. Bioeng.* **2011**, *108*, 1591–1602. [CrossRef] [PubMed]

27. Grainger, R.K.; James, D.C. Cho cell line specific prediction and control of recombinant monoclonal antibody n-glycosylation. *Biotechnol. Bioeng.* **2013**, *110*, 2970–2983. [CrossRef] [PubMed]

28. Surve, T.; Gadgil, M. Manganese increases high mannose glycoform on monoclonal antibody expressed in cho when glucose is absent or limiting: Implications for use of alternate sugars. *Biotechnol. Prog.* **2015**, *31*, 460–467. [CrossRef] [PubMed]

29. St. Amand, M.M.; Tran, K.; Radhakrishnan, D.; Robinson, A.S.; Ogunnaike, B.A. Controllability analysis of protein glycosylation in cho cells. *PLoS ONE* **2014**, *9*, e87973. [CrossRef] [PubMed]

30. St. Amand, M.M.; Radhakrishnan, D.; Robinson, A.S.; Ogunnaike, B.A. Identification of manipulated variables for a glycosylation control strategy. *Biotechnol. Bioeng.* **2014**, *111*, 1957–1970. [CrossRef] [PubMed]

31. Brühlmann, D.; Jordan, M.; Hemberger, J.; Sauer, M.; Stettler, M.; Broly, H. Tailoring recombinant protein quality by rational media design. *Biotechnol. Prog.* **2015**, *31*, 615–629. [CrossRef] [PubMed]

32. Ciucanu, I.; Kerek, F. A simple and rapid method for the permethylation of carbohydrates. *Carbohydr. Res.* **1984**, *131*, 209–217. [CrossRef]

33. Ciucanu, I.; Costello, C.E. Elimination of oxidative degradation during the per-o-methylation of carbohydrates. *J. Am. Chem. Soc.* **2003**, *125*, 16213–16219. [CrossRef] [PubMed]

34. Majid, F.A.A.; Butler, M.; Al-Rubeai, M. Glycosylation of an immunoglobulin produced from a murine hybridoma cell line: The effect of culture mode and the anti-apoptotic gene, bcl-2. *Biotechnol. Bioeng.* **2007**, *97*, 156–169. [CrossRef] [PubMed]

35. Kim, W.H.; Kim, J.-S.; Yoon, Y.; Lee, G.M. Effect of Ca^{2+} and Mg^{2+} concentration in culture medium on the activation of recombinant factor ix produced in chinese hamster ovary cells. *J. Biotechnol.* **2009**, *142*, 275–278. [CrossRef] [PubMed]

36. Kao, Y.-H.; Hewitt, D.P.; Trexler-Schmidt, M.; Laird, M.W. Mechanism of antibody reduction in cell culture production processes. *Biotechnol. Bioeng.* **2010**, *107*, 622–632. [CrossRef] [PubMed]

37. McAtee, A.G.; Templeton, N.; Young, J.D. Role of chinese hamster ovary central carbon metabolism in controlling the quality of secreted biotherapeutic proteins. *Pharm. Bioprocess.* **2014**, *2*, 63–74. [CrossRef]

38. Fan, Y.Z.; Del Val, I.J.; Muller, C.; Lund, A.M.; Sen, J.W.; Rasmussen, S.K.; Kontoravdi, C.; Baycin-Hizal, D.; Betenbaugh, M.J.; Weilguny, D.; et al. A multi-pronged investigation into the effect of glucose starvation and culture duration on fed-batch cho cell culture. *Biotechnol. Bioeng.* **2015**, *112*, 2172–2184. [CrossRef] [PubMed]

39. Del Val, I.J.; Nagy, J.M.; Kontoravdi, C. A dynamic mathematical model for monoclonal antibody n-linked glycosylation and nucleotide sugar donor transport within a maturing golgi apparatus. *Biotechnol. Prog.* **2011**, *27*, 1730–1743. [CrossRef] [PubMed]

antibodies

MDPI

Article

Tumor-Directed Blockade of CD47 with Bispecific Antibodies Induces Adaptive Antitumor Immunity

Elie Dheilly [†], Stefano Majocchi, Valéry Moine, Gérard Didelot, Lucile Broyer, Sébastien Calloud, Pauline Malinge, Laurence Chatel, Walter G. Ferlin, Marie H. Kosco-Vilbois, Nicolas Fischer and Krzysztof Masternak *

Novimmune S.A., 14 chemin des Aulx, CH-1228 Geneva, Switzerland; elie.dheilly@epfl.ch (E.D.); smajocchi@novimmune.com (S.M.); vmoine@novimmune.com (V.M.); gdidelot@novimmune.com (G.D.); lbroyer@novimmune.com (L.B.); scalloud@novimmune.com (S.C.); pmalinge@novimmune.com (P.M.); lchatel@novimmune.com (L.C.); wferlin@novimmune.com (W.G.F.); mkosco-vilbois@novimmune.com (M.H.K.-V.); nfischer@novimmune.com (N.F.)
* Correspondence: kmasternak@novimmune.com; Tel.: +41-22-839-41-51
† Present address: EPFL SV ISREC UPORICCHIO, SV 2824, CH-1015 Lausanne, Switzerland.

Received: 30 October 2017; Accepted: 22 December 2017; Published: 3 January 2018

Abstract: CD47 serves as an anti-phagocytic receptor that is upregulated by cancer to promote immune escape. As such, CD47 is the focus of intense immuno-oncology drug development efforts. However, as CD47 is expressed ubiquitously, clinical development of conventional drugs, e.g., monoclonal antibodies, is confronted with patient safety issues and poor pharmacology due to the widespread CD47 "antigen sink". A potential solution is tumor-directed blockade of CD47, which can be achieved with bispecific antibodies (biAbs). Using mouse CD47-blocking biAbs in a syngeneic tumor model allowed us to evaluate the efficacy of tumor-directed blockade of CD47 in the presence of the CD47 antigen sink and a functional adaptive immune system. We show here that CD47-targeting biAbs inhibited tumor growth in vivo, promoting durable antitumor responses and stimulating CD8[+] T cell activation in vitro. In vivo efficacy of the biAbs could be further enhanced when combined with chemotherapy or PD-1/PD-L1 immune checkpoint blockade. We also show that selectivity and pharmacological properties of the biAb are dependent on the affinity of the anti-CD47 arm. Taken together, our study validates the approach to use CD47-blocking biAbs either as a monotherapy or part of a multi-drug approach to enhance antitumor immunity.

Keywords: bispecific antibodies; CD47; antitumor immunity; immunotherapy; immunosurveillance; checkpoint inhibitors; phagocytosis

1. Introduction

CD47 is a ubiquitous anti-phagocytic receptor widely known as the "don't eat me" signal, as it inhibits phagocytosis by engaging a signal regulatory protein alpha (SIRPα) on macrophages and other phagocytes (reviewed in [1–4]). In addition, CD47 carries out other important physiological functions, such as regulation of cardiovascular homeostasis, neuronal development, bone remodeling, adaptive immunity, cellular response to stress, stem cell renewal, cell adhesion, motility, proliferation, and survival (reviewed in: [1,2,4–6]).

Cancer cells upregulate CD47 expression in order to evade antitumor immunity. In nearly all hematological and solid cancers, increased levels of CD47 expression correlate with more aggressive disease and poorer prognosis (reviewed in [7–9]). Blockade of CD47 increases phagocytosis of tumor cells and stimulates macrophage-mediated tumor elimination in various human xenograft tumor models (reviewed in [7–9]). In addition, when studied in tumor models in immunocompetent animals, the induction of adaptive immune responses in mediating therapeutic efficacy of CD47 neutralization has been observed [10–18]. These studies revealed that CD47 expressed on tumor cells is a key suppressor of antitumor immunity as well as a mediator of resistance to PD1 checkpoint blockade therapy [19,20]. However, it is not clear how CD47 blockade facilitates tumor immunosurveillance. Several mechanisms have been proposed, including enhanced immune cell infiltration [11,15,19], dendritic cell activation [17,21], and increased tumor antigen cross-presentation with activation of cytotoxic T cell responses due to enhanced phagocytosis of tumor cells [7,8,22,23]. Given the involvement of CD47 in various physiological and cellular functions, multiple non-mutually exclusive mechanisms may be involved. As such, a rationale for therapeutically targeting CD47 in cancer has evolved. Thus, several CD47-directed therapeutic strategies are being pursued (reviewed in [8,9]).

Nonetheless, ubiquitous CD47 expression on healthy cells presents a considerable hurdle for the development of CD47-blocking therapies. Molecules that block CD47 have been observed to induce hematotoxicity in mice and non-human primates. Furthermore, they have poor pharmacokinetic properties, which reflects the substantial target-mediated drug disposition due to the CD47 antigen sink [16,24–29]. Hematotoxicity was found to be dependent on the Fc region of the molecules as CD47-blocking mAbs and SIRPα-Fc fusion proteins bearing a functional Fc region show significant hematological toxicity, while Fc effector functionless CD47 blockers display a better safety profile [16,26,29].

Dual targeting bispecific antibodies (biAbs) co-engage two antigens at the cell surface, which stabilizes binding through avidity [30,31]. Therefore, creating anti-CD47/antitumor-associated antigen (TAA) biAbs to steer CD47 blockade towards TAA-positive malignant cells, and away from TAA-negative healthy cells expressing CD47 alone, provides a solution to poor pharmacology and safety issues related to ubiquitous CD47 expression [32–34]. We have previously demonstrated that selective targeting of CD47 on tumor cells can be achieved with anti-CD47/TAA biAbs with a low-affinity anti-CD47 arm and a high-affinity anti-TAA arm [34]. Such an "unbalanced affinity" design minimizes CD47 binding to TAA-negative cells while driving bivalent attachment of the biAb to double-positive cells via TAA/CD47 co-engagement. Furthermore, with an increased selectivity, a biAb can be safely endowed with an immunologically active Fc region, thus maximizing tumor cell phagocytosis via the engagement of FcγRs and a concurrent blockade of the "don't eat me" signal [34,35]. Indeed, the unbalanced affinity exemplified by our anti-human CD47/CD19 therapeutic biAb candidate, NI-1701, demonstrated potent antitumor activity in xenograft models. Moreover, in non-human primates, it displayed favorable pharmacokinetics and the absence of toxicity following multiple administrations of super-therapeutic doses [35].

However, the lack of cross-reactivity of anti-human therapeutic biAbs with mouse CD47 precludes in vivo evaluation in a physiologically relevant context, i.e., with tumors grown in syngeneic, immunocompetent mouse strains in the presence of the CD47 antigen sink. To overcome this limitation, we generated anti-mouse CD47-specific biAbs. Using these reagents and a syngeneic mouse model of B cell lymphoma, we show that the affinity of the anti-mouse CD47 biAb arm determines the safety, pharmacokinetic properties, and in vivo efficacy in the presence of the CD47 antigen sink. We observed that tumor-directed blockade of CD47 enhances tumor antigen cross-presentation and CD8$^+$ T cell activation in vitro. Finally, we show that tumor-directed blockade of CD47 via biAbs can increase the therapeutic effect of standard of care oncology treatments such as PD-L1-blocking antibodies and cyclophosphamide.

2. Materials and Methods

2.1. Antibodies

Generation of bispecific antibodies based on one heavy chain and two different light chains (one κ and one λ), including light chain library construction, phage display selection and screening, bispecific antibody production, purification, and analytical characterization, are described in detail in [36]. Briefly, anti-mouse CD47 antibody sequences were identified with the κλ body phage display platform [36] using a combination of recombinant mouse CD47 (produced in-house) or cells naturally expressing or transfected with mouse CD47. Screening was performed as described in [36], and positive hits were reformatted into hIgG1 monoclonal antibodies for further evaluation. A panel of anti-mouse CD47 antibody arms with different affinities was generated and two of these candidates were used for the generation of bispecific antibodies analyzed in this study. The human anti-mouse CD47 variable domains (κ-light chain) and the human variable domain of an anti-human CD19 antibody (λ-light chain) [36] were grafted on mouse heavy and light chain constant domains, giving rise to bispecific human-mouse chimeric mIgG2a molecules (Figure S1). Therefore, the two mouse CD47-targeting biAbs (biAb1 and biAb2) comprised two distinct anti-mouse CD47 arms with different affinities (Table S1), a common, previously described, high-affinity antitumor associated antigen (TAA) antibody arm recognizing the human CD19 antigen (hCD19), and a wild type mIgG2a Fc portion. For the construction of monovalent anti-mouse CD47 antibodies, the anti-human CD19 arm was replaced with a non-binding arm (non-cross-reactive anti-human CD47 arm, ref. [34]). In addition, we generated human-mouse chimeric anti-CD47 mAbs, consisting of human anti-mouse CD47 variable domains of biAb1 and biAb2 grafted on mouse IgG2aλ constant domains (CD47 mAb1 and CD47 mAb2). Finally, the human anti-human CD19 variable domain was also used to generate a chimeric CD19 mAb and a chimeric CD19 monovalent antibody, both with a mouse IgG2a constant region.

2.2. Cell Lines and Reagents

Mouse Diffuse Large B Cell Lymphoma cell line A20 (ATCC (Wesel, Germany) TIB208™, BALB/c) was cultured in RPMI-1840 medium and supplemented with 10% heat-inactivated fetal calf serum (FCS), 2 mM L-glutamine, 1 mM sodium pyruvate, 10 mM HEPES, 5 mM D-glucose (all from Sigma-Aldrich, Buchs, Switzerland) and 50 µM 2-mercaptoethanol (Gibco/Thermo Fisher Scientific, Reinach, Switzerland). Mouse melanoma cell line B16–F10 (ATCC® CRL6475™, C57Bl/6) was cultured in DMEM-Hi glucose medium and supplemented with 10% heat-inactivated FCS, 4 mM L-glutamine, and 1 mM sodium pyruvate. Cells were maintained at 37 °C in a water-jacketed incubator containing 5% CO_2. A20-hCD19, A20-hCD19-HA-GFP, A20-mCherry, A20-hCD19-GFP, and B16F10-hCD19 cell lines were obtained by transfecting with Lipofectamine 2000 (Thermo Fisher Scientific, Reinach, Switzerland) the parental wild type cell lines with the corresponding transgenes: human CD19 transmembrane and extracellular domains (UniProtKB-P15391), and hemagglutinin (HA) gene from influenza virus strain A/Puerto Rico/8/1934 H1N1 (UniProtKB-P03452). Proteins were addressed to the cell surface via a signal peptide. Stably expressing pools were enriched by successive Fluorescence-Activated Cell Sorting (FACS) of positive cells. Once stable pools were obtained, clones were generated by single cell FACS sorting in the absence of selective pressure. Cell surface mCD47 and hCD19 densities were determined by an indirect immunofluorescence assay (QIFIKIT) according to the manufacturer's instructions (Agilent Technologies, Basel, Switzerland). As the primary antibody, saturating concentrations of the high-affinity anti-mCD47 mAb1 and the anti-hCD19 mAb described in this manuscript were used.

2.3. Mice

Six- to eight-week-old female BALB/c mice and C57BL/6J mice were purchased from Charles River Laboratories (Ecully, France). All animal experiments were performed in accordance with the Swiss Federal Veterinary Office guidelines and as authorized by the Cantonal Veterinary Office.

2.4. Antibody Binding Assays

Harvested cells were washed with a FACS buffer (PBS, 2% BSA, and 0.1% sodium azide), incubated with test antibodies for 30 min at 4 °C in FACS buffer, washed once to remove unbound antibody, and stained with a fluorescently labeled anti-mouse IgG detection antibody (either an anti-mouse IgG (H + L) (A865, Life technologies/Thermo Fisher Scientific, Reinach, Switzerland) or an anti-mouse IgG (Fc) (115-606-072 Jackson ImmunoResearch, West Grove, PA, USA) for 15 min at 4 °C. A viability marker, SYTOX blue (Thermo Fischer Scientific, Reinach, Switzerland) or propidium iodide (Sigma-Aldrich,

Buchs, Switzerland), was added before acquisition to exclude dead cells. Antibody binding was measured by flow cytometry using either a Cytoflex flow cytometer (Beckman Coulter, Indianapolis, IN, USA) or a FACS Calibur flow cytometer (BD Biosciences, San Jose, CA, USA). FACS data were analyzed with FlowJo software (FlowJo LLC, Ashland, OR, USA).

For binding assays performed with A20 cells, cells were first incubated with anti-mouse Fc (same as secondary/detection antibody but unlabeled) to block the Fc portion of IgGs expressed at the surface of A20 cells. The staining procedure was as detailed above. To assess antibody selectivity to human CD19 positive A20 cells, A20-hCD19-GFP cells were mixed with A20-mCherry cells at a 1:1 ratio and stained and analyzed using the protocol described above. Data are presented as the ratio between Mean Fluorescent Intensity (MFI) measured with A20-hCD19-GFP and A20-mCherry cells.

Antibody binding to red blood cells (RBC) in whole blood was determined as follows. First, test antibody was pre-incubated with a fluorescently-labeled Fab Fragment Anti-Mouse IgG (H + L) (115-007-003, Jackson ImmunoResearch, West Grove, PA, USA) at a 2:1 ratio for 15 min at 4 °C to minimize interference arising from the secondary (detection) antibody binding to cell surface and serum IgG. This antibody mixture was prepared 20 times concentrated and was added to heparinized blood from BALB/c mice and incubated for 15 min at room temperature. Blood cells were then washed three times and antibody binding was analyzed by flow cytometry. RBC were identified and gated based on their SSC-FSC parameters.

2.5. CD47 Competition Assay

A20-hCD19 cells were incubated with CD47-blocking test antibodies for 30 min at room temperature (rt). Test antibodies were then subject to competition by a fluorescently-conjugated low affinity anti-mouseCD47-blocking mAb (mAb3, see Figure S1) on ice for 20 min. After that, cells were washed and the percentage of test antibody outcompeted by the indicator antibody was measured by quantifying the MFI of cells by flow cytometry.

2.6. Phagocytosis Assay

A single cell suspension of bone marrow cells was obtained from BALB/c or C57BL/6J mice and cultured for 7 days in complete macrophage medium (RPMI 1640, 10% heat-inactivated FCS, 2 mM L-glutamine, 1 mM sodium pyruvate, 10 mM HEPES buffer, 50 μM 2-mercaptoethanol, 25 μg/mL gentamicin (Sigma-Aldrich, Buchs, Switzerland)) supplemented with 20 ng/mL M-CSF (PeproTech, London, UK). On the day of the experiment, non-adherent cells were eliminated by washing the culture flasks with PBS. Adherent macrophages were detached using a non-enzymatic cell dissociation solution (Sigma-Aldrich, Buchs, Switzerland) and washed in complete macrophage medium. For phagocytosis to proceed, macrophages were mixed with CFSE-labeled target cells (effector to target ratio of 1:5) in ultra-low attachment plates (Corning/Sigma-Aldrich, Buchs, Switzerland) and incubated in the presence of the test antibody for 2.5 h at 37 °C. Macrophages were subsequently stained with an Alexa Fluor 647-labeled anti-mouse anti-F4/80 secondary antibody (Clone A3-1, Biorad, Cressier, Switzerland) and a viability marker (Sytox Blue by Thermo Fischer Scientific, Reinach, Switzerland) was added to exclude dead cells before acquisition using a Cytoflex flow cytometer (Beckman Coulter, Indianapolis, IN, USA). Double-positive events (positive for F4/80 and CFSE-labeled target cell) were identified as phagocytosis events. Phagocytosis is presented as the percentage of macrophages which have engulfed at least one target cell (double-positive events/total macrophages × 100).

2.7. Isolation of HA-Specific TCR CD8+ T Cells

HA-specific TCR CD8+ T cells were isolated from lymph nodes and spleens of CL4 transgenic mice kindly provided by Dr. Roland Liblau (Research Center Toulouse Purpan, CPTP-INSERM). The CL4 mice were bred and maintained in Novimmune's animal facility. Spleens and lymph nodes were digested using collagenase IV (Lubiosciences, Zürich, Switzerland) and DNase I (Sigma-Aldrich, Buchs, Switzerland) cocktail for 30 min at 37 °C. Single cell suspension was then processed through a negative

CD8 isolation kit (STEMCELL Technologies, Vancouver, BC, Canada) following the manufacturer's instructions. Purified CD8$^+$ T cells were subsequently stained with CellTrace Violet (Thermo Fisher Scientific, Reinach, Switzerland) before being used in cross-presentation experiments.

2.8. Cross-Presentation Assay

Phagocytosis was set up as described above except for the incubation time at 37 °C, which was extended from 2.5 h to overnight. Then 4.5×10^4 F4/80$^+$ macrophages were sorted for each condition with an Astrios FACS sorter (Beckman Coulter, Indianapolis, IN, USA) and subsequently co-cultured with 17.5×10^5 CellTrace Violet labeled, HA-specific TCR CD8$^+$ T cells for 48 h at 37 °C. At the end of the macrophage/T cell co-culture, supernatants were collected and analyzed for IFN-γ and IL-2 levels using Luminex bead-based multiplex assay (Thermo Fisher Scientific, Reinach, Switzerland). In parallel, the proliferation of T cells stained with an anti-mouse CD8a antibody (Clone 53-6.7, BD Biosciences, San Jose, CA, USA) was evaluated by quantifying CellTrace violet proliferation peaks by flow cytometry.

2.9. In Vivo Efficacy Experiments

In vivo, 3×10^6 A20-hCD19 cells were injected subcutaneously (sc) into the left flank of 6–8-week-old BALB/c mice (Charles River Laboratories). Tumors were measured every 2–3 days using a digital caliper and volumes were calculated using the formula (width × length × height × Pi)/6. Animals were euthanized when the tumor volume exceeded 1000 mm^3 or at the experimental endpoint. In biAb monotherapy experiments, the treatment began when the tumors reached 6 × 6 mm. Mice were either treated intraperitoneally with 400 μg every other day or intratumorally with 100 μg of antibody every three days. The number of injections varied between experiments and is specified in the corresponding figure legend. In combination therapy experiments, mice were recruited when the tumors reached 100 mm^3. For cyclophosphamide (CTX) combination therapy, we used a sequential combination therapy protocol as described by [12]. In brief, mice were treated with a single dose of 60 mg/kg CTX (Sandoz Pharmaceuticals AG, Rotkreuz, Switzerland) on day 6 followed by 100 μg antibody injected intratumorally on day 7, 10 and 13. For anti-PD-L1 combination therapy, mice received a mixture of 200 μg anti-PD-L1 (clone 10F.9G2, BioXcell, Lebanon, NH, USA) plus 400 μg biAb intraperitoneally on day 6, 8, 10, 12, 14. For rechallenge experiments, tumor-free mice and naive mice were injected (sc) with 3×10^6 of either A20 wild type (wt) or A20-hCD19 cells on the opposite flank 2 months after rejecting the primary tumor. For B16F10-hCD19 experiments, 6–8-week-old C57BL/6J mice were engrafted with 1×10^5 B16F10-hCD19 on the left flank and recruited once tumors became palpable. Mice received 1.2 mg antibody treatment on the day of recruitment and were subsequently treated every other day with 400 μg antibody.

2.10. Evaluation of biAb Bioavailability and Binding to RBC

BALB/c mice were injected intraperitoneally every other day with 400 μg antibody for a total of three injections. Mice were bled two days after the last injection to evaluate RBC concentration, antibody binding to RBC, and antibody bioavailability in the blood. RBC counts were determined using a HEMAVET 950 hematology analyzer (Drew Scientific Inc., Miami Lakes, FL, USA). To evaluate antibody binding to RBC, 2 μL of blood were washed three time with FACS buffer and subsequently stained with an anti-Mouse IgG (Life Technologies/Thermo Fisher Scientific, Reinach, Switzerland) for 20 min at room temperature. Cells were then washed three times in FACS Buffer and antibody binding was determined by flow cytometry.

For the assessment of blood bioavailability, blood was collected into serum separator tubes (BD Microtainer) and centrifuged at 10,000× *g* for 5 min. Resultant serum supernatant was stored at −20 °C until further analysis. Serum concentration of hCD19-specific antibody was determined by an ELISA, in which biotinylated hCD19 extracellular domain (produced in-house) was captured on streptavidin-coated plates. A dose range antibody spiked in serum from non-treated mice was used to calibrate the assay.

3. Results and Discussion

3.1. Generating Tumor Cells and Anti-CD47/TAA biAbs to Study Tumor-Specific CD47 Targeting In Vitro and In Vivo

To study immunosurveillance, and thus the ability to generate antitumor adaptive immune responses, we needed to work with immunocompetent mice and syngeneic tumor models. For the model tumor antigen, we chose to work with human CD19 (hCD19) transfected into an A20 cell line derived from a BALB/c diffuse large B cell lymphoma. In order to study biAb-mediated tumor-directed blockade of CD47, various anti-mouse CD47-targeting antibody sequences were identified using the κλ body biAb platform [36]. The two sequences analyzed in this study were initially expressed as mAbs and were shown to block mouse CD47-SIRPα interaction with different potency (CD47 mAb1 and CD47 mAb2, Figure S1). These sequences were then combined as biAb1 and biAb2 with the previously described high-affinity TAA sequence recognizing the human CD19 antigen, an anti-hCD19 sequence which is not cross-reactive with mouse CD19 [36]. Thus, biAb1 and biAb2 share the same anti-hCD19 arm but have different anti-mouse CD47 arms. These two anti-CD47 sequences bind overlapping epitopes on mouse CD47 but with significantly different affinities. The equilibrium dissociation constant (K_D) between biAb1 and mouse CD47 is in the nanomolar range (Table S1), while the affinity of biAb2 to CD47 was too weak to be reliably measured. The CD47/CD19 biAbs and the corresponding monovalent CD47 and CD19 antibodies used in this study are schematically depicted in Figure S2.

3.2. CD47/CD19 biAbs Selectively Target Mouse Lymphoma Cells Expressing Human CD19

After engineering, the A20-hCD19 cells displayed 30,000 copies of mouse CD47 per cell, similar to non-transfected A20 cells ("wild type", A20 wt), and 65,000 copies of hCD19. Figure 1 shows that the targeting of A20-hCD19 tumor cells is primarily driven by the anti-hCD19 antibody arm of the biAbs. Both biAb1 and biAb2 bind strongly to A20-hCD19 cells, but only weakly (biAb1) or negligibly (biAb2) to A20 wt cells (Figure 1a,b). Of note, the binding of the two biAbs is comparable to CD19 monovalent antibody binding to A20-hCD19 cells, with a slight EC50 shift in favor of the biAbs, reflecting a contribution of the CD47 arms to the aggregate binding affinity (Figure 1a,b). To test if CD47/CD19 biAbs are able to block mouse CD47 on double-positive cells upon hCD19 co-engagement, we used a CD47 competitive binding assay. Both biAbs efficiently out-competed fluorescently labeled CD47 ligand (Figure 1c) more potently than the corresponding CD47 monovalent antibodies (i.e., κλ bodies with a non-binding arm instead of the anti-hCD19 arm) as well as the 'bivalent' CD47 mAbs. Noticeably, biAb1 and biAb2 blocked CD47 with similar potency despite different affinities of their anti-CD47 arms (see Table S1 and compare binding to A20 wt cells on Figure 1a,b) illustrating the power of avidity in facilitating CD47/TAA co-engagement at the cell surface.

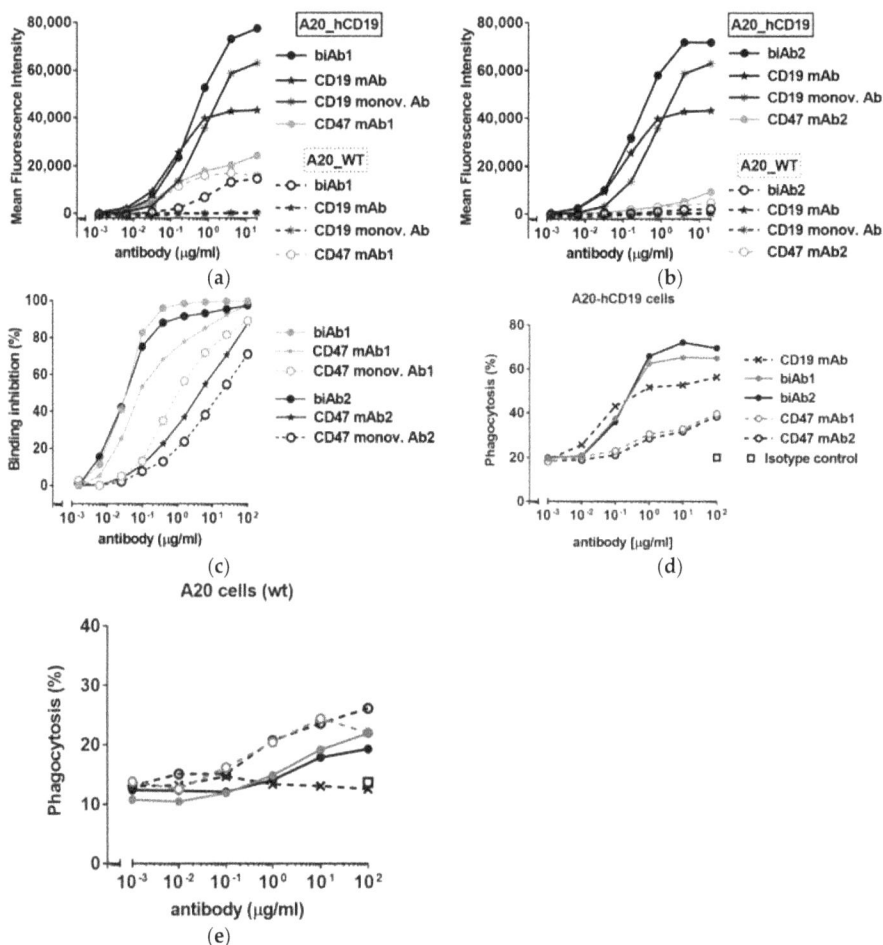

Figure 1. CD47/CD19 bispecific antibody (biAb) selectivity. (**a,b**) Binding, reported as mean fluorescence intensity values, to A20 mouse B cell lymphoma cells (A20 wild type (wt)) and A20 cells expressing human CD19 (A20-hCD19), of biAb1 (**a**) and biAb2 (**b**) as well as the anti-human CD19 mAb and the corresponding anti-mouse CD47 mAbs (mAb1 for biAb1 and mAb2 for biAb2). (**c**) Percent binding inhibition curves obtained by flow cytometry of a fluorescently-labeled CD47/ signal regulatory protein alpha (SIRPα) blocking mAb to A20-hCD19 cells by the biAbs and the corresponding anti-CD47 monovalent antibodies. (**d,e**) CD47/CD19 biAb-mediated antibody-dependent cellular phagocytosis (ADCP) as assessed by incubating fluorescently labeled A20-hCD19 (**d**) or A20 wt (**e**) cells with mouse macrophages (E:T ratio 1:5) in the presence of increasing concentrations of biAbs. Phagocytosis was assessed by flow cytometry and is expressed as the percentage of macrophages that ingested at least one target cell. The corresponding mAbs were tested for comparison.

3.3. Antitumor Efficacy of CD47-Blocking biAbs

The mouse IgG2a isotype binds activating Fcγ receptors with high affinity, mediating potent effector functions such as antibody-dependent cellular phagocytosis (ADCP). Phagocytosis of tumor cells can be further enhanced by blockade of CD47; the "don't eat me" signal [3,4,7–9,37]. Here we show that biAb-mediated tumor-directed blockade of CD47 led to enhanced phagocytosis of A20-hCD19 cells in vitro as compared to anti-CD47 and anti-CD19 mAbs, anti-CD47 and anti-CD19 monovalent antibodies, or a combination of both monovalent antibodies (Figures 1d and S3). We also show that the Fc region of CD47/CD19 biAbs is crucial for ADCP efficacy, as the F(ab')2 fragments of biAb1 and biAb2 did not trigger the phagocytosis of target cells (Figure S4). As expected, the biAbs induced only minimal phagocytosis of hCD19-negative (A20-wt) cells (Figure 1e). As in the CD47 blocking assay (Figure 1c), biAb1 and biAb2 showed comparable activity in the ADCP assay (Figure 1d), confirming that: (i) the affinity for binding hCD19 is the driving force of biAb activity; and (ii) the avidity generated upon hCD19 co-engagement on the cell surface of double-positive cells forces efficient CD47 blockade, compensating for differences in CD47 arm affinity. It is also likely that CD47 binding and blockade on the tumor cell is locally stabilized at the site of the phagocytic synapse, arising through the interactions of antibody Fc regions with Fcγ receptors on the macrophage.

To assess antitumor activity in vivo, BALB/c mice, engrafted subcutaneously (sc) with A20-hCD19 cells, received either intra-tumoral (it) or intra-peritoneal (ip) biAb treatment, initiated after the tumors reached a measurable volume. When administered locally, both biAbs were able to delay tumor growth (Figure 2a). However, upon ip administration, only biAb2 showed an antitumor effect (Figure 2b). This result was striking but not totally unexpected. While target cell selectivity of the CD47/CD19 biAbs is principally determined by the anti-hCD19 antibody arm, the higher-affinity anti-CD47 arm of biAb1 shows autonomous binding activity, in particular at higher antibody concentrations (Figure 1a). Thus, the higher affinity of the anti-CD47 arm in biAb1 resulted in a less stringent discrimination between double-positive and hCD19-negative cells as compared to biAb2 (Figure 1b). We hypothesized that the lack of in vivo efficacy of biAb1 administered systemically results from its interactions with CD47 expressed ubiquitously on healthy cells (i.e., the CD47 antigen sink), which, in turn, negatively impacts the pharmacokinetics, the level of toxicity, and eventually the efficacy of the biAb upon systemic administration.

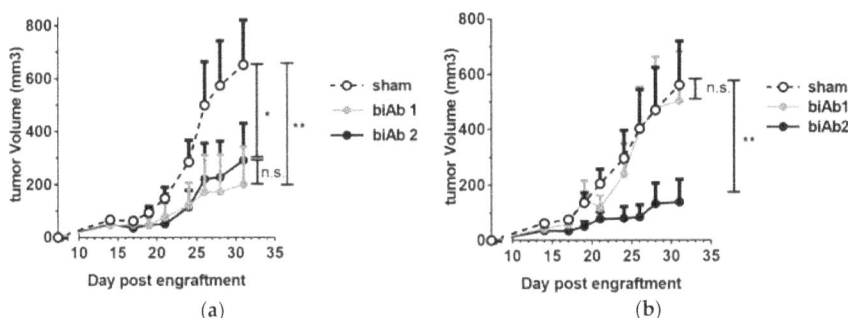

Figure 2. In vivo efficacy of CD47/CD19 bispecific antibodies. Treatment of BALB/c mice engrafted subcutaneously (sc) with A20-hCD19 cells began when the tumor size reached about 6×6 mm ($n = 7$ mice per group). (**a**) 100 μg of CD47/CD19 biAb was administered intra-tumorally on day 7, 10 and 13; (**b**) 400 μg of CD47/CD19 biAb was systemically administered intra-peritoneally on day 11, 13, 15 and 17. Tumor growth was measured three times a week and the average tumor volume per group +/− SEM (Standard Error of the Mean) reported. Statistical significance was determined using two way ANOVA comparing groups on d 31, *p*-value: * $p < 0.05$, ** $p < 0.01$; n.s., not significant.

3.4. The Affinity of the Anti-CD47 Arm Is Important for biAb Selectivity and Pharmacological Properties

To further assess the contribution of biAb anti-CD47 arms to cell binding, we incubated a 1:1 mix of fluorescently labeled A20-hCD19 cells and A20 wt cells with increasing concentrations of biAb and evaluated antibody binding by flow cytometry. Figure 3a shows relative binding, i.e., the ratio between MFI values obtained with A20-hCD19 and A20 wt cells. Contrary to the anti-CD47 mAbs, both biAbs showed a preference for double-positive cells (A20-hCD19), but biAb1 completely lost this binding selectivity at higher antibody concentrations (10 µg/mL or higher, Figure 3a).

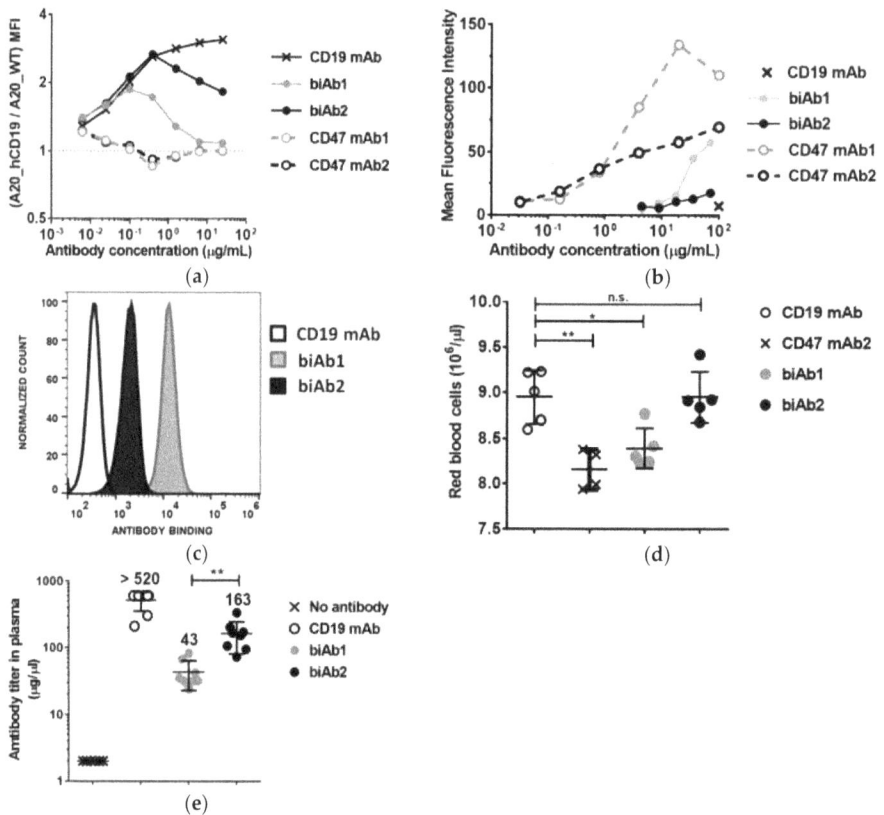

(a)

(b)

(c)

(d)

(e)

Figure 3. Affinity of the anti-CD47 arm influences biAb binding selectivity, hematological toxicity and bioavailability. (**a**) Binding selectivity was assessed using green fluorescent protein (GFP)-expressing A20-hCD19 cells and mCherry-expressing A20 wt cells at a 1:1 ratio, incubated with increasing concentrations of antibody and then analyzed by flow cytometry. The ratio between the Mean Fluorescence Intensity values obtained with A20-hCD19 cells and A20 wt cells is reported; (**b**) Antibody binding to erythrocytes from mouse (BALB/c) whole blood was determined by flow cytometry and the results reported as the Mean Fluorescence Intensity values obtained with biAbs and the corresponding mAbs; (**c,d**) To evaluate in vivo binding to RBC, BALB/c mice ($n = 8$/group) received three intraperitoneal 400 µg doses of anti-human CD19 antibody, biAb1 or biAb2 every 2 days. (**c**) Blood was collected 2 days after the last injection and RBC-bound antibody levels were determined by flow cytometry. Histogram plots show mean fluorescence intensity values obtained with the three antibodies tested; (**d**) Individual levels of RBC counts per mouse post antibody treatment; (**e**) Systemic bioavailability of the biAbs was assessed by obtaining the plasma and determining the antibody levels using a quantitative hCD19 ELISA assay. Normalized mean antibody titer +/− SD (standard deviation) is reported. Statistical significance was determined using one way ANOVA (**d**) or t-test (**e**); *p*-value * $p < 0.05$, ** $p < 0.01$; n.s., not significant.

We next assessed the binding of the CD47/CD19 biAbs to mouse red blood cells (RBC). RBCs displayed 20,000–25,000 copies of CD47 per cell. Given their abundance (between 6 and 10 billion cells per mL of blood), RBCs constitute potentially the most important antigen sink for anti-CD47 antibodies. To test binding to RBCs, antibodies were incubated with whole blood from BALB/c mice followed by flow cytometry analysis. BiAb1 bound mouse RBC relatively well at high concentrations, reaching binding levels comparable to anti-CD47 mAb2 at 100 μg/mL. In contrast, biAb2 showed only minimal binding (Figure 3b). Binding to RBC was also assessed in vivo two days after three ip injections of biAb at a dose of 20 mg/kg administered every other day. Consistent with the in vitro assay results, in vivo RBC binding levels were significantly higher with biAb1 than with biAb2 (Figure 3c). These results confirm once again the higher propensity of biAb1 to interact with the CD47 antigen sink. As anti-CD47 mAbs induce cytopenia and anemia in nonclinical and clinical studies, we assessed the effect of the two CD47/CD19 biAbs on RBC depletion. As shown on Figure 3d, RBCs declined following treatment with an anti-CD47 mAb (i.e., CD47 mAb2), to a lesser extent with biAb1, but, importantly, not with biAb2.

To assess how the ubiquitously expressed nature of CD47 affects biodistribution and pharmacokinetics of the biAbs, we determined the levels of circulating antibody two days after the last of three ip injections administered as described above. We observed a decreased bioavailability in the blood of both biAbs as compared to a non-binding control IgG (anti-hCD19 mAb) (Figure 3e). Consistent with the RBC binding results, biAb1 levels were significantly lower than biAb2 in circulation (Figure 3e).

These findings highlight the fact that the CD47 antigen sink could be a major challenge not only for monospecific anti-CD47 mAbs and SIRPα-Fc fusion proteins, but also for bispecific antibodies targeting CD47, depending on the affinity of the anti-CD47 arm. However, as illustrated by the biAb2, with a properly tuned affinity, negative effects of the CD47 antigen sink on bispecific drug pharmacology and safety can be minimized without compromising the efficacy of TAA-mediated, tumor-directed CD47 blockade.

3.5. CD47/CD19 biAb Induces Adaptive Antitumor Immunity

As it has been reported that CD47-SIRPα blockade enhances the T cell-dependent antitumor response, we also studied the induction of adaptive immunity. Using our system, we observed that biAb2 administered systemically was not only able to slow-down tumor progression (Figure 2b) but also could provoke complete tumor regression in some animals and generate long-term survival (Figure 4a), the latter suggesting creation of efficient tumor immunosurveillance. We hypothesized that animals that rejected the primary hCD19-positive tumors may have developed immunological memory against A20 tumor antigens other than the xenogeneic neo-antigen, hCD19. To test if indeed the antitumor memory responses were associated with epitope spreading, mice with complete responses were rechallenged with wild type A20 cells. In contrast to the control group (i.e., not treated with biAb nor previously challenged with tumor), none of the previously treated and then rechallenged mice developed tumors (Figure 4b), indicating the induction of long-lasting protective antitumor immunity as well as a broadened antigen specificity.

To further define how tumor-directed blockade of CD47 resulted in enhanced adaptive immune responses, an in vitro cross-priming system using hemagglutinin A (HA) CD8$^+$ transgenic T cells was established. The A20-hCD19 tumor cells were further transfected with HA (A20-hCD19-HA cells) to serve as the target tumor cell, and macrophages derived from mouse bone marrow (BMDMs) were serving as antigen-presenting cells. In the presence of biAb2, a substantially higher level of phagocytosis is observed as compared to the higher affinity anti-hCD19 mAb (Figure 4c). Furthermore, biAb2 promoted higher levels of T cell proliferation and cytokine production than with anti-hCD19 mAb-induced macrophages (Figure 4d–f). From these data, we conclude that enhanced tumor cell phagocytosis and antigen cross-presentation may contribute to in vivo efficacy of CD47-blocking biAbs. In any case, inhibition of CD47 on tumor cells promoted tumor control in vivo, as biAb2 demonstrated superior efficacy compared to the anti-hCD19 monovalent antibody and the anti-hCD19 mAb (Figure S5).

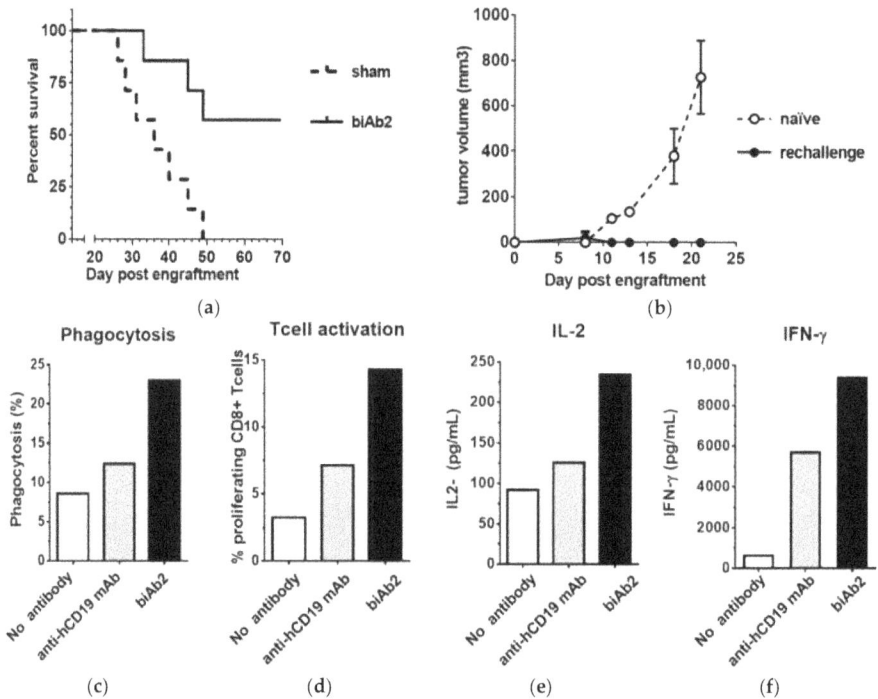

Figure 4. CD47/CD19 biAb promotes adaptive antitumor immunity. (**a**) Kaplan–Meier survival curves of animals treated with biAb2 in the experiment presented in Figure 2b; (**b**) Mice that rejected the A20-hCD19 primary tumor were rechallenged on the opposite flank with A20 wt cells and followed for tumor growth. The average tumor volume compared to naïve mice control group (*n* = 7) is reported. (**c**–**f**) biAb-mediated phagocytosis augments antigen cross-presentation and CD8+ T cell activation. Results from a representative experiment are presented; (**c**) Phagocytosis of HA-expressing tumor cells was assessed by incubating A20-hCD19-GFP-HA cells with bone marrow-derived macrophages from BALB/c mice and 5 µg/mL of antibody overnight. Phagocytosis was determined by flow cytometry and is expressed as the percentage of macrophages that ingested at least one target cell. The anti-human CD19 mAb was tested for comparison. (**d**) CD8+ T cell activation was assessed using macrophages from (c), co-cultured with cell tracer-labeled naive HA-specific TCR transgenic CD8+ T cells. The percentage of proliferating T cells was determined after 48 h of co-culture. (**e,f**) Cytokine secretion associated with T cell activation was assessed by quantifying the levels of IL-2 (**e**) and IFN-γ (**f**) released in the supernatant upon T cell activation.

3.6. CD47/CD19 biAbs Enhance the Therapeutic Effect of Immune Checkpoint Blockade and Chemotherapy

PD-1/PD-L1 checkpoint inhibitors, while curative in some patients, have clinically demonstrated the multifactorial primary and/or acquired resistance mechanisms that tumors use to thwart immunosurveillance [38,39]. Thus, drug combinations aimed at overcoming these resistance factors, are an area of intense clinical research. As our results demonstrate the ability to enhance antigen presentation and endogenous T cell immunity, we hypothesized that tumor-directed blockade of CD47 with biAbs offer yet another means of attack. Indeed, CD47 blockers have been shown to increase tumor control when combined with anti-PD-1/PD-L1 antibodies in preclinical models [13,15,16] and CD47 expressed on tumor cells has recently been identified as a key agent of resistance to PD-1 immunotherapy in mice [20]. Thus, we next examined the efficacy of a combination therapy involving PD-1/PD-L1 checkpoint inhibition with a CD47-blocking biAb. We observed that while an anti-PD-L1 mAb or biAb2 single agent treatment was able to delay A20-hCD19 tumor growth, the combination therapy was more potent (Figure 5a).

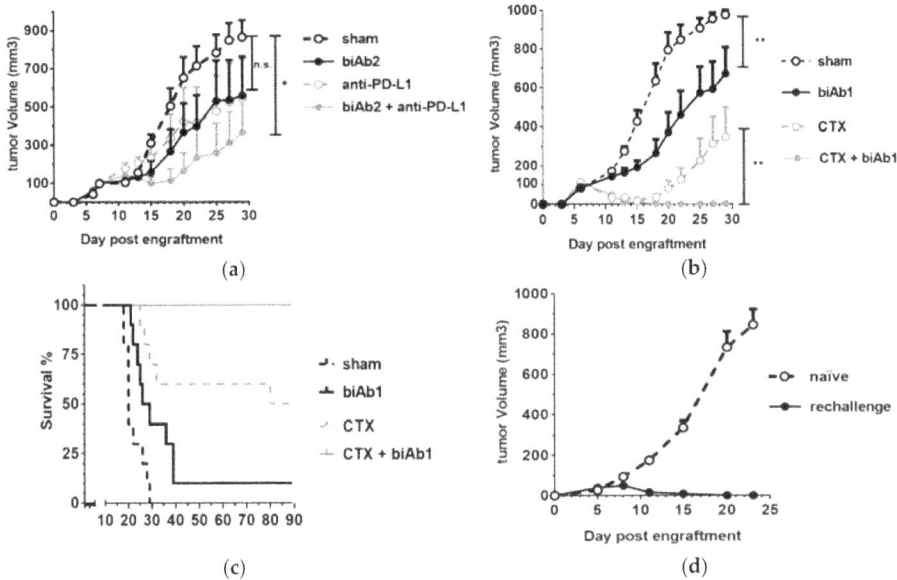

Figure 5. Combination therapies with anti-CD47 biAbs. (**a**) Combination of biAb2 and PD1/PD-L1 checkpoint blockade was assessed by engrafting BALB/c mice sc with A20-hCD19 cells and starting treatment when the tumor size reached 100 mm³ (intra-peritoneal doses of 400 μg for biAb2, or 200 μg doses for an anti-mouse PD-L1 antibody or a combination of the two antibodies on day 7, 9, 11, 13 and 15; *n* = 6 per group). (**b,c**) Combination of biAb2 and cyclophosphamide (CTX) was assessed by engrafting BALB/c mice sc with A20-hCD19 cells and starting treatment when the tumor size reached 100 mm³ (a single dose of CTX, 60 mg/kg ip, on d6, or three intra-tumoral 100 μg doses of biAb1 on day 7, 9 and 12, or a combination of both; *n* = 10 per group) (**b**) Tumor growth is shown as average tumor size per group +/− SEM. Statistical significance was determined using two-way ANOVA, *p*-value: * *p* < 0.05, ** *p* < 0.01; n.s., not significant; (**c**) Kaplan–Meier curves of tumor-bearing mice; (**d**) Mice that rejected the A20-hCD19 primary tumor were rechallenged on the opposite flank with A20-hCD19 cells and followed for tumor growth. The average tumor volume compared to naïve mice control group is reported.

We next studied the combination using cyclophosphamide (CTX). This chemotherapeutic agent promotes antitumor immunity by mediating selective immunodepletion of regulatory T cells [40]. We used a sequential combination therapy protocol to test the effect of tumor-directed blockade of CD47 in conjunction with CTX. A20-hCD19 tumors were allowed to grow to 100 mm³ and then the mice were treated with a single ip dose of CTX followed by three intra-tumoral injections of biAb1. The biAb, administered as monotherapy, was able to slow down tumor progression, whereas CTX led to a rapid tumor regression and durable responses in 50% of the animals (Figure 5b,c). Strikingly, the combination therapy allowed an even more potent effect, as 100% of the animals eliminated the tumor (Figure 5b,c) and developed long-term protective immunity, as demonstrated by resistance to tumor rechallenge (Figure 5d).

We used a second tumor type to confirm the antitumor efficacy of CD47-targeting biAbs. The B16F10 mouse melanoma cells that were stably transfected with human CD19 antigen (B16F10-hCD19 cells, Figure S6a) were engrafted sc in C57Bl/6 mice and treated intraperitoneally (ip) with biAb2. Similar to the results obtained with A20-hCD19 cells, we confirmed that binding of biAb2 to tumor cells is predominantly driven by its high-affinity anti-hCD19 arm (Figure S6b). We also demonstrated that tumor-directed blockade of CD47 with biAb2 administered systemically results in tumor growth inhibition and prolonged survival. (Figure S6c,d).

In conclusion, our study demonstrates that tumor-directed blockade of CD47 with dual targeting biAbs can inhibit tumor growth, promote tumor regression and induce durable antitumor memory responses. Our results further support the concept that CD47 blockade promotes T cell-dependent immunity via stimulating phagocytosis of tumor cells, antigen cross-presentation, and cytotoxic T cell cross-priming. We also show that tumor-directed blockage of CD47 can be combined with other therapeutic modalities, such as immune checkpoint inhibitors or chemotherapy, which should further improve antitumor responses and lift the proportion of curative therapies to offer to cancer patients. A key finding of this study is that using an in vivo system in which a CD47 antigen sink is present, the affinity of the anti-CD47 arm is important not only for drug safety and pharmacokinetics, but also for antitumor efficacy, in particular upon systemic administration. Therefore, the affinity of the CD47-blocking arm needs to be carefully considered when designing biAb antibodies targeting human CD47 for therapeutic intervention.

Supplementary Materials: The following are available online at http://www.mdpi.com/2073-4468/7/1/3/s1, Figure S1: Blockade of CD47-SIRPα interaction with anti-mouse CD47 mAbs. Binding of fluorescently labeled recombinant mouse SIRPα to CD47 expressed on the surface of murine pre-B cell lymphoma L1.2 cells in the presence of increasing concentrations of anti-mouse CD47 mAbs was assessed by FMAT and is represented as Mean Fluorescence Intensity (MFI) ± S.D. of four replicates, Figure S2: Schematic representation of CD47/CD19 biAbs and the corresponding monovalent antibodies, Figure S3: biAb2-induced phagocytosis of A20-hCD19 cells, Figure S4: CD47/CD19 bispecific antibody Fc region is required for efficient ADCP, Figure S5: Comparison of in vivo efficacy of biAb2 and anti-CD19 monospecific antibodies, Figure S6. Antitumor efficacy of biAb2 in the B16F10-hCD19 model, Table S1: Affinity of biAb1 and biAb2 antibody arms.

Acknowledgments: We would like to thank Nessie Costes and Nicolas Bosson for the identification of anti-mouse CD47 antibody sequences, Yves Poitevin and Guillemette Pontini for their guidance in bispecific antibody expression and purification, Giovanni Magistrelli for the plasmids expressing GFP-HA and human CD19, and Xavier Chauchet and Laura Cons for their help with conducting in vivo and antigen-cross-presentation experiments. The authors are grateful to Roland Liblau for the kind gift of CL4-TCR mice.

Author Contributions: E.D., S.M., W.F, M.H.K.-V., N.F. and K.M. conceived and designed research; L.B., E.D., S.M., V.M., G.D., P.M., S.C. and L.C. performed research and analyzed data; E.D., S.M., M.H.K.-V., N.F. and K.M. wrote the paper.

Conflicts of Interest: All authors are current or former employees of Novimmune SA. The authors declare no conflict of interest.

References

1. Oldenborg, P.A. CD47: A cell surface glycoprotein which regulates multiple functions of hematopoietic cells in health and disease. *ISRN Hematol.* **2013**, *2013*. [CrossRef] [PubMed]
2. Barclay, A.N.; Van den Berg, T.K. The interaction between signal regulatory protein alpha (SIRPalpha) and CD47: Structure, function, and therapeutic target. *Annu. Rev. Immunol.* **2014**, *32*, 25–50. [CrossRef] [PubMed]
3. Alvey, C.; Discher, D.E. Engineering macrophages to eat cancer: From "marker of self" CD47 and phagocytosis to differentiation. *J. Leukoc. Biol.* **2017**, *102*, 31–40. [CrossRef] [PubMed]
4. Murata, Y.; Kotani, T.; Ohnishi, H.; Matozaki, T. The CD47-SIRPalpha signalling system: Its physiological roles and therapeutic application. *J. Biochem.* **2014**, *155*, 335–344. [CrossRef] [PubMed]
5. Sick, E.; Jeanne, A.; Schneider, C.; Dedieu, S.; Takeda, K.; Martiny, L. CD47 update: A multifaceted actor in the tumour microenvironment of potential therapeutic interest. *Br. J. Pharmacol.* **2012**, *167*, 1415–1430. [CrossRef] [PubMed]
6. Soto-Pantoja, D.R.; Kaur, S.; Roberts, D.D. CD47 signaling pathways controlling cellular differentiation and responses to stress. *Crit. Rev. Biochem. Mol. Biol.* **2015**, *50*, 212–230. [CrossRef] [PubMed]
7. Chao, M.P.; Weissman, I.L.; Majeti, R. The CD47-SIRPalpha pathway in cancer immune evasion and potential therapeutic implications. *Curr. Opin. Immunol.* **2012**, *24*, 225–232. [CrossRef] [PubMed]

8. Matlung, H.L.; Szilagyi, K.; Barclay, N.A.; Van den Berg, T.K. The CD47-SIRPalpha signaling axis as an innate immune checkpoint in cancer. *Immunol. Rev.* **2017**, *276*, 145–164. [CrossRef] [PubMed]

9. Weiskopf, K. Cancer immunotherapy targeting the CD47/SIRPalpha axis. *Eur. J. Cancer* **2017**, *76*, 100–109. [CrossRef] [PubMed]

10. Maxhimer, J.B.; Soto-Pantoja, D.R.; Ridnour, L.A.; Shih, H.B.; DeGraff, W.G.; Tsokos, M.; Wink, D.A.; Isenberg, J.S.; Roberts, D.D. Radioprotection in normal tissue and delayed tumor growth by blockade of CD47 signaling. *Sci. Transl. Med.* **2009**, *1*. [CrossRef] [PubMed]

11. Soto-Pantoja, D.R.; Terabe, M.; Ghosh, A.; Ridnour, L.A.; DeGraff, W.G.; Wink, D.A.; Berzofsky, J.A.; Roberts, D.D. CD47 in the tumor microenvironment limits cooperation between antitumor T-cell immunity and radiotherapy. *Cancer Res.* **2014**, *74*, 6771–6783. [CrossRef] [PubMed]

12. Liu, X.; Pu, Y.; Cron, K.; Deng, L.; Kline, J.; Frazier, W.A.; Xu, H.; Peng, H.; Fu, Y.X.; Xu, M.M. CD47 blockade triggers T cell-mediated destruction of immunogenic tumors. *Nat. Med.* **2015**, *21*, 1209–1215. [CrossRef] [PubMed]

13. Sockolosky, J.T.; Dougan, M.; Ingram, J.R.; Ho, C.C.; Kauke, M.J.; Almo, S.C.; Ploegh, H.L.; Garcia, K.C. Durable antitumor responses to CD47 blockade require adaptive immune stimulation. *Proc. Natl. Acad. Sci. USA* **2016**, *113*, E2646–E2654. [CrossRef] [PubMed]

14. Yanagita, T.; Murata, Y.; Tanaka, D.; Motegi, S.I.; Arai, E.; Daniwijaya, E.W.; Hazama, D.; Washio, K.; Saito, Y.; Kotani, T.; et al. Anti-SIRPalpha antibodies as a potential new tool for cancer immunotherapy. *JCI Insight* **2017**, *2*. [CrossRef] [PubMed]

15. Tao, H.; Qian, P.; Wang, F.; Yu, H.; Guo, Y. Targeting CD47 enhances the efficacy of Anti-PD-1 and CTLA-4 in esophageal squamous cell cancer preclinical model. *Oncol. Res.* **2017**, *25*, 1579–1587. [CrossRef] [PubMed]

16. Ingram, J.R.; Blomberg, O.S.; Sockolosky, J.T.; Ali, L.; Schmidt, F.I.; Pishesha, N.; Espinosa, C.; Dougan, S.K.; Garcia, K.C.; Ploegh, H.L.; et al. Localized CD47 blockade enhances immunotherapy for murine melanoma. *Proc. Natl. Acad. Sci. USA* **2017**, *114*, 10184–10189. [CrossRef] [PubMed]

17. Xu, M.M.; Pu, Y.; Han, D.; Shi, Y.; Cao, X.; Liang, H.; Chen, X.; Li, X.D.; Deng, L.; Chen, Z.J.; et al. Dendritic cells but not macrophages sense tumor mitochondrial dna for cross-priming through signal regulatory protein alpha signaling. *Immunity* **2017**, *47*, 363–373. [CrossRef] [PubMed]

18. Wang, Y.; Xu, Z.; Guo, S.; Zhang, L.; Sharma, A.; Robertson, G.P.; Huang, L. Intravenous delivery of siRNA targeting CD47 effectively inhibits melanoma tumor growth and lung metastasis. *Mol. Ther.* **2013**, *21*, 1919–1929. [CrossRef] [PubMed]

19. Shuptrine, C.W.; Ajina, R.; Fertig, E.J.; Jablonski, S.A.; Kim, L.H.; Hartman, Z.C.; Weiner, L.M. An unbiased in vivo functional genomics screening approach in mice identifies novel tumor cell-based regulators of immune rejection. *Cancer Immunol. Immunother.* **2017**, *66*, 1529–1544. [CrossRef] [PubMed]

20. Manguso, R.T.; Pope, H.W.; Zimmer, M.D.; Brown, F.D.; Yates, K.B.; Miller, B.C.; Collins, N.B.; Bi, K.; LaFleur, M.W.; Juneja, V.R.; et al. In vivo CRISPR screening identifies Ptpn2 as a cancer immunotherapy target. *Nature* **2017**, *547*, 413–418. [CrossRef] [PubMed]

21. Yi, T.; Li, J.; Chen, H.; Wu, J.; An, J.; Xu, Y.; Hu, Y.; Lowell, C.A.; Cyster, J.G. Splenic dendritic cells survey red blood cells for missing self-CD47 to trigger adaptive immune responses. *Immunity* **2015**, *43*, 764–775. [CrossRef] [PubMed]

22. Tseng, D.; Volkmer, J.P.; Willingham, S.B.; Contreras-Trujillo, H.; Fathman, J.W.; Fernhoff, N.B.; Seita, J.; Inlay, M.A.; Weiskopf, K.; Miyanishi, M.; et al. Anti-CD47 antibody-mediated phagocytosis of cancer by macrophages primes an effective antitumor T-cell response. *Proc. Natl. Acad. Sci. USA* **2013**, *110*, 11103–11108. [CrossRef] [PubMed]

23. Liu, X.; Kwon, H.; Li, Z.; Fu, Y.X. Is CD47 an innate immune checkpoint for tumor evasion? *J. Hematol. Oncol.* **2017**, *10*. [CrossRef] [PubMed]

24. Willingham, S.B.; Volkmer, J.P.; Gentles, A.J.; Sahoo, D.; Dalerba, P.; Mitra, S.S.; Wang, J.; Contreras-Trujillo, H.; Martin, R.; Cohen, J.D.; et al. The CD47-signal regulatory protein alpha (SIRPa) interaction is a therapeutic target for human solid tumors. *Proc. Natl. Acad. Sci. USA* **2012**, *109*, 6662–6667. [CrossRef] [PubMed]

25. Majeti, R.; Chao, M.P.; Alizadeh, A.A.; Pang, W.W.; Jaiswal, S.; Gibbs, K.D., Jr.; van Rooijen, N.; Weissman, I.L. CD47 is an adverse prognostic factor and therapeutic antibody target on human acute myeloid leukemia stem cells. *Cell* **2009**, *138*, 286–299. [CrossRef] [PubMed]

26. Weiskopf, K.; Ring, A.M.; Ho, C.C.; Volkmer, J.P.; Levin, A.M.; Volkmer, A.K.; Ozkan, E.; Fernhoff, N.B.; van de Rijn, M.; Weissman, I.L.; et al. Engineered SIRPalpha variants as immunotherapeutic adjuvants to anticancer antibodies. *Science* **2013**, *341*, 88–91. [CrossRef] [PubMed]
27. Maute, R.L.; Gordon, S.R.; Mayer, A.T.; McCracken, M.N.; Natarajan, A.; Ring, N.G.; Kimura, R.; Tsai, J.M.; Manglik, A.; Kruse, A.C.; et al. Engineering high-affinity PD-1 variants for optimized immunotherapy and immuno-PET imaging. *Proc. Natl. Acad. Sci. USA* **2015**, *112*, E6506–E6514. [CrossRef] [PubMed]
28. Liu, J.; Wang, L.; Zhao, F.; Tseng, S.; Narayanan, C.; Shura, L.; Willingham, S.; Howard, M.; Prohaska, S.; Volkmer, J.; et al. Pre-clinical development of a humanized Anti-CD47 antibody with Anti-Cancer therapeutic potential. *PLoS ONE* **2015**, *10*, e0137345. [CrossRef] [PubMed]
29. Pietsch, E.C.; Dong, J.; Cardoso, R.; Zhang, X.; Chin, D.; Hawkins, R.; Dinh, T.; Zhou, M.; Strake, B.; Feng, P.H.; et al. Anti-leukemic activity and tolerability of anti-human CD47 monoclonal antibodies. *Blood Cancer J.* **2017**, *7*. [CrossRef] [PubMed]
30. Kaufman, E.N.; Jain, R.K. Effect of bivalent interaction upon apparent antibody affinity: Experimental confirmation of theory using fluorescence photobleaching and implications for antibody binding assays. *Cancer Res.* **1992**, *52*, 4157–4167. [PubMed]
31. Kontermann, R.E. Dual targeting strategies with bispecific antibodies. *MAbs* **2012**, *4*, 182–197. [CrossRef] [PubMed]
32. Piccione, E.C.; Juarez, S.; Liu, J.; Tseng, S.; Ryan, C.E.; Narayanan, C.; Wang, L.; Weiskopf, K.; Majeti, R. A bispecific antibody targeting CD47 and CD20 selectively binds and eliminates dual antigen expressing lymphoma cells. *MAbs* **2015**, *7*, 946–956. [CrossRef] [PubMed]
33. Piccione, E.C.; Juarez, S.; Tseng, S.; Liu, J.; Stafford, M.; Narayanan, C.; Wang, L.; Weiskopf, K.; Majeti, R. SIRPalpha-Antibody fusion proteins selectively bind and eliminate dual Antigen-Expressing tumor cells. *Clin. Cancer Res.* **2016**, *22*, 5109–5119. [CrossRef] [PubMed]
34. Dheilly, E.; Moine, V.; Broyer, L.; Salgado-Pires, S.; Johnson, Z.; Papaioannou, A.; Cons, L.; Calloud, S.; Majocchi, S.; Nelson, R.; et al. Selective blockade of the ubiquitous checkpoint receptor CD47 is enabled by dual-targeting bispecific antibodies. *Mol. Ther.* **2017**, *25*, 523–533. [CrossRef] [PubMed]
35. Buatois, V.; Johnson, Z.; Salgado Pires, S.; Papaioannou, A.; Hatterer, E.; Chauchet, X.; Richard, F.; Barba, L.; Cons, L.; Broyer, L.; et al. Preclinical development of a bispecific antibody that safely and effectively targets CD19 and CD47 for the treatment of B cell lymphomas and leukemias. Unpublished work. 2018.
36. Fischer, N.; Elson, G.; Magistrelli, G.; Dheilly, E.; Fouque, N.; Laurendon, A.; Gueneau, F.; Ravn, U.; Depoisier, J.F.; Moine, V.; et al. Exploiting light chains for the scalable generation and platform purification of native human bispecific IgG. *Nat. Commun.* **2015**, *6*. [CrossRef] [PubMed]
37. Van Bommel, P.E.; He, Y.; Schepel, I.; Hendriks, M.A.J.; Wiersma, V.R.; van Ginkel, R.J.; van Meerten, T.; Ammatuna, E.; Huls, G.; Samplonius, D.F.; et al. CD20-selective inhibition of CD47-SIRPα "don't eat me" signaling with a bispecific antibody-derivative enhances the anticancer activity of daratumumab, alemtuzumab and obinutuzumab. *Oncoimmunology* **2017**. [CrossRef]
38. O'Donnell, J.S.; Long, G.V.; Scolyer, R.A.; Teng, M.W.; Smyth, M.J. Resistance to PD1/PDL1 checkpoint inhibition. *Cancer Treat. Rev.* **2017**, *52*, 71–81. [CrossRef] [PubMed]
39. Sharma, P.; Hu-Lieskovan, S.; Wargo, J.A.; Ribas, A. Primary, adaptive, and acquired resistance to cancer immunotherapy. *Cell* **2017**, *168*, 707–723. [CrossRef] [PubMed]
40. Ahlmann, M.; Hempel, G. The effect of cyclophosphamide on the immune system: Implications for clinical cancer therapy. *Cancer Chemother. Pharmacol.* **2016**, *78*, 661–671. [CrossRef] [PubMed]

antibodies

MDPI

Review

Pharmacokinetic and Pharmacodynamic Considerations for the Use of Monoclonal Antibodies in the Treatment of Bacterial Infections

Shun Xin Wang-Lin and Joseph P. Balthasar *

Department of Pharmaceutical Sciences, University at Buffalo, State University of New York, Buffalo, NY 14214, USA; sl256@buffalo.edu
* Correspondence: jb@buffalo.edu; Tel.: +1-716-645-4807

Received: 8 December 2017; Accepted: 2 January 2018; Published: 4 January 2018

Abstract: Antibiotic-resistant bacterial pathogens are increasingly implicated in hospital- and community-acquired infections. Recent advances in monoclonal antibody (mAb) production and engineering have led to renewed interest in the development of antibody-based therapies for treatment of drug-resistant bacterial infections. Currently, there are three antibacterial mAb products approved by the Food and Drug Administration (FDA) and at least nine mAbs are in clinical trials. Antibacterial mAbs are typically developed to kill bacteria or to attenuate bacterial pathological activity through neutralization of bacterial toxins and virulence factors. Antibodies exhibit distinct pharmacological mechanisms from traditional antimicrobials and, hence, cross-resistance between small molecule antimicrobials and antibacterial mAbs is unlikely. Additionally, the long biological half-lives typically found for mAbs may allow convenient dosing and vaccine-like prophylaxis from infection. However, the high affinity of mAbs and the involvement of the host immune system in their pharmacological actions may lead to complex and nonlinear pharmacokinetics and pharmacodynamics. In this review, we summarize the pharmacokinetics and pharmacodynamics of the FDA-approved antibacterial mAbs and those are currently in clinical trials. Challenges in the development of antibacterial mAbs are also discussed.

Keywords: bacterial infections; monoclonal antibodies; pharmacokinetics; pharmacodynamics

1. Introduction

The clinical application of antibodies for the treatment of infectious diseases was first introduced in the form of serum therapy in early 1890s by Emil von Behring and Shibasaburo Kitasato [1]. Serum therapy was then widely applied to treat infections caused by several bacterial pathogens, including *Corynebacterium diphtheria*, *Streptococcus pneumoniae*, *Neisseria meningitides*, *Haemophilus influenzae*, Group A *Streptococcus*, and *Clostridium tetani* [2]. Although serum therapy became the standard-of-care for several infectious diseases in the pre-antibiotic era, treatment with polyclonal antisera poses several drawbacks, including "serum sickness" or immune complex hypersensitivity that can occur in 10–50% patients, lot-to-lot variation in efficacy, low content of specific antibodies, and potential hazards in the transmission of infectious diseases [2–5]. In 1937, the discovery of sulfonamides led to a boom in the development of antimicrobial chemotherapy [6], and the significant advantages associated with antimicrobials, such as less toxicity and cost, higher efficacy, and broad spectrum activity, resulted in the abandonment of serum therapy. Drug resistance, which has been a concern from the onset of antibacterial chemotherapy, has become a major clinical problem within the past few decades. It has been suggested that broad spectrum antimicrobial activity contributes to the widespread development of resistant strains, and specific mechanisms of resistance can either exist before, or emerge rapidly after, the clinical launch of new antibiotics [7].

In 2004/5, pan-drug resistant strains (i.e., resistant to all antimicrobials currently approved by FDA) of *Acinetobacter* and *Pseudomonas* were identified [7]. Unfortunately, as drug resistance has been increasing, the rate of approval of new antibiotics has been decreasing. The number of new antibacterial drugs approved has decreased from an average of five per year in the 1980s to less than one per year in the 2000s [7].

The discovery of hybridoma technology in 1975 and recent advances in monoclonal antibody (mAb) engineering, which make production of unlimited amount of human mAbs possible [8,9], have renewed interest in the development of antibacterial antibody therapies. Monoclonal antibodies are widely used to treat immune deficiencies, cancers, multiple sclerosis, rheumatoid arthritis, and psoriasis, but the application of mAbs for bacterial infections has progressed slowly. Currently, there are only three mAbs approved by the FDA for use in the treatment of bacterial infections (Table 1). All three mAbs are indicated as adjuvant therapies to antibiotics and do not have bactericidal activity. They are directed against bacterial exotoxins and protect host cells from toxin-mediated cytotoxicity through neutralization of exotoxin activities. As of December 2017, there are nine mAb products in clinical trials (Table 2). Of these, six are 'naked' mAbs, two are mAb cocktails containing two mAbs that bind to different targets (ASN100 and Shigamab), and one is an antibody-antibiotic conjugate (DSTA4637S) that kills intracellular bacteria through the intracellular delivery of a potent antibiotic. Of note, five of the nine mAb products in clinical testing bind to the bacterial cell surface and have shown bactericidal activity in preclinical studies (DSTA4637S, 514G3, MEDI3902, Aerumab, and Aerucin); the other four target exotoxins and protect against infections via toxin neutralization (MEDI4893, ASN100, Salvecin, and Shigamab).

In this review, we discuss important considerations of mAb-based therapies for the treatment of bacterial infections, including unique challenges, pharmacokinetic (PK) properties, and pharmacodynamic (PD) mechanisms of action. The PK/PD characteristics of FDA-approved antibacterial mAbs and those in clinical trial are also summarized.

2. Pharmacokinetic Considerations

All FDA-approved antibacterial mAbs and the majority of those in clinical trials are of the immune gamma globulin (IgG) isotype. IgG is the predominant immunoglobulin isotype, comprising approximately 80% of immunoglobulin in human serum. An intact IgG has a molecular weight of ~150 kDa, with two antigen binding domains and a highly conserved crystallizable region (Fc) that is responsible for binding to Fc gamma receptors (FcγRs) on immune cells and activating Fc-mediated effector functions. IgG typically exhibits linear pharmacokinetics (i.e., area under the drug concentration-time curve (AUC) is directly proportional to the dose) in healthy human subjects, with small volumes of distribution (3–9 L), relatively slow clearance (8–12 mL/h), and long half-lives (20–25 days) [10]. The long biological persistence of IgG is partially attributed to Brambell receptor (FcRn) mediated salvage of IgG from lysosomal catabolism [11]. In contrast to the pharmacokinetics of pooled endogenous IgG, therapeutic mAbs often demonstrate nonlinear PK (i.e., AUC is not proportional to the dose), depending on the total body load of the pharmacological target (i.e., the quantity of bacteria, in the case of antimicrobial mAb), the accessibility of the targets, mAb-target affinity, and mAb doses. Key PK considerations for antibacterial mAbs are summarized below, including a discussion of determinants of target mediated drug disposition (TMDD) and mAb distribution in infected organs.

Table 1. Food and Drug Administration (FDA)-approved monoclonal antibodies (mAbs) for use in bacterial infection.

Antibody	Company	Format	Pathogen/Target	First Approved Indication	Reported Mechanism of Action	Approval Year
Raxibacumab	GlaxoSmith Kline	Human IgG1(λ)	*Bacillus anthracis*/Protective antigen	Treatment and prophylaxis of inhalational anthrax	Toxin neutralization	2012
Obiltoxaximab	Elusys	Chimeric IgG1(κ)	*Bacillus anthracis*/Protective antigen	Treatment and prophylaxis of inhalational anthrax	Toxin neutralization	2016
Bezlotoxumab	Merck & Co.	Human IgG1	*Clostridium difficile*/Enterotoxin B	Prevention of *Clostridium difficile* infection recurrence	Toxin neutralization	2016

Table 2. mAbs currently in clinical trials.

Antibody	Sponsor	Format	Pathogen	Target	Reported Mechanism of Action	Current Status
MEDI4893	MedImmune	Human IgG1(κ)	*Staphylococcus aureus*	Alpha toxin	Toxin neutralization	Phase 2
ASN100	Arsanis	Human IgG1(κ)	*Staphylococcus aureus*	Alpha toxin and five leukocidins	Toxin neutralization	Phase 2
DSTA4637S	Genentech	Human IgG1	*Staphylococcus aureus*	β-O-linked N-acetylglucosamine on wall teichoic acids	Antibody-antibiotic conjugate	Phase 1
Salvecin (AR-301)	Aridis	Human IgG1	*Staphylococcus aureus*	Alpha toxin	Toxin neutralization	Phase 1/2a
514G3	XBiotech	Human IgG3	*Staphylococcus aureus*	Protein A	Opsonophagocytosis	Phase 1/2
MEDI3902	MedImmune	Human bispecific IgG1	*Pseudomonas aeruginosa*	Psl and PcrV	Opsonophagocytosis; inhibition of cell attachment and cytotoxicity	Phase 2
Aerumab (AR-101)	Aridis	Human IgM(κ)	*Pseudomonas aeruginosa*	O-antigen (serotype O11)	Opsonophagocytosis; complement-mediated bacterial killing	Phase 2b
Aerucin	Aridis	Human IgG1	*Pseudomonas aeruginosa*	Alginate (surface polysaccharide)	Opsonophagocytosis; complement-mediated bacterial killing	Phase 2
Shigamab	Bellus Health	Chimeric IgG1(κ)	*Escherichia coli*	Shiga toxin 1 and 2	Toxin neutralization	Phase 2

2.1. Target Mediated Drug Disposition

Target mediated drug disposition describes the phenomenon where binding of a high affinity drug to its pharmacological target affects the PK characteristics of the drug (i.e., kinetics of distribution and clearance) [12]. At relatively low doses (compared to the amount of target), high affinity mAb-target binding results in drug accumulation at the sites of action (i.e., target-expressing tissue or site of infection), which may lead to a large apparent volume of distribution of the mAb. With increased doses and increased mAb concentrations, the target sites become increasingly saturated, which may decrease tissue to plasma mAb concentration ratios, decreasing the apparent volume of distribution. Additionally, mAb-target binding may trigger receptor-mediated endocytosis, in which mAb-target complexes are engulfed and degraded in lysosomes. This target-mediated elimination accelerates the clearance of mAbs and thereby shortens their biological persistence and half-life. Due to the effect of the drug in enhancing the elimination of the target, the volume of distribution and clearance of the drug may decrease during the course of repeated dosing.

All marketed antibacterial mAbs and those in clinical trials were developed to kill bacteria or attenuate bacterial pathological activity via antibody-mediated effector functions or toxin neutralization (Tables 1 and 2). Antibody-toxin and antibody-bacteria complexes may be cleared by phagocytic cells through Fc engagement with FcγRs, and via subsequent endocytosis and catabolism in phagolysosomes. Thus, antibacterial mAbs may be expected to exhibit TMDD characteristics. However, the pharmacokinetics of antibacterial mAbs have not been well evaluated to date. Obiltoxaximab, a chimeric IgG1 targeting the protective antigen of *Bacillus anthracis*, showed ~2-fold faster clearances in rabbits and monkeys challenged with *B. anthracis* spores compared to those in non-infected animals [13]. Similarly, a mAb directed against *Staphylococcus aureus* also demonstrated significantly increased clearance (12.1–15.8 mL/day/kg) and decreased half-life (3.74–5.28 days) in *S. aureus* infected mice compared to those in non-infected mice (4.69–5.19 mL/day/kg and 16.4–18.0 days, respectively) [14]. In contrast, pulmonary infection with *Acinetobacter baumannii* did not impact the PK of an anti-K2 capsule mAb in mouse blood, although a substantial accumulation of mAb (5.64–36.1 fold higher amount) was observed in tissues (i.e., lung, liver, and spleen) with high bacterial load compared to values found in non-infected mice [15]. Further investigations are needed, as TMDD results in nonlinear PK, which impacts the design of efficacious dosing regimens, and contributes to potential intra- and inter-patient PK variability (e.g., due to differences in bacterial burden and immune status).

2.2. Distribution of mAbs in Infected Tissues

The tissue disposition of anti-bacterial mAbs may be complex, involving extravasation of mAb molecules from blood to tissue interstitial fluids, diffusion of the molecule in the interstitial fluids to bacterial targets, binding to bacterial targets, and elimination of mAbs from tissue via convective drainage through the lymphatics and via catabolism. Extravasation of IgG antibodies is typically thought to be governed by both diffusion and convection (i.e., bulk movement of molecules through paracellular pores in the vascular endothelium), but convection has been estimated to contribute more than 98% of the total transport [16–18]. Bacterial invasion and dissemination normally accompany disruption of the vascular endothelial integrity due to bacterial toxin-mediated cytotoxicity. This vascular damage may lead to increased antibody extravasation within infected tissues. Once bacterial cells seed in the tissue, rapid bacterial growth and release of exotoxins stimulate immune responses, including recruitment of effector cells (i.e., lymphocytes, polymorphonuclear leukocytes, and phagocytes) and massive release of cytokines and chemokines. These reactions on one hand result in increased fluid infiltration and increased vascular permeability that facilitate mAb extravasation; but, on the other hand, build up fluid pressure within tissue that hampers antibody distribution (by decreasing the hydrostatic pressure gradient driving convective transport of mAb from blood to tissue interstitial fluid) [19,20]. In addition, antibody-dependent phagocytosis accelerates the elimination of mAbs (i.e., target-mediated elimination) in the tissue. Bacterial infections in visceral organs may also result in formation of abscesses, which enclose bacteria by pseudocapsules and protect them from immune cells and mAbs. Furthermore, many pathogenic bacteria generate biofilms that are comparatively inaccessible to antibodies, immune cells, and even small molecule antibiotics. Another concern is that biofilms may mediate a near continuous release of virulence factors, such as exopolysaccharides of *Staphylococcus epidermis*, that act as decoys to reduce mAb molecules reaching the bacteria [21]. Therefore, formation of both abscesses and biofilms create barriers to mAb distribution, and hence adversely affect the antibody-mediated clearance of bacteria.

3. Pharmacodynamic Mechanisms of Action

Antibacterial mAbs have been developed against a variety of bacterial cell surface targets (i.e., proteins and polysaccharides) and soluble exotoxins (Tables 1 and 2). The potential pharmacodynamic mechanism of action depends on the nature of the target, its role in bacterial pathogenesis, and mAb isotype and structure (i.e., intact IgG mAb or IgG fragments, immunoconjugates, bispecific antibodies, etc.). Anti-exotoxin mAbs typically attenuate bacterial pathological activity via neutralization of exotoxins. Monoclonal antibodies targeting bacterial surface epitopes are expected to increase bacterial clearance through enhancing antibody-dependent phagocytosis, and/or complement-mediated bactericidal activity, or via immune system-independent bacterial killing. In addition, there has been increasing interest in the development of immunoconjugates and immunomodulatory mAbs that either carry potent antimicrobials or stimulate exhausted immune effector functions to augment bactericidal activity.

3.1. Toxin Neutralization

Antibacterial mAbs that act through the mechanism of neutralization are typically directed against exotoxins. mAb binding to soluble exotoxins leads to the formation of antibody-toxin complexes, which are primarily cleared by the reticuloendothelial system. All three marketed antibacterial mAbs achieve effects via toxin neutralization (Table 1). The efficacy of neutralizing mAbs has been shown to be directly correlated with mAb binding affinity. Anti-protective antigen (PA) mAb with higher binding affinity showed superior protection against anthrax lethal toxin challenge in macrophage cytotoxicity assays and in a rat infection model, when compared to mAbs with relatively low affinities [22]. Additionally, antibody-FcγR engagement was found to be required for anti-PA mAb neutralization activity, where mAb-mediated protection against anthrax infection was only shown in wild-type mice but not FcγR-deficient mice [23]. However, FcγR engagement may not be required for all anti-toxin mAb. For example, MAb166, an anti-PcrV (type III secretion injectisome) antibody, blocks the delivery of *Pseudomonas aeruginosa* type III toxins to host cells [24]. A single dose of 10 µg of MAb166 Fab fragments, which lack an Fc domain, was able to confer similar protection ($\geq$80% survival) as intact MAb166 against clinical isolates of *P. aeruginosa* in a mouse pneumonia infection model [24,25].

3.2. Opsonophagocytosis

Opsonophagocytosis has been considered as one of the key bactericidal mechanisms of the innate immune system. Antibody-mediated opsonophagocytosis involves antibody binding to bacterial surface antigens, followed by the engagement of FcγRs on the surface of professional phagocytes (i.e., monocytes/macrophages, neutrophils and dendritic cells), which in turn trigger actin-myosin driven endocytosis of antibody-bacteria complexes [26]. Phagosome vacuoles fuse with lysosomes, which leads to formation of phagolysosomes where bacteria are catabolized. Antibody-dependent phagocytosis is readily activated at the presence of phagocytes and antibody-bacteria complexes. It has been estimated that a surface density of only 5.33–26.7 antibodies/μm^2 (i.e., IgG density on the surface of targeted particles) is required to trigger antibody-dependent phagocytosis [27]. In one example of the significance of opsonophagocytosis for mAb treatment of bacterial infection, Russo et al. developed a mAb 13D6 against K1 capsular polysaccharide of *Acinetobacter baumannii*, which showed potent inhibitory effects on bacterial growth in a rat soft tissue infection model. Antibody-dependent phagocytic killing has been found to be the primary bactericidal mechanism in this study [28].

3.3. Complement-Dependent Cytotoxicity

Antibody-dependent (i.e., classical) complement activation is another important bactericidal mechanism of the innate immune system. Binding of antibodies on the bacterial surface enhances the recruitment and binding of soluble complement factors, including C1q, to the Fc domain of the mAb, which leads to the activation of the complement cascade (i.e., complement fixation), formation of the membrane attack complex, thus leading to bacterial killing. Activation of the antibody-dependent pathway requires interaction of C1q with at least two IgG molecules [29]. Based on the molecular size of C1q, it is estimated that the surface density of IgG must be such that IgG molecules are separated by no greater than ~40 nm to fix C1q [30]. In contrast, IgM is much more efficient in complement activation, as a single IgM molecule is able to fix C1q [29]. An anti-keratin antibody IgM (3B4) directed against Methicillin-resistant *Staphylococcus aureus* (MRSA) was generated by An et al. with strong binding to MRSA and mannose-binding lectin (MBL), which in turn activated the classical and MBL complement pathways and led to potent bactericidal activity [31]. Passive immunization with 3B4 significantly decreased bacterial burden in organs and improved animal survival in a mouse bacteremia model [31].

3.4. Direct Bactericidal mAbs

In addition to the mechanisms of action discussed above, antibacterial mAbs showing direct bactericidal activity have been occasionally identified. Binding of these mAbs may trigger lysis of bacterial cells directly (i.e., without requirement for fixation of complement of engagement of other components of the host immune system). For instance, LaRocca et al. developed an IgG1 mAb (CB2) that targets outer surface protein B (OspB) of *Borrelia burgdorferi* for the treatment of Lyme disease [32]. CB2 exhibited complement-independent pore forming when bound to *B. burgdorferi*, which resulted in osmotic lysis of bacterial cells. However, this bactericidal effect was not transferable to *Escherichia coli* expressing recombinant OspB, suggesting a unique interaction between CB2 and *B. burgdorferi* [32]. The underlying mechanism of action is unclear, but it was found to be correlated with cholesterol glycolipids in the *B. burgdorferi* outer membrane that exist as temperature-sensitive lipid raft-like microdomains [33].

3.5. Immunoconjugates

The use of mAbs to deliver highly potent payloads has been successfully applied in cancer treatment [34]. This strategy, on one hand, does not require the mAb itself to be protective; on the other hand, it increases the half-life and specificity of the payload and, hence, decreases off-target toxicity. However, application of this approach to bacterial infection is still in its infancy. In 2015, Lehar and colleagues for the first time adapted the immunoconjugate strategy to antimicrobials and developed a novel THIOMAB™ (Genentech, South San Francisco, CA, USA) antibody antibiotic conjugate (AAC) against *Staphylococcus aureus* [35]. The antibody module opsonizes *S. aureus* and mediates uptake into phagolysosomes, where a potent antibiotic payload is released, allowing efficient killing of intracellular bacteria [35]. This AAC strategy demonstrated promising bactericidal activity against vancomycin-resistant *S. aureus*, and it was especially efficacious for bacteria with an intracellular life cycle [36]. Antibody-antibiotic conjugates may be expected to demonstrate favorable pharmacokinetics (i.e., long half-lives), and decreased off-site toxicity. For example, the AAC strategy ameliorated antibiotic-mediated disruption to the normal flora, and may decrease selective pressure that enhances the development of cross resistance, due to the specificity provided by the antibody carrier. The advantages provided by antibody conjugation may allow for reconsideration of antimicrobials that failed in development due to unfavorable PK or toxicity.

Radioimmunoconjuates, which link radionuclides to mAbs, may allow targeted delivery of bactericidal radiation to bacteria. As a proof-of-principle, Dadachova et al. developed a radioimmunoconjugate with Bismuth-213 linked to a mAb (D11) targeting the pneumococcal capsular polysaccharide [37]. Administration of ^{213}Bi-D11 showed dose-dependent bacterial killing in vitro and in a mouse bacteremia model, without detectable hematological toxicity [37]. In addition, radioimmunotherapy has also shown to confer protection against human immunodeficiency virus (HIV) in severe combined immunodeficiency (SCID) mice, and to selectively kill HIV-infected human T cells and human peripheral blood mononuclear cells in vitro [38]. These data suggest that radioimmunoconjugates may provide protection in immunocompromised patients and may be efficacious against infected host cells that express bacterial antigens on cell surfaces, which could be a novel approach to clear latent intracellular bacteria.

3.6. Immunomodulatory mAbs

Immunomodulatory mAbs, such as T-cell engaging antibodies and antibodies targeting programmed cell death protein 1 (PD-1) or cytotoxic T lymphocyte-associated protein 4 (CRLA-4), have gained great success in the treatment of cancer. However, their application in bacterial infections has not been well explored. Akin to cancer, chronic exposure of antigens to T-cells during persistent infections leads to cellular exhaustion of effector functions [39]. Thus, immunomodulatory mAbs theoretically may aid in the clearance of bacteria through stimulating the host immune system. Recently, evidence from the literature supports the potential benefit of anti-PD-1 mAb for the treatment of tuberculosis (TB) infection. PD-1 and its ligands (PD-L1 and PD-L2) were found to be significantly decreased in $CD4^+$ and $CD8^+$ T cells in TB patients after standard-of-care therapy [40]. Treatment with anti-PD-1 mAb has been shown to restore cytokine secretion and antigen responsiveness of T cells isolated from TB patients ex vivo [41].

4. Challenges in the Development of mAbs for the Treatment of Bacterial Infections

Development of antibacterial mAbs has been progressing relatively slowly. Only three antibacterial mAbs have been marketed in the United States (Table 1), and nine mAb products are in Phase 1–2 clinical trials (Table 2). Although modest success in this area may be partly due to a real or perceived lack of economic incentive to the pharmaceutical industry, limited development of antibacterial mAbs may also relate to a host of scientific challenges. Some of the complexities include difficulties in the selection of accessible and conserved bacterial targets, risk for antibody-dependent enhancement of bacterial infection, and various bacterial countermeasures against antibodies.

4.1. Difficulties in Selection of Bacterial Targets

Antibacterial mAbs have been primarily developed to target either bacterial cell surface targets or secreted exotoxins. Although anti-exotoxin antibodies have been successful, these mAbs only attenuate bacterial pathological activity though neutralization of toxins, and are not expected to provide bactericidal activities. Anti-exotoxin mAbs, therefore, are typically indicated for prophylaxis or as adjunctive therapies to antibiotics. Antibacterial mAbs directed against bacterial surface epitopes have primarily been developed for binding to outer membrane proteins (OMPs) or exopolysaccharides (i.e., capsules or O-antigens of lipopolysaccharides) as shown in Table 2. OMPs are attractive therapeutic targets for vaccination and passive immunization due to the high conservation of OMPs among clinical isolates. Outer membrane protein A (OmpA)-like proteins, for instance, are conserved across all sequenced clinical isolates with high protein homology in many Gram-negative bacteria such as *Escherichia coli*, *Pseudomonas aeruginosa*, and *Acinetobacter baumannii* [42,43]. However, a concern for mAbs targeting OMPs are reports that exopolysaccharides mask these conserved targets, impede mAb binding, and hinder opsonization [44–48]. In contrast, exopolysaccharides are readily accessible to antibody binding. Several of the mAbs in development are directed against exopolysaccharides, such as DSTA4637S, MEDI3902, Aerumab, and Aerucin (Table 2). However, exopolysaccharide epitopes are not typically conserved. Large numbers of capsular serotypes have been identified for many bacterial pathogens. For example, at least 18 and 90 capsular serotypes have been described for *Staphylococcus aureus* and *Streptococcus pneumoniae*, respectively [49,50]. A single mAb thus may only provide protection against a specific capsular serotype. For instance, Aerumab binds only to serotype

O11 of *P. aeruginosa* (Table 2) and, therefore, mAb cocktails that recognize different serotypes may be required to confer meaningful antibacterial efficacy. In addition, exopolysaccharides may shed from bacterial cells and act as decoys, which may reduce the amount of unbound antibody reaching the bacterial surface. Capsular polysaccharides shed from *Klebsiella pneumoniae*, *Streptococcus pneumoniae*, and *P. aeruginosa* have been shown to increase the resistance to antimicrobial peptides (e.g., polymyxin B and neutrophil α-defensin 1) of an unencapsulated strains (i.e., 3-fold increase in minimum inhibitory concentrations). Incubation with these antimicrobial peptides also stimulated the release of capsular polysaccharide [51]. Shed capsular polysaccharides from *A. baumannii* appear to neutralize free anti-capsule mAb molecules and may contribute to the lack of efficacy of mAb treatment in an *A. baumannii* mouse pneumonia infection model [15].

4.2. Antibody-Dependent Enhancement of Infection

Although the role of immunoglobulin in host defense against invading microorganisms via activation of effector cells and complement is undisputed, accumulating evidence supports that, in some instances, antibody augments microorganism infection via assisting in their colonization and invasion to host cells. Antibody-dependent enhancement (ADE) of infection was first discovered in Murray Valley encephalitis virus by Hawkes [52], and it was found to also have relevance for other viruses, such as Dengue virus, human immunodeficiency virus, Ebola virus, and Zika virus [53–57]. ADE of viral infection is mainly due to the intracellular viability of viruses, where binding with mAb facilitates viral adherence and entry to host cells through interaction with Fc receptors or complement receptors [58,59]. ADE of bacterial infection has been reported less frequently; however, striking mechanisms have been identified. IgA1-dependent enhancement of pneumococcal adherence to Detroit 562 pharyngeal epithelial cells found by Weiser et al. is a particular example. IgA1 protease secreted by *Streptococcus pneumoniae* cleave the anti-capsule IgA1 mAb that is bound on the bacterial exopolysaccharide. Positive charges on the IgA1 Fab fragments neutralize the negative charges on the polysaccharide, which unmasks phosphorylcholine underneath the capsular polysaccharide, thus enhancing *S. pneumoniae* binding to epithelial cells [60]. The increase in *S. pneumoniae* adherence was found to be directly correlated with the isoelectric points of the IgA1 Fab fragments [60]. Recently, antibodies obtained from persons with latent tuberculosis were found to be protective, whereas antibodies obtained from those with active tuberculosis promote *Mycobacterium tuberculosis* infection of human lung epithelial cells and promote bacterial replication in macrophages [61]. Additionally, the activity of anti-*M. tuberculosis* mAbs (protective vs. non-protective) was shown to be correlated with both antibody isotype and glycosylation patterns [62,63]. Monoclonal antibodies directed against capsule epitopes of *Acinetobacter baumannii*, one of the three top priority pathogens listed by World Health Organization, also demonstrated ADE of infection. Our laboratory had shown that an anti-capsule mAb IgG3 enhances *A. baumannii* adherence/invasion to macrophages and human lung epithelial cells through IgG engagement of FcγRs, and this ADE of infection leads to a significant increase of animal mortality in a mouse pneumonia infection model [64].

4.3. Countermeasures against Antibacterial mAbs

Although antibacterial mAbs exploit pharmacological mechanisms that are distinct from those of antimicrobials and hence cross-resistance with small-molecule antibiotics is unlikely, host immunoglobulin has applied "selection pressure" to bacteria for millennia, which has led to the evolution of a variety of bacterial defense mechanisms. Countermeasures against host immunoglobulin may, in many cases, provide defense against therapeutic monoclonal antibacterial antibodies. Antibody neutralizing proteins, for instance, protein A of *S. aureus* and protein G of *Streptococcus*, are membrane proteins that bind antibody Fc domain and thus impede opsonophagocytosis and complement activation [65–67]. Additionally, binding of serum IgG via the Fc region decorates the bacterial surface, decreasing bacterial recognition by the immune system. Many bacteria also secrete proteinases that degrade antibodies and therefore inactivate antibody effector functions. *Streptococcus pyogenes*, for example, secrete IdeS (Immunoglobulin G-degrading enzyme of *S. pyogenes*) that specifically cleaves the γ-chain of human IgG in the hinge region, and SpeB (*Streptococcal* erythrogenic toxin B) that has a broad immunoglobulin-degrading activity toward IgG, IgA, IgM, IgD, and IgE [68,69]. Following IgG cleavage, the resultant antibody fragments (e.g., Fab or F(ab′)$_2$) may compete for binding with intact antibodies and further impede antibody-mediated bactericidal activities [70]. Similar proteinases are also found in *S. aureus*, *P. aeruginosa*, *Streptococcus pneumoniae*, and *Haemophilus influenzae* [71–74]. Antibody-based therapies that rely on antibody effector functions are likely to be affected by these antibody neutralizing proteins and proteinases. Additionally, some bacteria can survive and replicate inside phagocytes, which then turn into "Trojan horses" that contribute to the systemic dissemination of bacteria and recurrence of infection. *M. tuberculosis*, for instance, inhibits fusion of lysosomes with phagosomes in macrophages, which protects bacteria from killing mediated by lysosomal constituents [75]. *Rickettsia prowazekii* escapes from phagosome vacuoles before the phagosome-lysosome fusion, likely via phospholipase-mediated dissolution of phagosome membrane [76]. *S. aureus* is resistant to killing by phagolysosomal catabolism through neutralization of toxic oxygen radicals by released catalase, superoxide dismutase, and carotenoids [36,77,78]. Thus, passive immunization that depends on opsonophagocytosis alone may be ineffective against these bacteria, and antibody-antibiotic conjugates that release potent antibiotic molecules intracellularly may be an effective strategy to kill these intracellular bacteria [35].

5. Currently Marketed mAbs for the Treatment of Bacterial Infections

5.1. Raxibacumab

Raxibacumab (Abthrax) is an anti-protective antigen (PA) mAb that has been approved for the treatment of adult and pediatric patients with inhalational anthrax due to *Bacillus anthracis*. Raxibacumab is approved for use in combination with appropriate antibacterial drugs, and for prophylaxis of inhalational anthrax when alternative therapies are not available or are not appropriate. Raxibacumab binds free PA with a high affinity (equilibrium dissociation constant K_d = 2.78 nM), which inhibits engagement of PA to its cellular receptors on macrophages. The antibody impedes intracellular entry of anthrax lethal factor and edema factor, which contribute substantially to the pathogenic effects of anthrax toxin [79,80]. Raxibacumab demonstrated linear PK in the dose range of 1–40 mg/kg following single IV doses in healthy human volunteers with a half-life of 20–22 days [81,82]. Co-administration with ciprofloxacin, a standard-of-care (SoC) antibiotic for *B. anthracis* bacteremia, did not affect the PK of raxibacumab [82]. Likewise, raxibacumab did not alter the PK of ciprofloxacin.

Raxibacumab is the first biologic product that was developed and approved under the FDA Animal Rule that may be applied when it is not ethical or feasible to conduct controlled clinical trials in humans. The effectiveness of raxibacumab for treatment of inhalational anthrax thus is based on efficacy studies in rabbits and monkeys. Treatment with raxibacumab was initiated when PA was detected in serum (28–42 h) or when body temperature was sustained above baseline for 2 h in animals after challenge with aerosolized *B. anthracis* spores. Significantly improved survival was demonstrated in infected New Zealand White (NZW) rabbits and cynomolgus macaques (44% and 64% survival, respectively), when treated with 40 mg/kg raxibacumab compared to placebo groups (0% survival) [82]. In addition, the combination of raxibacumab and levofloxacin provided significantly enhanced protection (82% survival) compared to the antibiotic alone (65% survival) in *B. anthracis* challenged NZW rabbits [83]. Based on the observed and simulated systemic exposure of raxibacumab in animals versus humans, a single intravenous dose of 40 mg/kg was suggested to provide protection in humans.

5.2. Obiltoxaximab

Obiltoxaximab (Anthim) is also an anti-PA mAb that was approved for the same indication and usage as raxibacumab. However, premedication with diphenhydramine is recommended and close monitoring of individuals who receive obiltoxaximab is also required due to common adverse reactions including hypersensitivity (10.6%, 34/320 healthy subjects) and anaphylaxis (0.9% cases) observed in Phase 1 clinical trials [84]. Obiltoxaximab (K_d = 0.33 nM) protects against anthrax toxin through inhibition of PA binding to cellular receptors on host cells [85]. Obiltoxaximab demonstrated linear PK in dose range of 4–16 mg/kg following single IV administration in healthy humans. Although obiltoxaximab PK has not been studied in infected patients [84], infection of NZW rabbits and cynomolgus monkeys with *B. anthracis* led to significantly faster clearance (17.0 mL/day/kg and 8.6 mL/day/kg) compared to values observed in non-infected animals (8.7 and 4.2 mL/day/kg, respectively) [13]. These data are suggestive of target-mediated elimination upon binding of obiltoxaximab to PA; however, this hypothesis requires further investigation. The estimated half-life and volume of distribution of obiltoxaximab in healthy volunteers was 17–23 days and 6.3–7.5 L, respectively [84]. Low titers (1:20–1:320) of anti-obiltoxaximab antibodies were detected in eight subjects (2.5%) during phase 1 studies, but alterations in PK and toxicity profile were not observed in these individuals [84]. Further, the PK of obiltoxaximab was not affected by concomitant intravenous and oral doses of ciprofloxacin in healthy humans and vice versa [84].

Obiltoxaximab was also approved under the US FDA Animal Rule. Therapeutic efficacy of obiltoxaximab was assessed in animals challenged with aerosolized *B. anthracis* spores. The mAb was administered after animals exhibited clinical signs of systemic anthrax (i.e., presence of PA in serum or sustained elevation of body temperature above baseline), and intravenous obiltoxaximab at 16 mg/kg was able to significantly improve survival in NZW rabbits (62–93%) and cynomolgus macaques (31–47%) compared to 0–6% survival in placebo groups [86]. In prophylaxis studies, a single dose of obiltoxaximab (16 mg/kg) administered 24–72 h prior to *B. anthracis* infection provided full protection (i.e., 100% survival) in cynomolgus macaques versus 10% survival in control animals [13]. Furthermore, obiltoxaximab administered in combination with antibiotics such as levofloxacin, ciprofloxacin, and doxycycline resulted in higher survival rates than the antibiotic alone in *B. anthracis* infected animals [13].

5.3. Bezlotoxumab

Bezlotoxumab (Zinplava) is a human IgG1 that has been approved for use to reduce recurrence of *Clostridium difficile* infection (CDI) in patients ≥18 years of age who are receiving antibacterial drugs for CDI and are at high risk for CDI recurrence. Bezlotoxumab binds with high affinity (K_d < 1 nM) to toxin B, a pivotal virulence factor of *C. difficile*. The mAb inhibits toxin B binding to host cells and hence prevents toxin B-mediated inactivation of Rho GTPases and downstream signaling pathways in cells [87]. The PK of bezlotoxumab was assessed in *C. difficile*-infected patients in Phase 3 clinical trials, with estimated mean clearance, volume of distribution, and half-life of 0.317 L/day, 7.33 L, and 19 days, respectively [88]. Recurrence of CDI (i.e., development of a new episode of *C. difficile*-associated diarrhea following clinical cure of the presenting CDI episode) was significantly lower in patients receiving 10 mg/kg bezlotoxumab with SoC (17.4% and 15.7%) than the subjects receiving placebo with SoC therapy (27.6% and 25.7%) in two Phase 3 studies [89]. However, addition of bezlotoxumab to SoC did not improve clinical cure rate in *C. difficile*-infected patients compared to the SoC group [89]. Thus bezlotoxumab is indicated only for prevention of recurrence of CDI, but not for treatment of CDI.

6. Antibacterial mAbs in Clinical Trials

In addition to the three marketed mAb products, there are nine mAbs that are currently being investigated in clinical trials. Among the nine products listed in Table 2, five mAbs are developed against *Staphylococcus aureus*, three are targeting *Pseudomonas aeruginosa*, and one is for *Escherichia coli*. Released data from preclinical and clinical studies are summarized here to give a broad overview of the products that may be clinically available in the next few years.

6.1. MEDI4893

MEDI4893 is a human IgG1 mAb that specifically binds to and neutralizes alpha-toxin (AT) of *Staphylococcus aureus* and hence inhibits AT-mediated cytotoxic activity toward host cells [90]. AT is a 33-kDa pore-forming toxin that forms heptameric pores in host cells membranes and results in cell lysis [91]. Animal studies using isogenic AT negative mutants demonstrated that AT is a key virulence factor in *S. aureus* infections including sepsis, skin and soft tissue infection, and pneumonia [91–93]. AT was found to be expressed in 83% clinical isolates worldwide, and 91% of the isolates encoded AT subtypes that were neutralized by MEDI4893 [90]. In an acute pneumonia infection model, MEDI4893 was shown to provide both prophylactic and therapeutic effects in immunocompetent and immunocompromised mice. Further, sub-therapeutic MEDI4893 doses administered in combination with sub-therapeutic doses of antibiotics (vancomycin or linezolid) provided significantly improved survival rates compared to monotherapies [94,95]. In a recent phase 1 clinical trial, MEDI4893 exhibited linear PK in the dose range of 225–5000 mg/subject. YTE mutations (amino acid substitutions M252Y/S254T/T256E) in Fc region of the mAb, which increase binding affinity for FcRn, contribute to its favorable clearance of 42–50 mL/day and extended half-life of 80–112 days [96].

6.2. ASN100

ASN100 is a mAb combination of two human IgG1, ASN-1 and ASN-2, which is in development for the prevention of ventilator-associated *S. aureus* pneumonia (VASP). ASN-1 targets alpha-toxin and four leukocidins including gamma hemolysins (HlgAB and HlgCB), Panton-Valentine leukocidin (LukSF or PVL), and LukED. ASN-2 binds another leukocidin LukGH (LukAB) [97,98]. Leukocidins are pore-forming toxins that typically lyse human phagocytic cells and thus play a key role in bacterial evasion of the innate immune response [99–102]. Therefore, ASN100 binds six different toxin molecules to protect against lysis of multiple human cells, including polymorphonuclear leukocytes, monocytes, macrophages, red blood cells, T cells, epithelial, and endothelial cells. Among the five leukocidins, HlgAB, HlgCB, and LukGH are highly conserved in *S. aureus* clinical isolates. LukED is expressed in 50–75% isolates, while LukSF is only present in 5–10% isolates but is correlated with more severe infections [99]. ASN-1 and ASN-2 exhibit linear serum PK over the dose range of 200–4000 mg/subject either when administered alone or simultaneously in healthy human volunteers [103]. Estimated mean clearance, volume of distribution, and half-life of ASN-1 are 0.256 L/day, 7.14 L, and 23.5 days. Values for ASN-2 are 0.186 L/day, 6.45 L, and 26.7 days [103].

6.3. DSTA4637S

DSTA4637S is a THIOMAB™ (Genentech, South San Francisco, CA, USA) antibody antibiotic conjugate (AAC) that is comprised of an anti-*S. aureus* THIOMAB™ (Genentech, South San Francisco, CA, USA) antibody and a potent antibiotic, 4-dimethylamino piperidino-hydroxybenzoxazino rifamycin (dmDNA31), linked through a protease cleavable valine-citrulline linker [35]. It has been known for more than half-century that *S. aureus* can survive inside neutrophils and turn them into "Trojan horses", which assist in systemic bacterial dissemination and contribute to recurrence of infection following antibacterial therapy [104]. DSTA4637A (a preclinical formulation of DSTA4637S) demonstrated potent intracellular bactericidal activity against *S. aureus* both in vitro and in a mouse bacteremia model [35]. The antibody module of the AAC specifically targets the β-O-linked N-acetylglucosamine sugar modifications on cell wall teichoic acid residues of *S. aureus* and is responsible for opsonization of bacteria. Once the opsonized bacteria are taken up into phagolysosomes, proteases such as cathepsins cleave the linker and release the potent dmDNA31 antibiotic, which eradicates intracellular *S. aureus* [35]. Total concentrations of the DSTA4637A antibody (TAb) and antibody-conjugated dmDNA31 (ac-dmDNA31) were consistent with linear plasma PK over the dose range of 5–50 mg/kg in non-infected mice and 25–50 mg/kg in *S. aureus* bacteremia mice [14]. Infection with *S. aureus* had negligible impact on plasma PK of TAb and ac-dmDNA31 over the efficacious dose range of 25–50 mg/kg, with mean clearance of 4.95 vs. 6.08 mL/day/kg, volume of distribution of 94.9 vs. 119 mL/day/kg, and half-life of 14.3 vs. 13.9 days for TAb in non-infected and infected mice, respectively [14]. PK data for DSTA4637S in human subjects have not been published.

6.4. Salvecin

Salvecin (AR-301) is a mAb developed as an adjunctive therapy to SoC antibiotics for ventilator-associated *S. aureus* pneumonia [105]. It binds and neutralizes alpha-toxin and hence prevents AT-mediated lysis of host cells. Phase 1/2a study results met their primary endpoints, and showed that VASP patients who received Salvecin in combination with SoC antibiotics spent shorter time under mechanical ventilation than patients treated with placebo plus antibiotics [106]. In addition, blood bacterial burden was consistently lower in Salvecin-treated patients compared to the control group [106].

6.5. 514.G3

514G3 is a human mAb targeting *Staphylococcus* Protein A (SpA), a key virulence determinant of *S. aureus* that is expressed in all clinical isolates [107]. SpA is present in the *S. aureus* cell wall envelope and is released during bacterial growth [108]. SpA binds the Fc domain of human IgG and protects *S. aureus* from antibody-dependent phagocytic killing [109,110]. Additionally, released SpA triggers B cell superantigen activity through cross-linking of B cell receptors at V_H3 domain [111,112]. 514G3 displaces SpA-bound serum IgG on *S. aureus* surface and enhances opsonophagocytosis or other mechanisms of immune clearance of bacteria [107]. In a pilot Phase 2 study in patients hospitalized with *S. aureus* bacteremia, treatment with 40 mg/kg 514G3 led to 49% reduction in relative risk of overall incidence of serious adverse events (SAEs) and 56% relative risk reduction in *S. aureus* related SAEs compared to the placebo group [113]. More importantly, the duration of hospitalization was reduced by 33% in 514G3 treated patients compared to patients who received placebo (8.6 ± 7 days vs. 12.7 ± 9 days, respectively) [113].

6.6. MEDI3902

MEDI3902 is a bispecific mAb that targets both type III secretion injectisome PcrV anchored on bacterial cell wall and serotype-independent Psl exopolysaccharide of *Pseudomonas aeruginosa* [114]. PcrV is a critical component of the type III secretion system (T3SS) that delivers bacterial toxins and effector molecules into host cells in order to initiate infection. Psl exopolysaccharide is important for *P. aeruginosa* colonization/attachment to mammalian cells and formation of biofilms [115,116]. The majority of *P. aeruginosa* clinical isolates express Psl (89.8–91.2%) and PcrV (87.7–90.2%), and at least one of the targets were identified in 97.3–100% of isolates [114]. While binding of MEDI3902 to PcrV inhibits T3SS-mediated cytotoxicity, targeting Psl prevents *P. aeruginosa* attachment to host cells and enhances opsonophagocytic killing of bacteria. Intravenous administration of MEDI3902 (5 or 15 mg/kg) at 24 h before or 1 h after lethal *P. aeruginosa* challenge conferred 100% survival and significant reductions on tissue bacterial burdens in acute pneumonia and bacteremia animal models including mice, New Zealand rabbits, and pigs [114,117,118].

6.7. Aerumab

Aerumab (AR-101), previously known as panobacumab, is a human IgM mAb directed against the O-antigen of *P. aeruginosa* lipopolysaccharide serotype O11, which accounts for ~20% of clinical isolates. Aerumab is being developed as an adjunctive immunotherapy to SoC antibiotics for ventilator associated pneumonia caused by *P. aeruginosa* [119,120]. Binding of Aerumab to *P. aeruginosa* leads to enhanced bacterial clearance through either phagocytosis or complement-mediated bacterial killing [120]. Aerumab demonstrated linear PK over the dose range of 0.1–4 mg/kg in healthy human volunteers, with mean clearance of 0.039–0.120 L/h, volume of distribution of 4.75–5.47 L, and half-life of 70–95 h [121]. *P. aeruginosa* infection in patients did not affect the PK of Aerumab following IV doses of 1.2 mg/kg, where estimated clearance, volume of distribution, and half-life are 0.0579 L/h, 7.5 L, and 102 h, respectively [122]. Further, all 13 patients who received three doses of 1.2 mg/kg Aerumab as an adjunctive therapy given every 72 h survived, with a mean clinical resolution rate of 85% (11/13) in 8 days compared to a rate of 64% (9/14) in 18.5 days in patients who did not receive the mAb [122,123].

6.8. Aerucin

As indicated above, Aerumab can only recognize ~20% of *P. aeruginosa* clinical isolates. Aerucin is a second generation anti-*P. aeruginosa* mAb that binds to alginate (i.e., exopolysaccharide) in greater than 90% of clinical isolates [124]. Aerucin is also developed as an adjunctive therapy to SoC antibiotics for hospital-acquired and ventilator-associated pneumonia caused by *P. aeruginosa*, and binding of Aerucin is also expected to augment the opsonophagocytic killing and complement-dependent bactericidal activity against *P. aeruginosa* [124]. However, preclinical and clinical study results for Aerucin have not been published.

6.9. Shigamab

Shiga toxin (Stx)-producing *Escherichia coli* (STEC) is the major cause of hemorrhagic colitis by infectious agents in the United States. A serious consequence of STEC infection is hemolytic uremic syndrome (HUS) that can lead to renal failure and death in 5–15% of infected children [125]. Treatments for STEC infections are currently not available, and antibiotics may increase the risk of HUS [126]. Shigamab is a combination of two chimeric mabs, cαStx1 and cαStx2, which were developed to neutralize Stx1 and Stx2, respectively. Stx1 and Stx2 are the two major types of shiga toxin that are the key virulence factors contributing to the pathogenesis of HUS [127]. Treatment with 20 mg/kg of Shigamab conferred 90% survival in mice challenged with lethal doses of Stx1 and Stx2, whereas cαStx1 or cαStx2 alone did not protect mice against infection [128]. In addition, infection with STEC strain B2F1 in mice did not affect the PK of cαStx2 at 15 mg/kg [128]. Shigamab exhibited linear PK over the dose range of 1–3 mg/kg, and cαStx1 was shown to have greater clearance (0.38 ± 0.16 mL/h/kg) and shorter half-life (190 ± 140 h) than cαStx2 (0.20 ± 0.07 mL/h/kg and 261 ± 112 h, respectively) [129].

7. Conclusions

Pathogen-specific antibacterial mAbs have become an appealing therapeutic option due to recent advances in mAb production and engineering technologies. Antibodies kill bacteria or attenuate bacterial pathological activity via various mechanisms, including opsonophagocytosis, complement-mediated bactericidal activity, antibody-dependent cellular cytotoxicity, and neutralization of bacterial toxins. These pharmacodynamic mechanisms are distinct from those of small-molecule antimicrobials and therefore, such mAbs provide an attractive therapeutic option for antimicrobial resistant strains. The high specificity of mAbs may be expected to allow less disturbance to normal flora and less selective pressure for cross-resistance. Extended half-lives of mAbs may allow less frequent dosing and long-term prophylaxis. Antibacterial mAbs may also exhibit pharmacokinetic properties such as target-mediated drug disposition due to opsonophagocytosis or formation of antibody-toxin complexes, and there is some potential for complicated tissue distribution during the course of bacterial infection. These possible complexities require further study. Though there are only three mAbs marketed for prophylaxis or treatment of bacterial infection as of today, there is promise for a more prominent future role for antibacterial mAbs in view of their many advantages over traditional antimicrobials, and in view of the positive findings from clinical investigations of several mAbs in development.

Acknowledgments: This work was supported through funding provided by the Center of Protein Therapeutics (J.P.B).

Author Contributions: S.X.W.-L. performed the literature research. S.X.W.-L. and J.P.B. designed, wrote, and edited the review.

Conflicts of Interest: The authors declare no conflicts of interest.

References

1. Winau, F.; Winau, R. Emil von Behring and serum therapy. *Microbes Infect.* **2002**, *4*, 185–188. [CrossRef]
2. Casadevall, A. Antibody-based therapies for emerging infectious diseases. *Emerg. Infect. Dis.* **1996**, *2*, 200–208. [CrossRef] [PubMed]
3. Felton, L.D. The units of protective antibody in antipneumococcus serum and antibody solution. *J. Infect. Dis.* **1928**, *43*, 531–542. [CrossRef]
4. Weisman, L.E.; Cruess, D.F.; Fischer, G.W. Opsonic activity of commercially available standard intravenous immunoglobulin preparations. *Pediatr. Infect. Dis. J.* **1994**, *13*, 1122–1125. [CrossRef] [PubMed]
5. Slade, H.B. Human immunoglobulins for intravenous use and hepatitis C viral transmission. *Clin. Diagn. Lab. Immunol.* **1994**, *1*, 613–619. [PubMed]
6. Davies, J.; Davies, D. Origins and evolution of antibiotic resistance. *Microbiol. Mol. Biol. Rev.* **2010**, *74*, 417–433. [CrossRef] [PubMed]
7. Ventola, C.L. The antibiotic resistance crisis: Part 1: Causes and threats. *Pharm. Ther.* **2015**, *40*, 277–283.
8. Kohler, G.; Milstein, C. Continuous cultures of fused cells secreting antibody of predefined specificity. *Nature* **1975**, *256*, 495–497. [CrossRef] [PubMed]
9. Wright, A.; Shin, S.U.; Morrison, S.L. Genetically engineered antibodies: Progress and prospects. *Crit. Rev. Immunol.* **1992**, *12*, 125–168. [PubMed]
10. Waldmann, T.A.; Strober, W. Metabolism of immunoglobulins. *Prog. Allergy* **1969**, *13*, 1–110. [PubMed]
11. Junghans, R.P. Finally! The brambell receptor (FcRB). Mediator of transmission of immunity and protection from catabolism for IgG. *Immunol. Res.* **1997**, *16*, 29–57. [CrossRef] [PubMed]
12. Levy, G. Pharmacologic target-mediated drug disposition. *Clin. Pharmacol. Ther.* **1994**, *56*, 248–252. [CrossRef] [PubMed]
13. Greig, S.L. Obiltoxaximab: First global approval. *Drugs* **2016**, *76*, 823–830. [CrossRef] [PubMed]
14. Zhou, C.; Lehar, S.; Gutierrez, J.; Rosenberger, C.M.; Ljumanovic, N.; Dinoso, J.; Koppada, N.; Hong, K.; Baruch, A.; Carrasco-Triguero, M.; et al. Pharmacokinetics and pharmacodynamics of DSTA4637A: A novel THIOMAB antibody antibiotic conjugate against *staphylococcus aureus* in mice. *MAbs* **2016**, *8*, 1612–1619. [CrossRef] [PubMed]
15. Wang-Lin, S.X.; Russo, T.A.; Balthasar, J.P. Pharmacokinetics of a monoclonal anti-*acinetobacter baumannii* k2 capsule antibody in mice. Unpublished work. 2018.
16. Baxter, L.T.; Zhu, H.; Mackensen, D.G.; Jain, R.K. Physiologically based pharmacokinetic model for specific and nonspecific monoclonal antibodies and fragments in normal tissues and human tumor xenografts in nude mice. *Cancer Res.* **1994**, *54*, 1517–1528. [PubMed]
17. Flessner, M.F.; Lofthouse, J.; El Zakaria, R. In vivo diffusion of immunoglobulin G in muscle: Effects of binding, solute exclusion, and lymphatic removal. *Am. J. Physiol.* **1997**, *273*, H2783–H2793. [CrossRef] [PubMed]
18. Baxter, L.T.; Jain, R.K. Transport of fluid and macromolecules in tumors. I. Role of interstitial pressure and convection. *Microvasc. Res.* **1989**, *37*, 77–104. [CrossRef]
19. Lobo, E.D.; Hansen, R.J.; Balthasar, J.P. Antibody pharmacokinetics and pharmacodynamics. *J. Pharm. Sci.* **2004**, *93*, 2645–2668. [CrossRef] [PubMed]
20. Wang, W.; Wang, E.Q.; Balthasar, J.P. Monoclonal antibody pharmacokinetics and pharmacodynamics. *Clin. Pharmacol. Ther.* **2008**, *84*, 548–558. [CrossRef] [PubMed]
21. Cerca, N.; Jefferson, K.K.; Oliveira, R.; Pier, G.B.; Azeredo, J. Comparative antibody-mediated phagocytosis of staphylococcus epidermidis cells grown in a biofilm or in the planktonic state. *Infect. Immun.* **2006**, *74*, 4849–4855. [CrossRef] [PubMed]
22. Sawada-Hirai, R.; Jiang, I.; Wang, F.; Sun, S.M.; Nedellec, R.; Ruther, P.; Alvarez, A.; Millis, D.; Morrow, P.R.; Kang, A.S. Human anti-anthrax protective antigen neutralizing monoclonal antibodies derived from donors vaccinated with anthrax vaccine adsorbed. *J. Immune Based Ther. Vaccines* **2004**, *2*, 5. [CrossRef] [PubMed]

23. Abboud, N.; Chow, S.K.; Saylor, C.; Janda, A.; Ravetch, J.V.; Scharff, M.D.; Casadevall, A. A requirement for FcγR in antibody-mediated bacterial toxin neutralization. *J. Exp. Med.* **2010**, *207*, 2395–2405. [CrossRef] [PubMed]
24. Frank, D.W.; Vallis, A.; Wiener-Kronish, J.P.; Roy-Burman, A.; Spack, E.G.; Mullaney, B.P.; Megdoud, M.; Marks, J.D.; Fritz, R.; Sawa, T. Generation and characterization of a protective monoclonal antibody to pseudomonas aeruginosa PcrV. *J. Infect. Dis.* **2002**, *186*, 64–73. [CrossRef] [PubMed]
25. Baer, M.; Sawa, T.; Flynn, P.; Luehrsen, K.; Martinez, D.; Wiener-Kronish, J.P.; Yarranton, G.; Bebbington, C. An engineered human antibody Fab fragment specific for *Pseudomonas aeruginosa* PcrV antigen has potent antibacterial activity. *Infect. Immun.* **2009**, *77*, 1083–1090. [CrossRef] [PubMed]
26. Kuhn, D.A.; Vanhecke, D.; Michen, B.; Blank, F.; Gehr, P.; Petri-Fink, A.; Rothen-Rutishauser, B. Different endocytotic uptake mechanisms for nanoparticles in epithelial cells and macrophages. *Beilstein J. Nanotechnol.* **2014**, *5*, 1625–1636. [CrossRef] [PubMed]
27. Lewis, J.T.; Hafeman, D.G.; McConnell, H.M. Kinetics of antibody-dependent binding of haptenated phospholipid vesicles to a macrophage-related cell line. *Biochemistry* **1980**, *19*, 5376–5386. [CrossRef] [PubMed]
28. Russo, T.A.; Beanan, J.M.; Olson, R.; MacDonald, U.; Cox, A.D.; St Michael, F.; Vinogradov, E.V.; Spellberg, B.; Luke-Marshall, N.R.; Campagnari, A.A. The K1 capsular polysaccharide from *Acinetobacter baumannii* is a potential therapeutic target via passive immunization. *Infect. Immun.* **2013**, *81*, 915–922. [CrossRef] [PubMed]
29. Sompayrac, L. *How the Immune System Works*, 4th ed.; Wiley-Blackwell: Chichester, UK; Hoboken, NJ, USA, 2012; p. 1.
30. Salvador-Morales, C.; Sim, R.B. Complement activation. In *Frontiers in Nanobiomedical Research*; World Scientific Singapore: Singapore, 2012.
31. An, J.; Li, Z.; Dong, Y.; Wu, J.; Ren, J. Complement activation contributes to the anti-methicillin-resistant staphylococcus aureus effect of natural anti-keratin antibody. *Biochem. Biophys. Res. Commun.* **2015**, *461*, 142–147. [CrossRef] [PubMed]
32. LaRocca, T.J.; Holthausen, D.J.; Hsieh, C.; Renken, C.; Mannella, C.A.; Benach, J.L. The bactericidal effect of a complement-independent antibody is osmolytic and specific to *Borrelia*. *Proc. Natl. Acad. Sci. USA* **2009**, *106*, 10752–10757. [CrossRef] [PubMed]
33. LaRocca, T.J.; Crowley, J.T.; Cusack, B.J.; Pathak, P.; Benach, J.; London, E.; Garcia-Monco, J.C.; Benach, J.L. Cholesterol lipids of *Borrelia burgdorferi* form lipid rafts and are required for the bactericidal activity of a complement-independent antibody. *Cell Host Microbe* **2010**, *8*, 331–342. [CrossRef] [PubMed]
34. Beck, A.; Reichert, J.M. Antibody-drug conjugates: Present and future. *MAbs* **2014**, *6*, 15–17. [CrossRef] [PubMed]
35. Lehar, S.M.; Pillow, T.; Xu, M.; Staben, L.; Kajihara, K.K.; Vandlen, R.; DePalatis, L.; Raab, H.; Hazenbos, W.L.; Morisaki, J.H.; et al. Novel antibody-antibiotic conjugate eliminates intracellular *S. Aureus*. *Nature* **2015**, *527*, 323–328. [CrossRef] [PubMed]
36. Gresham, H.D.; Lowrance, J.H.; Caver, T.E.; Wilson, B.S.; Cheung, A.L.; Lindberg, F.P. Survival of *Staphylococcus aureus* inside neutrophils contributes to infection. *J. Immunol.* **2000**, *164*, 3713–3722. [CrossRef] [PubMed]
37. Dadachova, E.; Burns, T.; Bryan, R.A.; Apostolidis, C.; Brechbiel, M.W.; Nosanchuk, J.D.; Casadevall, A.; Pirofski, L. Feasibility of radioimmunotherapy of experimental pneumococcal infection. *Antimicrob. Agents Chemother.* **2004**, *48*, 1624–1629. [CrossRef] [PubMed]
38. Dadachova, E.; Patel, M.C.; Toussi, S.; Apostolidis, C.; Morgenstern, A.; Brechbiel, M.W.; Gorny, M.K.; Zolla-Pazner, S.; Casadevall, A.; Goldstein, H. Targeted killing of virally infected cells by radiolabeled antibodies to viral proteins. *PLoS Med.* **2006**, *3*, e427. [CrossRef] [PubMed]
39. Wherry, E.J.; Kurachi, M. Molecular and cellular insights into T cell exhaustion. *Nat. Rev. Immunol.* **2015**, *15*, 486–499. [CrossRef] [PubMed]
40. Hassan, S.S.; Akram, M.; King, E.C.; Dockrell, H.M.; Cliff, J.M. PD-1, PD-l1 and PD-l2 gene expression on T-cells and natural killer cells declines in conjunction with a reduction in PD-1 protein during the intensive phase of tuberculosis treatment. *PLoS ONE* **2015**, *10*, e0137646. [CrossRef] [PubMed]

41. Bandaru, A.; Devalraju, K.P.; Paidipally, P.; Dhiman, R.; Venkatasubramanian, S.; Barnes, P.F.; Vankayalapati, R.; Valluri, V. Phosphorylated STAT3 and PD-1 regulate IL-17 production and IL-23 receptor expression in mycobacterium tuberculosis infection. *Eur. J. Immunol.* **2014**, *44*, 2013–2024. [CrossRef] [PubMed]
42. Beher, M.G.; Schnaitman, C.A.; Pugsley, A.P. Major heat-modifiable outer membrane protein in gram-negative bacteria: Comparison with the ompa protein of *Escherichia coli*. *J. Bacteriol.* **1980**, *143*, 906–913. [PubMed]
43. Luo, G.; Lin, L.; Ibrahim, A.S.; Baquir, B.; Pantapalangkoor, P.; Bonomo, R.A.; Doi, Y.; Adams, M.D.; Russo, T.A.; Spellberg, B. Active and passive immunization protects against lethal, extreme drug resistant-*Acinetobacter baumannii* infection. *PLoS ONE* **2012**, *7*, e29446. [CrossRef] [PubMed]
44. Hyams, C.; Camberlein, E.; Cohen, J.M.; Bax, K.; Brown, J.S. The *Streptococcus pneumoniae* capsule inhibits complement activity and neutrophil phagocytosis by multiple mechanisms. *Infect. Immun.* **2010**, *78*, 704–715. [CrossRef] [PubMed]
45. Russo, T.A.; Beanan, J.M.; Olson, R.; MacDonald, U.; Cope, J.J. Capsular polysaccharide and the O-specific antigen impede antibody binding: A potential obstacle for the successful development of an extraintestinal pathogenic *Escherichia coli* vaccine. *Vaccine* **2009**, *27*, 388–395. [CrossRef] [PubMed]
46. Pluschke, G.; Mayden, J.; Achtman, M.; Levine, R.P. Role of the capsule and the O antigen in resistance of O18:K1 *Escherichia coli* to complement-mediated killing. *Infect. Immun.* **1983**, *42*, 907–913. [PubMed]
47. Van der Ley, P.; Kuipers, O.; Tommassen, J.; Lugtenberg, B. O-antigenic chains of lipopolysaccharide prevent binding of antibody molecules to an outer membrane pore protein in *Enterobacteriaceae*. *Microb. Pathog.* **1986**, *1*, 43–49. [CrossRef]
48. Wang-Lin, S.X.; Olson, R.; Beanan, J.M.; MacDonald, U.; Balthasar, J.P.; Russo, T.A. The capsular polysaccharide of *Acinetobacter baumannii* is an obstacle for therapeutic passive immunization strategies. *Infect. Immun.* **2017**, *85*, e00591-17. [CrossRef] [PubMed]
49. O'Riordan, K.; Lee, J.C. *Staphylococcus aureus* capsular polysaccharides. *Clin. Microbiol. Rev.* **2004**, *17*, 218–234. [CrossRef] [PubMed]
50. Henrichsen, J. Six newly recognized types of Streptococcus pneumoniae. *J. Clin. Microbiol.* **1995**, *33*, 2759–2762. [PubMed]
51. Llobet, E.; Tomas, J.M.; Bengoechea, J.A. Capsule polysaccharide is a bacterial decoy for antimicrobial peptides. *Microbiology* **2008**, *154*, 3877–3886. [CrossRef] [PubMed]
52. Hawkes, R.A. Enhancement of the infectivity of arboviruses by specific antisera produced in domestic fowls. *Aust. J. Exp. Biol. Med. Sci.* **1964**, *42*, 465–482. [CrossRef] [PubMed]
53. Sasaki, T.; Setthapramote, C.; Kurosu, T.; Nishimura, M.; Asai, A.; Omokoko, M.D.; Pipattanaboon, C.; Pitaksajjakul, P.; Limkittikul, K.; Subchareon, A.; et al. Dengue virus neutralization and antibody-dependent enhancement activities of human monoclonal antibodies derived from dengue patients at acute phase of secondary infection. *Antiviral Res.* **2013**, *98*, 423–431. [CrossRef] [PubMed]
54. Bardina, S.V.; Bunduc, P.; Tripathi, S.; Duehr, J.; Frere, J.J.; Brown, J.A.; Nachbagauer, R.; Foster, G.A.; Krysztof, D.; Tortorella, D.; et al. Enhancement of Zika virus pathogenesis by preexisting antiflavivirus immunity. *Science* **2017**, *356*, 175–180. [CrossRef] [PubMed]
55. Halstead, S.B. Pathogenesis of dengue: Challenges to molecular biology. *Science* **1988**, *239*, 476–481. [CrossRef] [PubMed]
56. Takada, A.; Feldmann, H.; Ksiazek, T.G.; Kawaoka, Y. Antibody-dependent enhancement of Ebola virus infection. *J. Virol.* **2003**, *77*, 7539–7544. [CrossRef] [PubMed]
57. Toth, F.D.; Mosborg-Petersen, P.; Kiss, J.; Aboagye-Mathiesen, G.; Zdravkovic, M.; Hager, H.; Aranyosi, J.; Lampe, L.; Ebbesen, P. Antibody-dependent enhancement of HIV-1 infection in human term syncytiotrophoblast cells cultured in vitro. *Clin. Exp. Immunol.* **1994**, *96*, 389–394. [CrossRef] [PubMed]
58. Tirado, S.M.; Yoon, K.J. Antibody-dependent enhancement of virus infection and disease. *Viral. Immunol.* **2003**, *16*, 69–86. [CrossRef] [PubMed]
59. Takada, A.; Kawaoka, Y. Antibody-dependent enhancement of viral infection: Molecular mechanisms and in vivo implications. *Rev. Med. Virol.* **2003**, *13*, 387–398. [CrossRef] [PubMed]
60. Weiser, J.N.; Bae, D.; Fasching, C.; Scamurra, R.W.; Ratner, A.J.; Janoff, E.N. Antibody-enhanced pneumococcal adherence requires IgA1 protease. *Proc. Natl. Acad. Sci. USA* **2003**, *100*, 4215–4220. [CrossRef] [PubMed]

61. Casadevall, A. Antibodies to mycobacterium tuberculosis. *N. Engl. J. Med.* **2017**, *376*, 283–285. [CrossRef] [PubMed]

62. Lu, L.L.; Chung, A.W.; Rosebrock, T.R.; Ghebremichael, M.; Yu, W.H.; Grace, P.S.; Schoen, M.K.; Tafesse, F.; Martin, C.; Leung, V.; et al. A functional role for antibodies in tuberculosis. *Cell* **2016**, *167*, 433–443. [CrossRef] [PubMed]

63. Zimmermann, N.; Thormann, V.; Hu, B.; Kohler, A.B.; Imai-Matsushima, A.; Locht, C.; Arnett, E.; Schlesinger, L.S.; Zoller, T.; Schurmann, M.; et al. Human isotype-dependent inhibitory antibody responses against *Mycobacterium tuberculosis*. *EMBO Mol. Med.* **2016**, *8*, 1325–1339. [CrossRef] [PubMed]

64. Wang-Lin, S.X.; Olson, R.; Beanan, J.M.; MacDonald, U.; Russo, T.A.; Balthasar, J.P. Antibody dependent enhancement of *acinetobacter baumannii* infection through immunoglobulin g engagement of fc gamma receptors. *J. Immunol.* **2017**. submitted.

65. Bjorck, L.; Kronvall, G. Purification and some properties of streptococcal protein G, a novel IgG-binding reagent. *J. Immunol.* **1984**, *133*, 969–974. [PubMed]

66. Akerstrom, B.; Brodin, T.; Reis, K.; Bjorck, L. Protein G: A powerful tool for binding and detection of monoclonal and polyclonal antibodies. *J. Immunol.* **1985**, *135*, 2589–2592. [PubMed]

67. Falugi, F.; Kim, H.K.; Missiakas, D.M.; Schneewind, O. Role of protein A in the evasion of host adaptive immune responses by *Staphylococcus aureus*. *MBio* **2013**, *4*, e00575-13. [CrossRef] [PubMed]

68. Von Pawel-Rammingen, U.; Johansson, B.P.; Bjorck, L. Ides, a novel streptococcal cysteine proteinase with unique specificity for immunoglobulin G. *EMBO J.* **2002**, *21*, 1607–1615. [CrossRef] [PubMed]

69. Collin, M.; Olsen, A. Effect of SpeB and EndoS from *Streptococcus pyogenes* on human immunoglobulins. *Infect. Immun.* **2001**, *69*, 7187–7189. [CrossRef] [PubMed]

70. Fick, R.B., Jr.; Naegel, G.P.; Squier, S.U.; Wood, R.E.; Gee, J.B.; Reynolds, H.Y. Proteins of the cystic fibrosis respiratory tract. Fragmented immunoglobulin g opsonic antibody causing defective opsonophagocytosis. *J. Clin. Investig.* **1984**, *74*, 236–248. [CrossRef] [PubMed]

71. Karlsson, A.; Arvidson, S. Variation in extracellular protease production among clinical isolates of *Staphylococcus aureus* due to different levels of expression of the protease repressor sarA. *Infect. Immun.* **2002**, *70*, 4239–4246. [CrossRef] [PubMed]

72. Rooijakkers, S.H.; van Wamel, W.J.; Ruyken, M.; van Kessel, K.P.; van Strijp, J.A. Anti-opsonic properties of staphylokinase. *Microbes Infect.* **2005**, *7*, 476–484. [CrossRef] [PubMed]

73. Fick, R.B., Jr.; Baltimore, R.S.; Squier, S.U.; Reynolds, H.Y. IgG proteolytic activity of *Pseudomonas aeruginosa* in cystic fibrosis. *J. Infect. Dis.* **1985**, *151*, 589–598. [CrossRef] [PubMed]

74. Mulks, M.H.; Kornfeld, S.J.; Plaut, A.G. Specific proteolysis of human IgA by *Streptococcus pneumoniae* and *Haemophilus influenzae*. *J. Infect. Dis.* **1980**, *141*, 450–456. [CrossRef] [PubMed]

75. Hart, P.D.; Young, M.R.; Gordon, A.H.; Sullivan, K.H. Inhibition of phagosome-lysosome fusion in macrophages by certain mycobacteria can be explained by inhibition of lysosomal movements observed after phagocytosis. *J. Exp. Med.* **1987**, *166*, 933–946. [CrossRef] [PubMed]

76. Whitworth, T.; Popov, V.L.; Yu, X.J.; Walker, D.H.; Bouyer, D.H. Expression of the *Rickettsia prowazekii* pld or tlyC gene in *Salmonella enterica* serovar typhimurium mediates phagosomal escape. *Infect. Immun.* **2005**, *73*, 6668–6673. [CrossRef] [PubMed]

77. Flannagan, R.S.; Heit, B.; Heinrichs, D.E. Intracellular replication of staphylococcus aureus in mature phagolysosomes in macrophages precedes host cell death, and bacterial escape and dissemination. *Cell. Microbiol.* **2016**, *18*, 514–535. [CrossRef] [PubMed]

78. Voyich, J.M.; Braughton, K.R.; Sturdevant, D.E.; Whitney, A.R.; Said-Salim, B.; Porcella, S.F.; Long, R.D.; Dorward, D.W.; Gardner, D.J.; Kreiswirth, B.N.; et al. Insights into mechanisms used by staphylococcus aureus to avoid destruction by human neutrophils. *J. Immunol.* **2005**, *175*, 3907–3919. [CrossRef] [PubMed]

79. Chen, Z.; Moayeri, M.; Purcell, R. Monoclonal antibody therapies against anthrax. *Toxins (Basel)* **2011**, *3*, 1004–1019. [CrossRef] [PubMed]

80. Mazumdar, S. Raxibacumab. *MAbs* **2009**, *1*, 531–538. [CrossRef] [PubMed]

81. Subramanian, G.M.; Cronin, P.W.; Poley, G.; Weinstein, A.; Stoughton, S.M.; Zhong, J.; Ou, Y.; Zmuda, J.F.; Osborn, B.L.; Freimuth, W.W. A phase 1 study of Pamab, a fully human monoclonal antibody against *Bacillus anthracis* protective antigen, in healthy volunteers. *Clin. Infect. Dis.* **2005**, *41*, 12–20. [CrossRef] [PubMed]

82. Migone, T.S.; Subramanian, G.M.; Zhong, J.; Healey, L.M.; Corey, A.; Devalaraja, M.; Lo, L.; Ullrich, S.; Zimmerman, J.; Chen, A.; et al. Raxibacumab for the treatment of inhalational anthrax. *N. Engl. J. Med.* **2009**, *361*, 135–144. [CrossRef] [PubMed]

83. Corey, A.; Migone, T.S.; Bolmer, S.; Fiscella, M.; Ward, C.; Chen, C.; Meister, G. *Bacillus anthracis* protective antigen kinetics in inhalation spore-challenged untreated or levofloxacin/raxibacumab-treated New Zealand white rabbits. *Toxins (Basel)* **2013**, *5*, 120–138. [CrossRef] [PubMed]

84. Nagy, C.F.; Leach, T.S.; Hoffman, J.H.; Czech, A.; Carpenter, S.E.; Guttendorf, R. Pharmacokinetics and tolerability of obiltoxaximab: A report of 5 healthy volunteer studies. *Clin. Ther.* **2016**, *38*, 2083–2097. [CrossRef] [PubMed]

85. Nagy, C.F.; Mondick, J.; Serbina, N.; Casey, L.S.; Carpenter, S.E.; French, J.; Guttendorf, R. Animal-to-Human Dose Translation of Obiltoxaximab for Treatment of Inhalational Anthrax Under the US FDA Animal Rule. *Clin. Transl. Sci.* **2017**, *10*, 12–19. [CrossRef] [PubMed]

86. Yamamoto, B.J.; Shadiack, A.M.; Carpenter, S.; Sanford, D.; Henning, L.N.; O'Connor, E.; Gonzales, N.; Mondick, J.; French, J.; Stark, G.V.; et al. Efficacy projection of obiltoxaximab for treatment of inhalational anthrax across a range of disease severity. *Antimicrob. Agents Chemother.* **2016**, *60*, 5787–5795. [CrossRef] [PubMed]

87. Orth, P.; Xiao, L.; Hernandez, L.D.; Reichert, P.; Sheth, P.R.; Beaumont, M.; Yang, X.; Murgolo, N.; Ermakov, G.; DiNunzio, E.; et al. Mechanism of action and epitopes of *Clostridium difficile* toxin B-neutralizing antibody bezlotoxumab revealed by X-ray crystallography. *J. Biol. Chem.* **2014**, *289*, 18008–18021. [CrossRef] [PubMed]

88. Markham, A. Bezlotoxumab: First global approval. *Drugs* **2016**, *76*, 1793–1798. [CrossRef] [PubMed]

89. Wilcox, M.H.; Gerding, D.N.; Poxton, I.R.; Kelly, C.; Nathan, R.; Birch, T.; Cornely, O.A.; Rahav, G.; Bouza, E.; Lee, C.; et al. Bezlotoxumab for prevention of recurrent clostridium difficile infection. *N. Engl. J. Med.* **2017**, *376*, 305–317. [CrossRef] [PubMed]

90. Tabor, D.E.; Yu, L.; Mok, H.; Tkaczyk, C.; Sellman, B.R.; Wu, Y.; Oganesyan, V.; Slidel, T.; Jafri, H.; McCarthy, M.; et al. *Staphylococcus aureus* alpha-toxin is conserved among diverse hospital respiratory isolates collected from a global surveillance study and is neutralized by monoclonal antibody MEDI4893. *Antimicrob. Agents Chemother.* **2016**, *60*, 5312–5321. [CrossRef] [PubMed]

91. Bubeck Wardenburg, J.; Bae, T.; Otto, M.; Deleo, F.R.; Schneewind, O. Poring over pores: Alpha-hemolysin and panton-valentine leukocidin in *Staphylococcus aureus* pneumonia. *Nat. Med.* **2007**, *13*, 1405–1406. [CrossRef] [PubMed]

92. Powers, M.E.; Kim, H.K.; Wang, Y.; Bubeck Wardenburg, J. Adam10 mediates vascular injury induced by *Staphylococcus aureus* alpha-hemolysin. *J. Infect. Dis.* **2012**, *206*, 352–356. [CrossRef] [PubMed]

93. Tkaczyk, C.; Hamilton, M.M.; Datta, V.; Yang, X.P.; Hilliard, J.J.; Stephens, G.L.; Sadowska, A.; Hua, L.; O'Day, T.; Suzich, J.; et al. *Staphylococcus aureus* alpha toxin suppresses effective innate and adaptive immune responses in a murine dermonecrosis model. *PLoS ONE* **2013**, *8*, e75103. [CrossRef] [PubMed]

94. Hua, L.; Hilliard, J.J.; Shi, Y.; Tkaczyk, C.; Cheng, L.I.; Yu, X.; Datta, V.; Ren, S.; Feng, H.; Zinsou, R.; et al. Assessment of an anti-alpha-toxin monoclonal antibody for prevention and treatment of *Staphylococcus aureus*-induced pneumonia. *Antimicrob. Agents Chemother.* **2014**, *58*, 1108–1117. [CrossRef] [PubMed]

95. Hua, L.; Cohen, T.S.; Shi, Y.; Datta, V.; Hilliard, J.J.; Tkaczyk, C.; Suzich, J.; Stover, C.K.; Sellman, B.R. MEDI4893* promotes survival and extends the antibiotic treatment window in a *Staphylococcus aureus* immunocompromised pneumonia model. *Antimicrob. Agents Chemother.* **2015**, *59*, 4526–4532. [CrossRef] [PubMed]

96. Yu, X.Q.; Robbie, G.J.; Wu, Y.; Esser, M.T.; Jensen, K.; Schwartz, H.I.; Bellamy, T.; Hernandez-Illas, M.; Jafri, H.S. Safety, tolerability, and pharmacokinetics of MEDI4893, an investigational, extended-half-life, anti-*Staphylococcus aureus* alpha-toxin human monoclonal antibody, in healthy adults. *Antimicrob. Agents Chemother.* **2017**, *61*, e01020-16. [CrossRef] [PubMed]

97. Badarau, A.; Rouha, H.; Malafa, S.; Battles, M.B.; Walker, L.; Nielson, N.; Dolezilkova, I.; Teubenbacher, A.; Banerjee, S.; Maierhofer, B.; et al. Context matters: The importance of dimerization-induced conformation of the lukgh leukocidin of *Staphylococcus aureus* for the generation of neutralizing antibodies. *MAbs* **2016**, *8*, 1347–1360. [CrossRef] [PubMed]

98. Rouha, H.; Badarau, A.; Visram, Z.C.; Battles, M.B.; Prinz, B.; Magyarics, Z.; Nagy, G.; Mirkina, I.; Stulik, L.; Zerbs, M.; et al. Five birds, one stone: Neutralization of alpha-hemolysin and 4 bi-component leukocidins of *Staphylococcus aureus* with a single human monoclonal antibody. *MAbs* **2015**, *7*, 243–254. [CrossRef] [PubMed]

99. Vandenesch, F.; Lina, G.; Henry, T. *Staphylococcus aureus* hemolysins, bi-component leukocidins, and cytolytic peptides: A redundant arsenal of membrane-damaging virulence factors? *Front. Cell. Infect. Microbiol.* **2012**, *2*, 12. [CrossRef] [PubMed]

100. Alonzo, F., III; Torres, V.J. The bicomponent pore-forming leucocidins of *Staphylococcus aureus*. *Microbiol. Mol. Biol. Rev.* **2014**, *78*, 199–230. [CrossRef] [PubMed]

101. DuMont, A.L.; Torres, V.J. Cell targeting by the *Staphylococcus aureus* pore-forming toxins: It's not just about lipids. *Trends Microbiol.* **2014**, *22*, 21–27. [CrossRef] [PubMed]

102. DeLeo, F.R.; Diep, B.A.; Otto, M. Host defense and pathogenesis in *Staphylococcus aureus* infections. *Infect. Dis. Clin. North Am.* **2009**, *23*, 17–34. [CrossRef] [PubMed]

103. Magyarics, Z.; Leslie, F.; Luperchio, S.; Bartko, J.; Schorgenhofer, C.; Schwameis, M.; Derhaschnig, U.; Lagler, H.; Stiebellehner, L.; Jilma, B.; et al. Safety and pharmacokinetics of ASN100, a monoclonal antibody combination for the prevention and treatment of *Staphylococcus aureus* infections, from a single ascending dose phase 1 clinical study in healthy adult volunteers. In Proceedings of the European Congress of Clinical Microbiology and Infectious Diseases, Vienna, Austria, 22–25 April 2017.

104. Rogers, D.E. Studies on bacteriemia. I. Mechanisms relating to the persistence of bacteriemia in rabbits following the intravenous injection of staphylococci. *J. Exp. Med.* **1956**, *103*, 713–742. [CrossRef] [PubMed]

105. Health, N.I.O. Safety, Pharmacokinetics and Efficacy of KBSA301 in Severe Pneumonia (*S. Aureus*). Available online: https://clinicaltrials.gov/ct2/show/NCT01589185?recrs=abdefghim&cond=Staphylococcus+Aureus&intr=Antibodies%2C+Monoclonal&rank=2 (accessed on 22 July 2017).

106. Aridis Pharmaceuticals. Ar-301: Fully Human mAb against *Straphylococcus aureus*. Available online: http://www.aridispharma.com/ar-301/ (accessed on 22 July 2017).

107. Huynh, T.; Stecher, M.; McKinnon, J.; Jung, N.; Rupp, M. Safety and tolerability of 514G3, a ture human anti-protein a monoclonal antibody for the treatment of *S. Aureus* bacteremia. In *Open Forum Infectious Diseases*; Oxford University Press: Oxford, UK, 2016.

108. Ton-That, H.; Liu, G.; Mazmanian, S.K.; Faull, K.F.; Schneewind, O. Purification and characterization of sortase, the transpeptidase that cleaves surface proteins of *Staphylococcus aureus* at the LPXTG motif. *Proc. Natl. Acad. Sci. USA* **1999**, *96*, 12424–12429. [CrossRef] [PubMed]

109. Sjodahl, J. Repetitive sequences in protein a from *Staphylococcus aureus*. Arrangement of five regions within the protein, four being highly homologous and Fc-binding. *Eur. J. Biochem.* **1977**, *73*, 343–351. [CrossRef] [PubMed]

110. Forsgren, A.; Quie, P.G. Effects of staphylococcal protein a on heat labile opsonins. *J. Immunol.* **1974**, *112*, 1177–1180. [PubMed]

111. Cary, S.; Krishnan, M.; Marion, T.N.; Silverman, G.J. The murine clan V_h III related 7183, J606 and S107 and DNA4 families commonly encode for binding to a bacterial B cell superantigen. *Mol. Immunol.* **1999**, *36*, 769–776. [CrossRef]

112. Goodyear, C.S.; Silverman, G.J. Death by a B cell superantigen: In vivo VH-targeted apoptotic supraclonal B cell deletion by a staphylococcal toxin. *J. Exp. Med.* **2003**, *197*, 1125–1139. [CrossRef] [PubMed]

113. Otero, A. Patients Receiving 514G3 Therapy Had Reduced Hospitalization and Fewer Infection-Related Serious Adverse Events. Available online: http://investors.xbiotech.com/phoenix.zhtml?c=253990&p=irol-newsArticle&ID=2259222 (accessed on 22 July 2017).

114. DiGiandomenico, A.; Keller, A.E.; Gao, C.; Rainey, G.J.; Warrener, P.; Camara, M.M.; Bonnell, J.; Fleming, R.; Bezabeh, B.; Dimasi, N.; et al. A multifunctional bispecific antibody protects against *Pseudomonas aeruginosa*. *Sci. Transl. Med.* **2014**, *6*, 262ra155. [CrossRef] [PubMed]

115. DiGiandomenico, A.; Warrener, P.; Hamilton, M.; Guillard, S.; Ravn, P.; Minter, R.; Camara, M.M.; Venkatraman, V.; Macgill, R.S.; Lin, J.; et al. Identification of broadly protective human antibodies to *Pseudomonas aeruginosa* exopolysaccharide Psl by phenotypic screening. *J. Exp. Med.* **2012**, *209*, 1273–1287. [CrossRef] [PubMed]

116. Warrener, P.; Varkey, R.; Bonnell, J.C.; DiGiandomenico, A.; Camara, M.; Cook, K.; Peng, L.; Zha, J.; Chowdury, P.; Sellman, B.; et al. A novel anti-PcrV antibody providing enhanced protection against *Pseudomonas aeruginosa* in multiple animal infection models. *Antimicrob. Agents Chemother.* **2014**, *58*, 4384–4391. [CrossRef] [PubMed]

117. Li Bassi, G.; Aguilera, E.; Senussi, T.; Iodone, F.A.; Motos, A.; Chiurazzi, C.; Travierso, C.; Amaro, R.; Hua, Y.; Bobi, J.; et al. MEDI3902 targeting *P. Aeruginosa* virulence factors PcrV and Psl for the prevention of pulmonary colonization during mechanical ventilation. In Proceedings of the American Thoracic Society 2017 International Conference, Washington, DC, USA, 19–24 May 2017.

118. DiGiandomenico, A.; Le, H.; Pinheiro, M.G.; Le, V.T.M.; Aguiar-Alves, F.; Quetz, J.; Tran, V.G.; Stover, C.K.; Diep, B.A. Protective activity of MEDI3902 for the prevention or treatment of lethal pneumonia and bloodstream infection caused by *pseudomonas aeruginosa* in rabbits. In Proceedings of the American Thoracic Society 2017 International Conference, Washington, DC, USA, 19–24 May 2017.

119. Secher, T.; Fas, S.; Fauconnier, L.; Mathieu, M.; Rutschi, O.; Ryffel, B.; Rudolf, M. The anti-*Pseudomonas aeruginosa* antibody panobacumab is efficacious on acute pneumonia in neutropenic mice and has additive effects with meropenem. *PLoS ONE* **2013**, *8*, e73396. [CrossRef] [PubMed]

120. Secher, T.; Fauconnier, L.; Szade, A.; Rutschi, O.; Fas, S.C.; Ryffel, B.; Rudolf, M.P. Anti-*Pseudomonas aeruginosa* serotype O11 LPS immunoglobulin M monoclonal antibody panobacumab (KBPA101) confers protection in a murine model of acute lung infection. *J. Antimicrob. Chemother.* **2011**, *66*, 1100–1109. [CrossRef] [PubMed]

121. Lazar, H.; Horn, M.P.; Zuercher, A.W.; Imboden, M.A.; Durrer, P.; Seiberling, M.; Pokorny, R.; Hammer, C.; Lang, A.B. Pharmacokinetics and safety profile of the human anti-*Pseudomonas aeruginosa* serotype O11 immunoglobulin M monoclonal antibody KBPA-101 in healthy volunteers. *Antimicrob. Agents Chemother.* **2009**, *53*, 3442–3446. [CrossRef] [PubMed]

122. Lu, Q.; Rouby, J.J.; Laterre, P.F.; Eggimann, P.; Dugard, A.; Giamarellos-Bourboulis, E.J.; Mercier, E.; Garbino, J.; Luyt, C.E.; Chastre, J.; et al. Pharmacokinetics and safety of panobacumab: Specific adjunctive immunotherapy in critical patients with nosocomial *Pseudomonas aeruginosa* O11 pneumonia. *J. Antimicrob. Chemother.* **2011**, *66*, 1110–1116. [CrossRef] [PubMed]

123. Que, Y.A.; Lazar, H.; Wolff, M.; Francois, B.; Laterre, P.F.; Mercier, E.; Garbino, J.; Pagani, J.L.; Revelly, J.P.; Mus, E.; et al. Assessment of panobacumab as adjunctive immunotherapy for the treatment of nosocomial *Pseudomonas aeruginosa* pneumonia. *Eur. J. Clin. Microbiol. Infect. Dis.* **2014**, *33*, 1861–1867. [CrossRef] [PubMed]

124. Aridis Pharmaceuticals. Aerucin: Broadly Active Human IgG Mab against *P. Aeruginosa*. Available online: http://www.aridispharma.com/aerucin/ (accessed on 23 July 2017).

125. Tarr, P.I.; Gordon, C.A.; Chandler, W.L. Shiga-toxin-producing *Escherichia coli* and haemolytic uraemic syndrome. *Lancet* **2005**, *365*, 1073–1086. [CrossRef]

126. Ahn, C.K.; Holt, N.J.; Tarr, P.I. Shiga-toxin producing *Escherichia coli* and the hemolytic uremic syndrome: What have we learned in the past 25 years? *Adv. Exp. Med. Biol.* **2009**, *634*, 1–17. [PubMed]

127. Melton-Celsa, A.R.; Smith, M.J.; O'Brien, A.D. Shiga toxins: Potent poisons, pathogenicity determinants, and pharmacological agents. *EcoSal Plus* **2005**, *1*. [CrossRef] [PubMed]

128. Melton-Celsa, A.R.; Carvalho, H.M.; Thuning-Roberson, C.; O'Brien, A.D. Protective efficacy and pharmacokinetics of human/mouse chimeric anti-stx1 and anti-stx2 antibodies in mice. *Clin. Vaccine Immunol.* **2015**, *22*, 448–455. [CrossRef] [PubMed]

129. Bitzan, M.; Poole, R.; Mehran, M.; Sicard, E.; Brockus, C.; Thuning-Roberson, C.; Riviere, M. Safety and pharmacokinetics of chimeric anti-shiga toxin 1 and anti-shiga toxin 2 monoclonal antibodies in healthy volunteers. *Antimicrob. Agents Chemother.* **2009**, *53*, 3081–3087. [CrossRef] [PubMed]

antibodies

MDPI

Review

Infusion Reactions Associated with the Medical Application of Monoclonal Antibodies: The Role of Complement Activation and Possibility of Inhibition by Factor H

Tamás Fülöp [1,2,*], Tamás Mészáros [1,2], Gergely Tibor Kozma [1,2], János Szebeni [1,2,3] and Mihály Józsi [4,5,*]

1. Nanomedicine Research and Education Center, Semmelweis University, 1089 Budapest, Hungary; tmeszaros@seroscience.com (T.M.); kozmalak@gmail.com (G.T.K.); jszebeni2@gmail.com (J.S.)
2. SeroScience Ltd., 1089 Budapest, Hungary
3. Department of Nanobiotechnology and Regenerative Medicine, Faculty of Health, Miskolc University, 3515 Miskolc, Hungary
4. Complement Research Group, Department of Immunology, ELTE Eötvös Loránd University, 1117 Budapest, Hungary
5. MTA-ELTE Immunology Research Group, Department of Immunology, ELTE Eötvös Loránd University, 1117 Budapest, Hungary
*   Correspondence: fulopgyulatamas@gmail.com (T.F.); mihaly.jozsi@gmx.net (M.J.); Tel.: +36-1-381-2175 (M.J.)

Received: 24 December 2017; Accepted: 8 March 2018; Published: 14 March 2018

Abstract: Human application of monoclonal antibodies (mAbs), enzymes, as well as contrast media and many other particulate drugs and agents referred to as "nanomedicines", can initiate pseudoallergic hypersensitivity reactions, also known as infusion reactions. These may in part be mediated by the activation of the complement system, a major humoral defense system of innate immunity. In this review, we provide a brief outline of complement activation-related pseudoallergy (CARPA) in general, and then focus on the reactions caused by mAb therapy. Because the alternative pathway of complement activation may amplify such adverse reactions, we highlight the potential use of complement factor H as an inhibitor of CARPA.

Keywords: CARPA; complement; complement activation; factor H; hypersensitivity; infusion reaction; monoclonal antibody therapy; pseudoallergic reaction

---

1. Introduction: Monoclonal Antibodies and Hypersensitivity Reactions

Monoclonal antibodies (mAbs) are made by identical immune cells that are all clones of a unique parent B cell, and are widely used both in basic research and the therapy of various diseases. For the latter purpose, one of the main goals of scientists became to create "fully" human products to reduce the side effects of humanized or chimeric therapeutic antibodies. These side effects include the induction of hypersensitivity reactions (HSRs), also known as infusion reactions (IRs) [1]. A selected list of anticancer and anti-inflammatory mAbs that cause such HSRs with various incidence and severity is shown in Table 1 [2–6].

Table 1. Information on hypersensitivity reactions to marketed monoclonal antibodies.

Brand Name (Manufacturer)	INN, Isotype (Target Antigen)	Indication	Incidence	Symptoms	References
		Anticancer use			
Avastin (Genentech, San Francisco, CA, USA; Roche, Basel, Switzerland)	bevacizumab, humanized IgG1 (VEGF-A)	combination chemotherapy of metastatic colon, lung, and kidney cancer, and glioblastoma	<3%, severe: 0.2%	chest pain, diaphoresis, headache, hypertension, neurologic signs and symptoms, oxygen desaturation, rigors, wheezing	[2,4]
Campath (Genzyme, Cambridge, MA, USA)	alemtuzumab–IH, humanized IgG1κ (CD52 on T and B cells)	B cell chronic lymphocytic leukemia (B-CLL)	4–7%	bronchospasm, chills, dyspnea, emesis, fever, hypotension, nausea, pyrexia, rash, rigors, tachycardia, urticaria	[2,5]
Erbitux (Bristol-Myers Squibb, New York, NY, USA; Eli Lilly, Indianapolis, IN, USA)	cetuximab, chimeric IgG1κ (EGFR)	metastatic colorectal cancer, head and neck cancer, squamous cell carcinomas	<3%, fatal < 0.1%	anaphylaxis, angioedema, bronchospasm, cardiac arrest, chills, dizziness, dyspnea, fever, hoarseness, hypotension, pruritus, rash, rigor, stridor, urticaria, wheezing	[1–3]
Herceptin (Genentech, San Francisco, CA, USA)	trastuzumab, humanized IgG1κ (EGFR receptor 2, HER2/neu/erbB2)	metastatic breast and gastric cancer	<1%	asthenia, bronchospasm, chills, death within hours, dizziness, dyspnea, further pulmonary complications, headache, hypotension, hypoxia, nausea, pain, rash, severe hypotension, vomiting	[1–3]
Rituxan (Genentech, San Francisco, CA, USA)	rituximab, chimeric IgG1κ (CD20 on B cells)	B cell leukemias, rheumatoid arthritis and non-Hodgkin's B-cell lymphoma	>80%, severe: <10%	ARDS, bronchospasm, cardiogenic shock, flushing, hypotension, hypoxia, itching, myocardial infarction, pain (at the site of the tumor), pulmonary infiltrates, runny nose, swelling of the tongue or throat, ventricular fibrillation, vomiting	[1–3,6]
		Anti-inflammatory use			
Remicade (Janssen Biotech. Inc., Horsham, PA, USA)	infliximab, chimeric IgG1κ (TNF alpha)	Crohn's disease, rheumatoid arthritis, spondylitis ankylopoetica, arthritis psoriatica, ulcerative colitis	18%	bronchospasm, laryngeal edema, pharyngeal edema, dyspnea, hypotension, urticaria, serum sickness-like reactions	[3]
Xolair (Genentech, San Francisco, CA, USA)	omalizumab, humanized IgG4 (IgE)	atopia, asthma	39%, Severe: 0.2%	anaphylaxis, bronchospasm, hypotension, syncope, urticaria, and/or angioedema of the throat or tongue, delayed anaphylaxis (with onset two to 24 h or even longer) beyond one year after beginning regularly administered treatment	[1]

INN: international nonproprietary names; ARDS: acute respiratory distress syndrome.

HSRs have been traditionally categorized in four groups, from I to IV, according to Coombs and Gell. This concept defined Type I reactions as IgE-mediated acute reactions, while the rest of the categories included subacute or chronic immune changes triggered or mediated by IgG, immune complexes, or lymphocytes [7]. However, it has increasingly been recognized that a substantial portion of acute allergic reactions, whose symptoms fit in Coombs and Gell's Type I category, are actually not initiated or mediated by pre-existing IgE antibodies. These reactions are known to be "pseudoallergic" or "anaphylactoid". There are estimates that pseudoallergy may represent as high as 77% of all immune-mediated immediate HSRs [8], implying hundreds of thousands of reactions and numerous fatalities every year [9]. Many of these reactions involve the activation of the complement system, an essential humoral arm of innate immunity. Complement activation-related pseudoallergy (CARPA) is linked to adverse events evoked by several liposomal and micellar formulations, nanoparticles, radiocontrast agents, and therapeutic antibodies [9].

Intravenous application of numerous drugs and medical agents, including therapeutic mAbs, enzymes, radiocontrast media, and many other particulate drugs with physical size in the upper nano $(10^{-8}–10^{-7}$ m) dimension (nanomedicines), can elicit HSRs with symptoms listed in Table 2.

Table 2. Symptoms of pseudoallergy. The most life-threatening symptoms are highlighted in bold [10]. Reprinted from Molecular Immunology, Vol. 61, Szebeni J., Complement activation-related pseudoallergy: A stress reaction in blood triggered by nanomedicines and biologicals, Pages 163–173, Copyright (2014), with permission from Elsevier.

Cardiovascular	Broncho-Pulmonary	Hematological	Mucocutaneous	Gastrointestinal	Neuro-Psycho-Somatic	Systemic
Angioedema	Apnea	Granulopenia	Cyanosis	Bloating	Back pain	Chills
Arrhythmia	Bronchospasm	Leukopenia	Erythmea	Cramping	Chest pain	Diaphoresis
Cardiogenic shock	Coughing	Lymphopenia	Flushing	Diarrhea	Chest tightness	Feeling of warmth
Edema	Dyspnea	Rebound leukocytosis	Nasal congestion	Metallic taste	Confusion	Fever
Hypertension	Hoarsness	Rebound granulocytosis	Rash	Nausea	Dizziness	Loss of consciousness
Hypotension	Hyperventillation	Trombocytopenia	Rhinitis	Vomiting	Feeling of imminent death	Rigors
Hypoxia	Laryngospasm		Swelling		Fright	Sweating
Myocardial infarction	Respiratory distress		Tearing		Headache	Wheezing
Tachycardia	Shortness of breath		Urticaria		Panic	
Ventricular fibrillation	Sneezing					
Syncope	Stridor					

2. The Consequences of Complement Activation for the Activator and the Host

One of the major tasks of the complement system is to mark and dispose of potentially dangerous particles, such as pathogenic microbes and altered host cells. This is achieved by targeted activation on foreign surfaces as well as on modified host targets, such as apoptotic cells. The classical pathway is activated by immunoglobulins bound to their target antigens, and the classical and lectin complement pathways are activated upon the recognition of certain molecular patterns associated with microbes or altered self, while the alternative pathway is activated constantly at a low rate and in an indiscriminative manner [11]. The activation can result in the deposition of opsonic molecules on the target cells or particles, thus labeling them for phagocytosis and, if not inhibited, allowing the initiation of the terminal pathway that may generate lytic complexes in the target cell's membrane. The three pathways merge at the activation of the central C3 molecule, which is cleaved into the anaphylatoxin and inflammatory mediator C3a and the larger, opsonic fragment C3b. C3b feeds back to the alternative pathway because it is part of the enzyme complex that cleaves additional C3 molecules. Thus, the alternative pathway can amplify complement activation initiated by any of the three pathways. Importantly, complement regulators expressed in body fluids and on cell surfaces protect the host from bystander damage [11].

Complement activation by liposomes can easily be rationalized on the basis of their resemblance to pathogenic viruses. In fact, both are phospholipid-coated vesicles in the same size range (60–200 nm), with the difference being that liposomes do not express surface proteins as viruses do. In the case of viruses, some of these surface proteins inhibit complement activation just as complement receptor type 1 (CR1), decay accelerating factor (DAF), and membrane cofactor protein (MCP) do on the surface of host blood cells and other cells. One may therefore conclude that liposomal nanomedicines activate complement because the immune system considers them as pathogenic viruses, and liposomes do not have a shield that protects them against complement attack [12]. The mechanism of complement activation by smaller nanoparticles (d < 10 nm), such as PEGylated polyethylene-imine polymers (PEG is polyethylene glycol) [13] or micelles formed from Cremophor EL (CrEL) and other polyethoxylated surfactants (PS-80 and PS-20, also known as Tween-20 and Tween-80) [14] is more difficult to explain. In those cases, complement activation may involve unconventional direct interaction with complement proteins, or, as it was suggested for CrEL, prior interaction with plasma lipoproteins that can lead to the formation of large(r) aggregates [9].

Furthermore, it is already shown in vitro that the aggregation of proteins during the preparation of mAbs can induce the activation of human monocyte-derived dendritic cells as well as T cell responses [15]. Complement activation is also possible in such conditions.

3. Therapeutic mAbs, Complement Activation, and CARPA

Antibodies are well known to activate the classical complement pathway upon binding to their target antigen, which allows for the binding of C1q, the recognition molecule of the activation initiator C1 complex, to the Fc part of the antibodies. Therapeutic mAbs may exploit this feature and can be engineered to enhance the effectiveness of the treatment while circumventing certain (e.g., Fc-receptor-mediated) adverse effects [16,17].

The role that complement plays in mAb therapy is exemplified well by the prototypic mAb rituximab. Rituximab, a murine-human chimera type anti-CD20, has been used since 1997 in clinical practice to treat malignant and autoimmune disorders related to the disfunction of B cells [18,19]. Besides the direct downregulation of CD20-related cell functions, both complement-dependent and complement-independent immune reactions participate in the elimination of CD20 highly positive B cells (Figure 1). Complement-dependent mechanisms include complement-dependent cytotoxicity (CDC), initiated upon C1q binding, through the classical complement activation cascade [20], and complement-enhanced antibody-dependent cell-mediated phagocytosis (ADCP). The most important complement-independent mechanism is antibody-dependent cell-mediated cytotoxicity (ADCC), which is performed mainly by natural killer (NK) cells (and macrophages). Programmed cell death (PCD) seems to be less important in the case of rituximab, but it may have more prominent role in the action of Type II-anti-CD20 antibodies, like tositumomab and GA101 [18,19]. However, it is likely that the complement-activating capacity of rituximab is also responsible for the high frequency of CARPA associated with this mAb [21].

Human IgG1 and IgG3 are particularly effective at fixing complement to the target cell surface, and many of the currently approved therapeutic mAbs, like rituximab, are indeed of the IgG1 isotype. A variety of cell-based assays have demonstrated the ability of mAbs to recruit complement components in vitro, but the efficiency of CDC to kill tumor cells in vivo is less clear, particularly for solid tumors, in part because tumor cells themselves express membrane-bound complement regulators as well as the soluble regulator factor H [22–24]. Since most of these mAbs work against cancer cells with the help of complement activation, a clear distinction has to be made between complement activation on the target cell surface with the help of the cell-bound mAb (i.e., CDC) and adverse hypersensitivity reaction related to complement activation in serum caused by the therapeutic antibody itself. This means that the same mechanisms are involved in the beneficial effects and hypersensitivity.

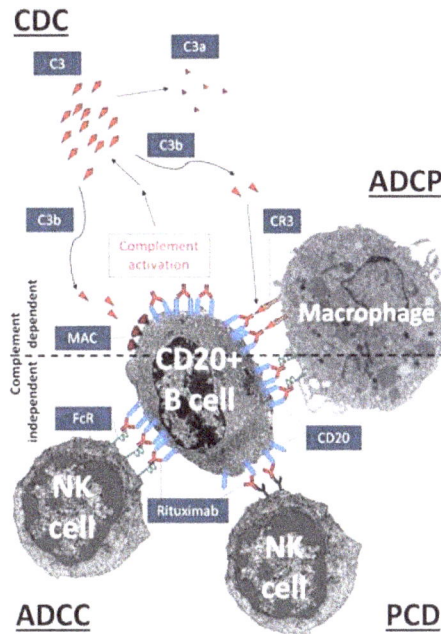

Figure 1. Complement activation as an essential mechanism of the therapeutic action of rituximab, an anti-CD20 antibody. Rituximab recognizes CD20 on the surface of pre- and mature B cells. After binding, the complement activation cascade is initiated by the classical pathway leading to the cleavage of C3 into C3a and C3b. C3b can cause complement-dependent cytotoxicity (CDC) by promoting the assembly of the membrane attack complex (MAC), while complement receptors on phagocytic cells, such as complement receptor type 3 (CR3) on macrophages, can mediate complement-enhanced antibody-dependent cell-mediated phagocytosis (ADCP). Surface-bound rituximab can trigger natural killer (NK) cells and macrophages by complement-independent mechanisms, via antibody-dependent cell-mediated cytotoxicity (ADCC), ADCP and, to a lesser degree, the induction of programmed cell death (PCD).

All currently available or publicly known mAbs can be considered to be potentially direct immunogens, as their molecular size is large enough and their structure is different from endogenous proteins. Despite current efforts to produce highly humanized or "human-like" mAbs, immunogenicity is not yet totally eradicated. Treatment of human patients with mAbs can be associated with the development of specific antibodies against these therapeutic antibodies (anti-drug antibodies, ADAs). These neutralizing ADAs can block the biological activity of the drug either by binding directly to the epitope(s) within their active site, or by steric hindrance due to binding to epitope(s) in close proximity to the active site. The presence of neutralizing ADAs may not result in adverse clinical effect, except that it decreases the efficacy of the therapeutic mAb, requiring its administration at higher doses. Furthermore, the presence of specific ADAs against mAbs can be associated in some cases with hypersensitivity reactions identical to the CARPA phenomenon delineated above for the case of liposomes and other nano-pharmaceuticals. The rare anaphylactic reactions associated with mAbs including cetuximab, infliximab, or basiliximab represent typical CARPA [25].

True allergic reactions, which are mediated by anti-drug IgE, require prior exposure to the mAb and, consequently, do not occur on the first infusion, except in rare cases where patients have pre-existing antibodies that cross-react with the drug. However, pseudoallergic reactions (IgE-independent reactions possibly mediated by direct immune cell and complement activation) and cytokine release syndrome (CRS) both occur primarily on the first infusion of the drug, although they can also occur on subsequent administrations. The symptoms of all three types of immunologically-mediated infusion reactions (IRs) overlap, making it difficult to identify the cause without additional laboratory work [26].

Rituximab and trastuzumab induce the highest incidence of IRs. In general, the incidence of mAb-induced IRs varies from ~15–20% for cetuximab (including 3% more severe, grade 3, and life threatening, grade 4 reactions) and 40% for trastuzumab first infusion (<1% grades 3–4) to 77% for rituximab first infusion (10% grades 3–4). Even after the fourth infusion, 30% of cancer patients react to rituximab, and the incidence of IRs remains 14% after the eighth infusion. Approximately 80% of fatal reactions occur after the first rituximab infusion. The incidence of IRs to the humanized mAb bevacizumab and the fully humanized panitumumab is significantly lower [27].

Thrombocytopenia, neutropenia, and anemia can occur in some patients treated with mAbs as part of anticancer immunotherapy, but the mechanisms of these potentially severe side effects frequently remain unexplored. Interestingly, these symptoms are also characteristic of liposome-induced CARPA. Late-onset neutropenia, especially after rituximab treatment, has been examined in a growing number of reports; however, with each of the three cytopenias seen during mAb therapy, it is frequently unclear whether the depletion of cells is due to an immunological mechanism. Type III hypersensitivities, such as serum sickness-like reactions and vasculitis, are also known to occur in response to mAbs. Some pulmonary events, including mAb-induced lung diseases, are hypersensitivity reactions that result from the interaction of the drug with the immune system and involve drug-specific antibodies or T cells [2].

Although it remains to be shown in humans, it is hypothesized that mAbs could stimulate anti-mAb IgGs bound to Fc-gamma-receptors on macrophages, basophils, and neutrophils, triggering the release of platelet-activating factor, as shown in the mouse model of IgG-dependent anaphylaxis [28]. In addition, the complement system could be activated by the formation of large immune complexes, thereby generating anaphylatoxins (C3a and C5a). It is also important to point out that patients with anti-infliximab IgGs are at increased risk of immediate HSRs compared with patients without such antibodies [1]. Thus, in addition to the preferred complement activation induced by the binding of therapeutic mAbs to their targets, complement activation can also arise as a consequence of the binding of naturally forming ADAs against the therapeutic mAbs. The molecular background of mAb-induced CARPA is yet to be studied in more detail.

4. Potential Role of Factor H in Mitigating Complement Activation

The use of natural or engineered complement inhibitors may represent an attractive way to prevent CARPA-mediated HSRs. Early approaches used the complement-regulatory domains of the natural complement inhibitor CR1 linked to a myristoyl group that mediated incorporation in liposomal membranes [29]. A recent study suggested that factor H could be also employed to reduce or eliminate complement activation triggered by liposomes, micelles, or therapeutic mAbs [30]. Factor H is the main soluble inhibitor of the alternative pathway and the amplification loop of complement [31,32]. It was shown that liposomal Amphotericin B, CrEL, and rituximab caused less complement activation in serum in vitro when factor H was added to the serum in excess, as compared with the serum without exogenous factor H [30]. Moreover, the artificial inhibitor, recombinant mini-factor H [33], which unites the N-terminal complement-regulatory domains and the C-terminal host surface recognition domains of the natural molecule, was even more effective in inhibiting such complement activation compared with factor H [30]. These data suggest that factor H-based complement inhibition could be a viable strategy to prevent or mitigate CARPA induced by nanomedicines, including therapeutic mAbs.

5. Conclusions and Outlook

The prevention of IRs induced by mAbs can be addressed the same way as the prevention of similar adverse reactions occurring upon nanomedicine treatments. The surface modification of liposomes and other therapeutic proteins can lead to prevention of the aggregation of these agents and reduction of immunogenicity and antigenicity. Recently, more and more antibodies and, predominantly, antibody fragments designed for therapeutic purposes use the covalent attachment of polyethylene glycol (PEG). PEGylation generally prolongs the half-life in the circulation and prevents the immunogenicity of many liposomal drugs and mAb molecules [34]. However, in some cases the generation of an IR event could be connected to the presence of PEGylation on the surfaces of liposomes. The formation of anti-PEG IgMs against PEG molecules on liposomes are observed in CARPA studies with animal models [35].

Another possible approach is the administration of complement inhibitors together with the therapeutic agents to reduce the chance of a possible adverse reaction. Even though this could be a good option as a prevention measure, most patients may not even need such an action if they are not prone to IRs, and this approach would just elevate the costs of the therapies. The best scenario would be to pre-screen each patient for proneness to any adverse reaction, using an in vitro test that could predict from a blood sample if any CARPA event could arise during introduction of a therapeutic agent, such as mAbs.

Acknowledgments: The work of the authors was supported in part by the Hungarian Academy of Sciences (grant LP2012-43 to M.J.), the European Union Seventh Framework Program grants No. NMP-2012-309820 (NanoAthero) and FP-7/2013-Innovation-1-602923-2 (TheraGlio), and the Applied Materials and Nanotechnology "Center of Excellence", Miskolc University.

Author Contributions: T.F. prepared the initial draft of the paper. T.F., T.M., and G.T.K. prepared the tables and the figure. T.F., G.T.K., J.S. and M.J. wrote the paper. All authors revised and approved the paper.

Conflicts of Interest: The authors declare no conflict of interest. The founding sponsors had no role in the design of the study; in the collection, analyses, or interpretation of data; in the writing of the manuscript, and in the decision to publish the results.

References

1. Picard, M.; Galvão, V.R. Current Knowledge and Management of Hypersensitivity Reactions to Monoclonal Antibodies. *J. Allergy Clin. Immunol. Pract.* **2017**, *5*, 600–609. [CrossRef] [PubMed]
2. Baldo, B.A. Adverse events to monoclonal antibodies used for cancer therapy: Focus on hypersensitivity responses. *Oncoimmunology* **2013**, *2*, e26333. [CrossRef] [PubMed]
3. Hong, D.I.; Bankova, L.; Cahill, K.N.; Kyin, T.; Castells, M.C. Allergy to monoclonal antibodies: Cutting-edge desensitization methods for cutting-edge therapies. *Expert Rev. Clin. Immunol.* **2012**, *8*, 43–52. [CrossRef] [PubMed]
4. Choueiri, T.K.; Mayer, E.L.; Je, Y.; Rosenberg, J.E.; Nguyen, P.L.; Azzi, G.R.; Bellmunt, J.; Burstein, H.J.; Schutz, F.A. Congestive Heart Failure Risk in Patients With Breast Cancer Treated With Bevacizumab. *J. Clin. Oncol.* **2011**, *29*, 632–638. [CrossRef] [PubMed]
5. Keating, M.J.; Flinn, I.; Jain, V.; Binet, J.L.; Hillmen, P.; Byrd, J.; Albitar, M.; Brettman, L.; Santabarbara, P.; Wacker, B.; et al. Therapeutic role of alemtuzumab (Campath-1H) in patients who have failed fludarabine: Results of a large international study. *Blood* **2002**, *99*, 3554–3561. [CrossRef] [PubMed]
6. Kimby, E. Tolerability and safety of rituximab (MabThera®). *Cancer Treat. Rev.* **2005**, *31*, 456–473. [CrossRef] [PubMed]
7. Coombs, R.R.A.; Gell, P.G.H. Classification of allergic reactions responsible for drug hypersensitivity reactions. In *Clinical Aspects of Immunology*; Coombs, R.R.A., Gell, P.G.H., Eds.; Davis: Philadelphia, PA, USA, 1968; pp. 575–596.
8. Demoly, P.; Lebel, B.; Messaad, D.; Sahla, H.; Rongier, M.; Daurès, J.P.; Godard, P.; Bousquet, J. Predictive capacity of histamine release for the diagnosis of drug allergy. *Allergy* **1999**, *54*, 500–506. [CrossRef] [PubMed]

9. Szebeni, J. Complement activation-related pseudoallergy caused by liposomes, micellar carriers of intravenous drugs, and radiocontrast agents. *Crit. Rev. Ther. Drug Carrier Syst.* **2001**, *18*, 567–606. [CrossRef] [PubMed]

10. Szebeni, J. Complement activation-related pseudoallergy: A stress reaction in blood triggered by nanomedicines and biologicals. *Mol. Immunol.* **2014**, *61*, 163–173. [CrossRef] [PubMed]

11. Ricklin, D.; Hajishengallis, G.; Yang, K.; Lambris, J.D. Complement: A key system for immune surveillance and homeostasis. *Nat. Immunol.* **2010**, *11*, 785–797. [CrossRef] [PubMed]

12. Szebeni, J.; Muggia, F.; Gabizon, A.; Barenholz, Y. Activation of complement by therapeutic liposomes and other lipid excipient-based therapeutic products: Prediction and prevention. *Adv. Drug Deliv. Rev.* **2011**, *63*, 1020–1030. [CrossRef] [PubMed]

13. Merkel, O.M.; Urbanics, R.; Bedocs, P.; Rozsnyay, Z.; Rosivall, L.; Toth, M.; Kissel, T.; Szebeni, J. In vitro and in vivo complement activation and related anaphylactic effects associated with polyethylenimine and polyethylenimine-graft-poly (ethylene glycol) block copolymers. *Biomaterials* **2011**, *32*, 4936–4942. [CrossRef] [PubMed]

14. Weiszhar, Z.; Czúcz, J.; Révész, C.; Rosivall, L.; Szebeni, J.; Rozsnyay, Z. Complement activation by polyethoxylated pharmaceutical surfactants: Cremophor-EL, Tween-80 and Tween-20. *Eur. J. Pharm. Sci.* **2012**, *45*, 492–498. [CrossRef] [PubMed]

15. Rombach-Riegraf, V.; Karle, A.C.; Wolf, B.; Sordé, L.; Koepke, S.; Gottlieb, S.; Krieg, J.; Djidja, M.C.; Baban, A.; Spindeldreher, S.; et al. Aggregation of human recombinant monoclonal antibodies influences the capacity of dendritic cells to stimulate adaptive T-cell responses in vitro. *PLoS ONE* **2014**, *9*, e86322. [CrossRef] [PubMed]

16. Cook, E.M.; Lindorfer, M.A.; van der Horst, H.; Oostindie, S.; Beurskens, F.J.; Schuurman, J.; Zent, C.S.; Burack, R.; Parren, P.W.; Taylor, R.P. Antibodies That Efficiently Form Hexamers upon Antigen Binding Can Induce Complement-Dependent Cytotoxicity under Complement-Limiting Conditions. *J. Immunol.* **2016**, *197*, 1762–1775. [CrossRef] [PubMed]

17. Lee, C.H.; Romain, G.; Yan, W.; Watanabe, M.; Charab, W.; Todorova, B.; Lee, J.; Triplett, K.; Donkor, M.; Lungu, O.I.; et al. IgG Fc domains that bind C1q but not effector Fcγ receptors delineate the importance of complement-mediated effector functions. *Nat. Immunol.* **2017**, *18*, 889–898. [CrossRef] [PubMed]

18. Lim, S.H.; Beers, S.A.; French, R.R.; Johnson, P.W.; Glennie, M.J.; Cragg, M.S. Anti-CD20 monoclonal antibodies: Historical and future perspectives. *Haematologica* **2010**, *95*, 135–143. [CrossRef] [PubMed]

19. Boross, P.; Leusen, J.H. Mechanisms of action of CD20 antibodies. *Am. J. Cancer Res.* **2012**, *2*, 676–690. [PubMed]

20. Zhou, X.; Hu, W.; Qin, X. The role of complement in the mechanism of action of rituximab for B-cell lymphoma: Implications for therapy. *Oncologist* **2008**, *13*, 954–966. [CrossRef] [PubMed]

21. Van der Kolk, L.E.; Grillo-Lopez, A.J.; Baars, J.W.; Hack, C.E.; van Oers, M.H. Complement activation plays a key role in the side-effects of rituximab treatment. *Br. J. Haematol.* **2001**, *115*, 807–811. [CrossRef] [PubMed]

22. Junnikkala, S.; Jokiranta, T.S.; Friese, M.A.; Jarva, H.; Zipfel, P.F.; Meri, S. Exceptional resistance of human H2 glioblastoma cells to complement-mediated killing by expression and utilization of factor H and factor H-like protein 1. *J. Immunol.* **2000**, *164*, 6075–6081. [CrossRef] [PubMed]

23. Ajona, D.; Castaño, Z.; Garayoa, M.; Zudaire, E.; Pajares, M.J.; Martinez, A.; Cuttitta, F.; Montuenga, L.M.; Pio, R. Expression of complement factor H by lung cancer cells: Effects on the activation of the alternative pathway of complement. *Cancer Res.* **2004**, *64*, 6310–6318. [CrossRef] [PubMed]

24. Rogers, L.M.; Veeramani, S.; Weiner, G.J. Complement in Monoclonal Antibody Therapy of Cancer. *Immunol. Res.* **2014**, *59*, 203–210. [CrossRef] [PubMed]

25. Descotes, J. Immunotoxicity of monoclonal antibodies. *mAbs* **2009**, *1*, 104–111. [CrossRef] [PubMed]

26. Brennan, F.R.; Morton, L.D.; Spindeldreher, S.; Kiessling, A.; Allenspach, R.; Hey, A.; Muller, P.Y.; Frings, W.; Sims, J. Safety and immunotoxicity assessment of immunomodulatory monoclonal antibodies. *mAbs* **2010**, *2*, 233–255. [CrossRef] [PubMed]

27. Chung, C.H. Managing premedications and the risk for reactions to infusional monoclonal antibody therapy. *Oncologist* **2008**, *13*, 725–732. [CrossRef] [PubMed]

28. Strait, R.T.; Morris, S.C.; Yang, M.; Qu, X.W.; Finkelman, F.D. Pathways of anaphylaxis in the mouse. *J. Allergy Clin. Immunol.* **2002**, *109*, 658–668. [CrossRef] [PubMed]

29. Smith, G.P.; Smith, R.A. Membrane-targeted complement inhibitors. *Mol. Immunol.* **2011**, *38*, 249–255. [CrossRef]

30. Mészáros, T.; Csincsi, Á.I.; Uzonyi, B.; Hebecker, M.; Fülöp, T.G.; Erdei, A.; Szebeni, J.; Józsi, M. Factor H inhibits complement activation induced by liposomal and micellar drugs and the therapeutic antibody rituximab in vitro. *Nanomedicine* **2016**, *12*, 1023–1031. [CrossRef] [PubMed]

31. Kopp, A.; Hebecker, M.; Svobodová, E.; Józsi, M. Factor H: A complement regulator in health and disease, and a mediator of cellular interactions. *Biomolecules* **2012**, *2*, 46–75. [CrossRef] [PubMed]

32. Parente, R.; Clark, S.J.; Inforzato, A.; Day, A.J. Complement factor H in host defense and immune evasion. *Cell. Mol. Life Sci.* **2017**, *74*, 1605–1624. [CrossRef] [PubMed]

33. Hebecker, M.; Alba-Domínguez, M.; Roumenina, L.T.; Reuter, S.; Hyvärinen, S.; Dragon-Durey, M.A.; Jokiranta, T.S.; Sánchez-Corral, P.; Józsi, M. An engineered construct combining complement regulatory and surface-recognition domains represents a minimal-size functional factor H. *J. Immunol.* **2013**, *191*, 912–921. [CrossRef] [PubMed]

34. Milla, P.; Dosio, F.; Cattel, L. PEGylation of proteins and liposomes: A powerful and flexible strategy to improve the drug delivery. *Curr. Drug Metab.* **2012**, *13*, 105–119. [CrossRef] [PubMed]

35. Hashimoto, Y.; Shimizu, T.; Abu Lila, A.S.; Ishida, T.; Kiwada, H. Relationship between the concentration of anti-polyethylene glycol (PEG) immunoglobulin M (IgM) and the intensity of the accelerated blood clearance (ABC) phenomenon against PEGylated liposomes in mice. *Biol. Pharm. Bull.* **2015**, *38*, 417–424. [CrossRef] [PubMed]

MDPI

St. Alban-Anlage 66

4052 Basel, Switzerland

Tel. +41 61 683 77 34

Fax +41 61 302 89 18

http://www.mdpi.com

Antibodies Editorial Office

E-mail: antibodies@mdpi.com

http://www.mdpi.com/journal/antibodies

www.ingramcontent.com/pod-product-compliance
Lightning Source LLC
Chambersburg PA
CBHW051837210326
41597CB00033B/5681